CONTAMINANT EFFECTS ON FISHERIES

Volume

16

in the Wiley Series in

Advances in Environmental Science and Technology

JEROME O. NRIAGU, Series Editor

CONTAMINANT EFFECTS ON FISHERIES

Edited by

Victor W. Cairns
Peter V. Hodson
Jerome O. Nriagu

A WILEY-INTERSCIENCE PUBLICATION

JOHN WILEY & SONS

New York ● Chichester ● Brisbane ● Toronto ● Singapore

Library of Congress Cataloging in Publication Data:
Main entry under title:

Contaminant effects on fisheries.

(Advances in environmental science and technology;
v. 16)
"A Wiley-Interscience publication."
Includes index.
1. Fishes—Effect of water pollution on.   2. Water—
Pollution—Measurement.   I. Cairns, V. W.   II. Hodson,
Peter V.   III. Nriagu, Jerome O.   IV. Series.

TD180.A38 vol. 16   [SH174]   628s   84-7488
ISBN 0-471-88014-0   [597'.05222]

Printed in the United States of America

10 9 8 7 6 5 4 3 2 1

CONTRIBUTORS

ADDISON, R. F., Marine Ecology Laboratory, Bedford Institute of Oceanography, Dartmouth, Nova Scotia

BERGMAN, HAROLD L., Fish Physiology and Toxicology Laboratory, Department of Zoology and Physiology, University of Wyoming, Laramie, Wyoming

BLUNT, BEVERLY R., Great Lakes Fisheries Research Branch, Department of Fisheries and Oceans, Canada Centre for Inland Waters, Burlington, Ontario

BOUCK, GERALD R., Division of Fish and Wildlife, Bonneville Power Administration, U.S. Department of Energy, Portland, Oregon

BRECK, J. E., Environmental Sciences Division, Oak Ridge National Laboratory, Oak Ridge, Tennessee

CAIRNS, VICTOR W., Great Lakes Fisheries Research Branch, Canada Center for Inland Waters, Burlington, Ontario.

COLBY, PETER J., Ontario Ministry of Natural Resources, Fisheries Research Section, Thunder Bay, Ontario

DEANGELIS, D. L., Environmental Sciences Division, Oak Ridge National Laboratory, Oak Ridge, Tennessee

DICKIE, L. M., Marine Ecology Laboratory, Bedford Institute of Oceanography, Dartmouth, Nova Scotia

DONALDSON, EDWARD M., West Vancouver Laboratory, Fisheries Research Branch, Department of Fisheries and Oceans, West Vancouver, British Columbia

DUNCAN, D. A., Department of Fisheries and Oceans, Freshwater Institute, Winnipeg, Manitoba

FAGERLUND, ULF H. M., West Vancouver Laboratory, Fisheries Research Branch, Department of Fisheries and Oceans, West Vancouver, British Columbia

FREEMAN, H. C., Fisheries Research Branch, Department of Fisheries and Oceans, Halifax, Nova Scotia

GOODYEAR, C. PHILLIP, U.S. Fish and Wildlife Service, National Fishery Center, Leetown, West Virginia

HODSON, PETER V., Great Lakes Fisheries Research Branch, Department of Fisheries and Oceans, Canada Centre for Inland Waters, Burlington, Ontario

HOHREITER, DAVID W., School of Forest Resources, University of Georgia, Athens, Georgia

JOHNSON, M. G., Great Lakes Fishery Commission, Great Lakes Fisheries Research Branch, Owen Sound Laboratory, Owen Sound, Ontario

JOHNSON, RODNEY D., Fish Physiology and Toxicology Laboratory, Department of Zoology and Physiology, University of Wyoming, Laramie, Wyoming

KERR, S. R., Marine Ecology Laboratory, Bedford Institute of Oceanography, Dartmough, Nova Scotia

KLAVERKAMP, J. F., Department of Fisheries and Oceans, Freshwater Institute, Winnipeg, Manitoba

KLONTZ, GEORGE W., Department of Fish and Wildlife Resources, University of Idaho, Moscow, Idaho

LEATHERLAND, J. F., Department of Zoology, University of Guelph, Guelph, Ontario

LOCKHART, W. L., Department of Fisheries and Oceans, Freshwater Institute, Winnipeg, Manitoba

MACDONALD, W. A., Department of Fisheries and Oceans, Freshwater Institute, Winnipeg, Manitoba

MCBRIDE, J. R., West Vancouver Laboratory, Fisheries Research Branch, Department of Fisheries and Oceans, West Vancouver, British Columbia

MCLEAY, DONALD J., D. McLeay and Associates Ltd., Vancouver, British Columbia

METNER, D. A., Department of Fisheries and Oceans, Freshwater Institute, Winnipeg, Manitoba

PASSINO, DORA R. MAY, Great Lakes Fishery Laboratory, U.S. Fish and Wildlife Service, Ann Arbor, Michigan

PEARCE, JOHN B., National Marine Fisheries Service, Highlands, New Jersey

RAPPORT, D. J., Statistics Canada, Tunneys Pasture, Ottawa, Ontario

REINERT, ROBERT E., School of Forest Resources, University of Georgia, Athens, Georgia

SANGALANG, G. B., Fisheries Research Branch, Department of Fisheries and Oceans, Halifax, Nova Scotia

SONSTEGARD, R. A., Departments of Biology and Pathology, McMaster University, Hamilton, Ontario

UTHE, J. F., Fisheries Research Branch, Department of Fisheries and Oceans, Halifax, Nova Scotia

VAUGHAN, D. S., National Marine Fisheries Service, Beaufort Laboratory, Beaufort, North Carolina

WAGEMANN, R., Department of Fisheries and Oceans, Freshwater Institute, Winnipeg, Manitoba

WEDEMEYER, GARY A., U.S. Fish and Wildlife Service, National Fishery Research Center, Seattle, Washington

WHITTLE, D. MICHAEL, Great Lakes Fisheries Research Branch, Department of Fisheries and Oceans, Canada Centre for Inland Waters, Burlington, Ontario

YOSHIYAMA, R. M., Environmental Sciences Division, Oak Ridge National Laboratory, Oak Ridge, Tennessee

INTRODUCTION
TO THE SERIES

The deterioration of environmental quality, which began when humankind first congregated into villages, has existed as a serious problem since the industrial revolution. In the second half of the twentieth century, under the ever-increasing impacts of exponentially growing population and of industrializing society, environmental contamination of the air, water, soil, and food has become a threat to the continued existence of many plant and animal communities of various ecosystems and may ultimately threaten the very survival of the human race. Understandably, many scientific, industrial, and governmental communities have recently committed large resources of money and humanpower to the problems of environmental pollution and pollution abatement by effective control measures.

Advances in Environmental Sciences and Technology deals with creative reviews and critical assessments of all studies pertaining to the quality of the environment and to the technology of its conservation. The volumes published in the series are expected to serve several objectives: (1) stimulate interdisciplinary cooperation and understanding among the environmental scientists; (2) provide scientists with a periodic overview of environmental developments that are of general concern or of relevance to their own work or interests; (3) provide the graduate student with a critical assessment of past accomplishment that may help stimulate interest in career opportunities in this vital area; and (4) provide the research manager and the legislative or administrative official with an assured awareness of newly developing research work on the critical pollutants and with the background information important to their responsibility.

As the skills and techniques of many scientific disciplines are brought to bear on the fundamental and applied aspects of the environmental issues, there is a heightened need to draw together the numerous threads and to present a coherent picture of the various research endeavors. This need and the recent tremendous growth in the field of environmental studies have clearly made some editorial adjustments necessary. Apart from the changes in style and format, each future volume in the series will focus on one particular theme or timely topic, starting with Volume 12. The author(s) of each pertinent section will be expected to critically review the literature and the most important recent developments in the particular field; to critically evaluate new concepts, methods, and data; and to focus attention on important unresolved or controversial questions and on probable future trends.

Monographs embodying the results of unusually extensive and well-rounded investigations will also be published in the series. The net result of the new editorial policy should be more integrative and comprehensive volumes on key environmental issues and pollutants. Indeed, the development of realistic standards of environmental quality for many pollutants often entails such a holistic treatment.

JEROME O. NRIAGU, Series Editor

PREFACE

In April 1970, the Lake St. Clair fishery was closed because levels of mercury in several species of fish exceeded the recommended safe level for human consumption. This was the first documented evidence that chemical contamination could affect the fisheries of the Great Lakes. Concern for mercury was soon followed by reports of elevated tissue burdens of DDT, PCBs, and Mirex, and more recently, dibenzo-*p*-dioxins, dibenzofurans, and toxaphene have been added to the expanding list of toxic substances found in Great Lakes fish.

Obvious effects of these chemicals include the loss of livelihood for commercial fishermen and a substantial reduction in revenue from commercial and recreational fisheries. Less obvious is the possibility that chemicals may have adverse effects on the health, reproduction, growth, or behavior of exposed fish.

There are some data indicating that Great Lakes fish are responding to chemical stress. As early as 1964 biologists from New York State reported that DDT in lake trout eggs was affecting lake trout reproduction. Since that time, strong circumstantial evidence has implicated chemical contaminants with fish and herring gull reproductive failures, physiological dysfunction, and epizootics of benign and malignant tumors in some fish species.

Cause and effect relationships are difficult to establish in wild populations and many of the stress responses, particularly at the population and community levels, are common to both chemical and nonchemical stressors. The absence of hard data describing the effects of contaminants on Great Lakes fish reflects the current state of knowledge. Many diagnostic tools that have demonstrated utility as sensitive, specific indicators of contaminant effects in laboratory experiments have not yet been field tested. Some promising techniques have been developed, such as the pathological, biochemical, and physiological approaches recommended by the International Council for Exploration of the Seas. However, Great Lakes fisheries agencies have failed to incorporate them into their fish assessment programs. This may be due to the apparent lack of communication between fisheries managers and laboratory technologists. Fisheries biologists are frequently criticized for failing to acknowledge the potential impacts of contaminants on fish populations. Similarly, fish toxicologists are reluctant to extrapolate the results of laboratory studies to estimate the effects of contaminants on fish populations and communities.

The Great Lakes Fishery Commission recognized these problems and hosted a workshop to discuss suitable methods for assessing the effects of contaminants on wild fish populations. Toxicologists and fisheries biologists from Canada and the United States were invited to present their own work and recommend promising avenues of research. Some of their presentations and suggestions have been incorporated into this book.

The first two chapters provide background to Great Lakes chemical and fisheries problems and introduce the basic concepts of epidemiology. These are followed by descriptions of histological, biochemical, and physiological responses to chemical exposure. The final chapters discuss the utility of population and community indicators of chemical and natural stressors.

The workshop provided a unique forum for laboratory toxicologists and fisheries managers to discuss opportunities for cooperative monitoring programs. We hope that this book will help fisheries agencies develop an integrated approach to contaminant effects monitoring on individuals and populations.

Many people contributed to the success of the workshop and to this publication. We are especially grateful to Drs. Clay Edwards, Dave Evans, George Dixon, and Murray Johnson, and Messrs. Wayne Willford and Michael Gilbertson, for their assistance as session chairmen and editors. Special thanks also to Miss Margaret Ross, Dr. Robert Campbell, Mr. John Fitzsimons, and Mr. Michael Whittle. Finally, we wish to thank all of the contributors for their excellent cooperation, and the Great Lakes Fishery Commission for their interest and encouragement.

VICTOR W. CAIRNS
PETER V. HODSON
JEROME O. NRIAGU

Burlington, Ontario
May 1984

CONTENTS

CONTAMINANT EFFECTS
ON FISHERIES

1

GREAT LAKES FISHERIES AND ENVIRONMENTAL ISSUES

M. G. Johnson

Great Lakes Fishery Commission
and
Great Lakes Fisheries Research Branch
Department of Fisheries and Oceans
Owen Sound, Ontario

1. INTRODUCTION

We urgently need a scientifically sound strategy to assess impacts of environmental stresses, including those posed by toxic substances, on aquatic communities in the Great Lakes, which comprise one of the most perturbed of the world's freshwater ecosystems. I would like to thank the editors and the contributors of this book, on behalf of the Great Lakes Fishery Commission (GLFC) and our associated fisheries agencies, for giving their time and experience toward this endeavor.

The Great Lakes ecosystem is home to 37 million people, 33% of the population of Canada and 14% of the population of the United States; 5% of the land area is urban. The population is expected to rise to 60 million people by 2020 (PLUARG, 1978). Industrial and municipal growth is projected to rise accordingly, mostly centered on the largest urban areas—the south end of Lake Michigan, western Lake Erie and Lake St. Clair from Sarnia to Cleveland, and western Lake Ontario from the Niagara Frontier to Oshawa. The Great Lakes Water Quality Board Annual Report for 1977 (Water Quality Board, 1978) listed 825 major municipal and industrial wastewater dischargers in the basin. There are approximately 4000 liquid and solid waste disposal sites (PLUARG, 1977). The Water Quality Board, in 1978, listed 381 organic and heavy metal contaminants that have been identified as present in the Great Lakes, and 38 additional contaminants were identified in 1979. At present, there are 89 thermal electric generating stations, rated at 54,000 MW, that use water directly from the Great Lakes, equivalent to about one-third of the flow of the St. Lawrence River. Seventeen new plants with a combined capacity of 31,000 MW are planned (Kelso and Milburn, 1979).

Pollution of Canada-U.S. boundary waters of this ecosystem has been documented in several reports to the International Joint Commission (IJC) (Upper Lakes Reference Group, 1976; PLUARG, 1978; Water Quality Board, 1979). The issue of long-range transport of air pollutants was described by the United States-Canada Research Consultation Group (Environment Canada, 1979). Reports to the GLFC on Great Lakes Ecosystem Rehabilitation (Frances et al., 1979), special symposia papers (Loftus and Regier, 1972; Colby, 1977), and other sources (Beeton, 1965; Smith, 1972; Christie, 1974) describe changes in limnological features, especially fish communities, and in fisheries in the Great Lakes in relation to stresses of overexploitation, invasion of exotic fish species, and pollution.

A historical description of Great Lakes fisheries is vivid testimony to the general depreciation of Great Lakes resources over the last several decades. Degradation has accelerated at an alarming rate since 1950. Valuable species such as Atlantic salmon, lake trout, lake whitefish, sauger, walleye, and blue pike (respectively, *Salmo salar*, *Salvelinus namaycush*, *Coregonus clupeaformis*, *Stizostedion canadense*, *Stizostedion vitreum vitreum*, and *S. v. glaucum*), which comprised 70–80% of early fisheries, now make up less than 30% and, in Lake Ontario, only 5% of the catch (Great Lakes Fishery Commission, 1975). Furthermore, with the exception of Lake Erie, the total catch of fish in each of the Great Lakes has been declining steadily over the last 30 years. Peak production (Canada and U.S.) occurred in the period 1891–1910, averaging 60 million kg valued at 38 million dollars at 1979 prices. The commercial catch recently has averaged only about 45 million kg valued at 25 million dollars at 1979 prices (Talhelm et al., 1979). With full- and part-time employment for 7000 people, including those engaged in processing, commercial fishing is still a significant, albeit struggling, industry in the Great Lakes. In fact, some port towns are primarily dependent on fishing. In spite of some setbacks, recreational fishing continues to grow. Recreational fishing activity totals about 25 million angler days per year. A large annual sum, estimated at 600 million dollars, was put into the economy through sport fisheries.

Stresses on aquatic communities attributable to human activities include toxic metals and organic compounds, limnological changes due to excessive nutrient enrichment, acid rain, mortalities at water intakes by impingement and entrainment, losses of natural littoral habitat, depreciation of streams and bays in many places, pressures from exotic species, and overexploitation.

In the face of effects from these stresses, combined with variability due to natural factors, it becomes extremely difficult to attain tolerable fluctuations in quantity and quality of fish yields. Instability does not favor public and private investment in the resource, and, to add insult to injury, funds available for management of the stocks and fisheries may be diverted to programs aimed at understanding and dealing with stresses imposed by the activites of other water users, often on a too-short-term, fire-fighting basis. I am of the opinion that fisheries agencies cannot avoid dealing with these issues, not only because we must try to understood major stresses, whatever their origin, but also because we have the expertise and the conviction to protect and, if necessary, to rehabilitate aquatic communities. However, it is important to control *how* we shall perform this task (that is, with a rational, scientifically sound, pragmatic approach, which is a central concern of this book).

However, I believe we shall have to rekindle the interest of many fisheries managers in environmental issues. The reasons for the postwar sag in the level of participation by fisheries agencies in processes to deal with environmental issues are varied and not at all obvious. In early years fisheries agencies appeared to be in the forefront of efforts to reduce pollution by sewage, sawdust, mine wastes, and paper mill wastes. To say that they are now overextended in dealing with problems of allocation and control of harvest denies the obvious and increasing importance of habitat protection in Great Lakes fisheries policy. The fundamental reason appears to be that current environmental management philosophies have moved away from origins in conservation-minded resource management, to become preoccupied with regulatory measures as ends, rather than only means. The field is jealously guarded by those with this "regulation viewpoint."

That the environmental protection mandate is unique and an end in itself has been so oversold that even some resource managers appear to believe it, and they rely almost entirely on environmental agencies to protect fisheries resources. There are, for those fisheries managers who try to become involved, the frustrations of basic, often irreconcilable differences in concept between them and environmental quality managers. The former often are accused of being unrealistic in seeking pristine purity, of quibbling over fine points of data, of indecisiveness, and of being negatively critical. The latter are thought to be consumed with producing water quality numbers regardless of their significance and precision, in dealing with symptoms rather than root causes, in searching for nuts-and-bolts recipes rather than holistic strategies. The impasse is maintained to the extent that the artifical distinction is perpetuated between resource management and environmental quality management, and it will continue as long as monopolies within public policy fields are allowed to persist. It appears that one of the most significant jobs for the GLFC (and IJC) is to continue to break down this impasse and develop productive dialogue among resource management, environmental protection, and other agencies (Johnson, 1980).

GLFC's interest in toxic substances and other environmental issues has matured over the last 10 years, and it continues to strengthen. Early reaction, for example, to DDT and mercury in fish was one of skepticism and indignation. Now the Commission's interest is more constructive and comprehensive in several ways. We have discussed with IJC the possibility of combining aspects of surveillance programs to serve our common information needs. We are considering the Commission's role in intervening with or on behalf of fisheries agencies on environment issues, our role in assisting the agencies in tactical planning in relation to habitat, and our role in securing fisheries agency input to IJC and other basin-wide activities.

2. RESEARCH NEEDS

In 1978, spurred on by some of our Lake Committees, we asked the Scientific Advisory Committee (SAC) to determine, "If current research on contaminants was adequate to assess effects of contaminants on Great Lakes fish." The approach taken was by interviews (many by mail based on a set of questions). Of 25 experts interviewed, 17 responded. The synthesized responses were circulated for further comment. In the most general terms, the expert opinion was that there were few field examples of effects of contaminants, mainly because we have not yet learned how to look for good evidence. On the other hand, we were cautioned that, even if we knew exactly how to measure the health of aquatic organisms, populations, and communities, we would be assessing the cumulative impact of a myriad of stresses in retroactive fashion. Also, to focus on toxics that bioaccumulate in the field in fish tissues is to miss many fugitive chemicals that may be equally harmful.

To be more specific and constructive, the following five points attempt to capture research needs most frequently cited by these experts (Great Lakes Fishery Commission, 1979):

1.   Agencies responsible for fisheries management must pursue a strategy of "ecosystem epidemiology" based on clinical fish health and improved methods of fish stock and forage base assessments. The approach to ecosystem health has utility as an early warning system to recognize deteriorating conditions *before* they reach crisis proportions and as a means of monitoring the effectiveness of contaminant regulations. Failure to develop clinical fish health and improved methods of stock assessment will deprive fisheries agencies of the only tools available for measuring the integrated response of the ecosystem to contaminants.

2.   Contaminants research is reactive by nature, responding to issues on a crisis basis. This approach must be balanced with more anticipatory research to identify locations and contaminants of future concern. As part of this anticipatory program, fisheries and research groups must develop or adopt a hazard-assessment scheme, employing accepted protocols and reflecting Great Lakes fishery interests, to assess the potential adverse effects of new chemicals proposed for use in the Great Lakes basin and to establish research priorities for contaminants known to accumulate in Great Lakes biota.

3.   Research is needed to develop protocols for integrating laboratory and field studies and to establish correlations between adverse effects in the laboratory and adverse effects in the field. Researchers agreed that neither laboratory nor field studies alone are sufficient to address all phases of the contaminants issue. Yet, very few studies have attempted to confirm laboratory observations

in the field. Respondents suggested laboratory studies could be linked to field programs at the cellular (biochemical-histological) level and the relevance of the response extrapolated from the individual to the population to the community. Laboratory studies should establish relevance by simulating, wherever possible, realistic environmental conditions and should attempt to correlate observed adverse effects with exposure, tissue residues and biochemical and histological perturbations.

4. More research is needed to identify chemical fate and construct mass balances of contaminants in the Great Lakes ecosystem. Data on source, quantity, degradation, translocation, transformation, and removal of contaminants is necessary to estimate environmental concentrations and develop estimates of total contaminant loadings to the Great Lakes.

5. More research is needed describing the effects of contaminants on Great Lakes invertebrates, particularly zooplankton which are more diverse, easier to test and in many cases more sensitive than fish to environmental contaminants. Research describing the effects of contaminants on fish could be made more relevant to the Great Lakes situation by testing fish indigenous to the Great Lakes as opposed to surrogate species such as the flagfish or zebrafish.

We are led, therefore, toward a double-barreled strategy, that is, toward two lines of defence. Initially we must be involved in protocols to prevent and/or control the use of hazardous substances (or processes, or facilities). There are strategies to make best use of dollars and expertise by careful selection of chemicals, tests, test animals, and so on. But we must not lose sight of the fact that effectiveness of all of our policies and programs to protect the environment are assessed, in time and in the aggregate, only by assessing the health and well-being of organisms, populations, and communities in the field. This implicitly would evaluate our own hazard assessment protocols. There is a need for substrategies that deal with how to perform this task, and further substrategies that deal with research needs, institutional arrangements, and public information needs to ensure sound and useful results. However, it seems to me that, if GLFC and its associated fisheries agencies want to have an active, credible, and responsible role, they will have to develop an "ecosystem epidemiology" and some very compelling arguments why "sick ecosystems" mean trouble for people.

What does the GLFC expect to happen as a result of this book? Hopefully, some of the following, considering the mix of interests and experience that are involved here. We should try to strengthen our understanding of effects and especially interactions among stresses, and consider the ways stresses may be integrated and effects measured at the level of the organism, the population, and the community, all in the context of success of species and communities of species. We should help clarify purposes and linkages between field assessment and laboratory assays and experiments (for example, to know residues of a toxic material in lab measurement of response would aid in interpreting residues found in fish in the field). It is important to stimulate awareness of the kinds of ancillary information that are important in understanding environmental stresses and to explore tests and applications of research findings that would be most useful to this end. We should encourage and help agencies in broadening their field assessment programs, by stimulating interest in other yardsticks, including some clinical measures. We have an opportunity to assist GLFC and fisheries agencies in development of fish habitat and environmental protection strategies, and to encourage them toward a more active role in using information from assessment programs in support of habitat protection and environmental programs.

ACKNOWLEDGMENTS

The Great Lakes Fishery Commission gratefully acknowledges the role of Niles Kevern (Michigan State University) and Vic Cairns (Department of Fisheries and Oceans) in bringing together expert advice to GLFC on the toxic contaminant issue. We thank the following contributors to the 1979 task of our Scientific Advisory Committee: John Allin and Doug Dodge (Ontario Ministry of Natural Resources), John Hesse (Michigan Department of Natural Resources), Edward Horn (New York State Department of Environmental Conservation), Gary Chapman, Wayland Swain, and Don Mount (Environmental Protection Agency), Blake Grant, Joe Kutkuhn, Fred Meyer, Richard Schoettger, Jim Seelye, and Wayne Willford (U.S. Fish and Wildlife Service), Peter Hodson, Jack Klaverkamp, Denis Rivard, Mike Waldichuk, and Vladimir Zitko (Canada Department of Fisheries and Oceans), John Cairns (Virginia Polytechnic Institute), Howard Johnson (Michigan State University), and Dwight Webster (Cornell University).

REFERENCES

Beeton, A. M. (1965). Eutrophication of the St. Lawrence Great Lakes. *Limnol. Oceanogr.* **10**, 240–254.

Christie, W. J. (1974). Changes in the fish species composition of the Great Lakes. *J. Fish. Res. Board Can.* **31**, 827–854.

Colby, P. J. (1977). Percid International Symposium (PERCIS). *J. Fish. Res. Board Can.* **34**, 1445–1999.

Environment Canada (1979). The LRTAP problem in North America. Rep. U.S. Canada Research Consultation Group on the long-range transport of air pollutants, Ottawa, Ont., 131 pp.

Frances, G. R., Magnuson, J. J., Regier, H. A., and Talhelm, D. R. (1979). Rehabilitating Great Lakes ecosystems. Rep. to the Great Lakes Fishery Commission, Ann Arbor, MI, 140 pp.

Great Lakes Fishery Commission (1975). Environmental quality and fishery resources of the Great Lakes. A brief to the Int. Joint Commission, Ann Arbor, MI, September, 1975, 14 pp.

Great Lakes Fishery Commission (1979). Contaminant research needs in the Great Lakes. Scientific Advisory Committee rep., June 28, 1979, 55 pp.

Johnson, M. G. (1980). Great Lakes environmental protection policies from a fisheries perspective. *Can. J. Fish. Aquat. Sci.* **37**, 1196–1204.

Kelso, J. R. M. and Milburn, G. S. (1979). Entrainment and impingement of fish by power plants in the Great Lakes which use the once-through cooling process. *J. Great Lakes Res.* **5**, 182–194.

Loftus, K. H. and Regier, H. A. (1972). Proceedings 1971 symposium on salmonid communities in oligotrophic lakes. *J. Fish. Res. Board Can.* **29**, 611–986.

PLUARG (Pollution from Land Use Activities Reference Group) (1977). Land use and land use practices in the Great Lakes basin. Int. Joint Commission, Windsor, Ont., 13 pp.

PLUARG (Pollution from Land Use Activities Reference Group) (1978). Environmental management strategy for the Great Lakes system. Int. Joint Commission, Windsor, Ont., 113 pp.

Smith, S. H. (1972). The future of salmonid communities in the Laurentian Great Lakes. *J. Fish. Res. Board Can.* **29**, 951–957.

Talhelm, D. R., Bishop, R. C., Cox, I. W., Smith, N. W., Steinnes, D. N., and Tuomi, A. L. W. (1979). Current estimates of Great Lakes fisheries values. Great Lakes Fishery Commission, Ann Arbor, MI, 10 pp.

Upper Lakes Reference Group (1976). The waters of Lake Huron and Lake Superior. Vol. 1. Summary and Recommendations. Int. Joint Commission, Windsor, Ont., 236 pp.

Water Quality Board (1978). Great Lakes Water Quality 1977 Annual Report. Int. Joint Commission, Windsor, Ont., 89 pp.

Water Quality Board (1979). Great Lakes Water Quality 1978 Annual Report. Int. Joint Commission, Windsor, Ont., 110 pp.

2

EPIDEMIOLOGY OF DISEASES IN WILD FISH POPULATIONS

George W. Klontz

Department of Fish and Wildlife Resources
University of Idaho
Moscow, Idaho

1. INTRODUCTION

Epidemiology is that branch of medicine that describes the occurrence, distribution, and types of diseases in populations of animals (Austin and Werner, 1974). The term has traditionally been applied to the study of disease episodes or epidemics in human populations. Its counterpart, epizootiology, has been applied to episodes of diseases in confined and free-living populations of mammals, birds, reptiles, amphibians, and fish. However, during the past two–three decades, the term epizootiology virtually has been dropped from common usage in favor of the term epidemiology, thus giving rise to the more inclusive definition of epidemiology as "the medical aspect of ecology" (Schwabe, 1969).

The art and science of epidemiology have been practiced for more than 2000 years. The process reportedly began with an awareness that disease episodes were associated with certain causal factors. For example, years ago, people living near a river had a higher rate of disease than did those living on the plains. It became apparent to the thinkers of those times that there were causal relationships between the incidence of disease and living near a river (Hippocrates, 377 B.C.).

Over the years, since those early cognitions, the application of the basic concepts of epidemiology have enabled the human medical profession to eradicate smallpox world-wide and measles on a locale-by-locale basis. Likewise, the veterinary medical profession has been able to locally eradicate brucellosis from cattle and rabies from canids.

2. CONCEPTS OF EPIDEMIOLOGY

An epidemiological investigation of a disease episode begins with establishing that a problem does exist. The next step is to define the problem and to identify the population at risk. This could be a small discrete group of individuals in a defined geographical area (a disease episode in this case would be termed an epidemic) or a very large area (a disease episode in this case could be termed a pandemic). The population at risk is the total number of individuals that the defined events affected or could have affected.

By comparing the number and nature of affected individuals within the population at risk, several *rates* can be established. The *incidence rate* is a measurement of the number of new cases of a disease or event (birth, marriage, etc.) occurring within a specific period of time (week, month, or year). The *prevalence rate* is a measurement of all cases of a disease or other event prevalent during a specified period of time. The *mortality rate* is a measurement of the deaths due to a specific disease occurring within the population at risk during a specified period of time. The *morbidity rate* is a measurement of the occurrence of the specific clinical manifestations of disease within the defined population at risk during a specified period of time (Russell, 1977).

Then, by comparing the derived rates, decisions may be made to establish and to characterize the epidemiological factors involved in the episode in question. The epidemiological factors serve to further characterize the events according to individuals, population, place, and time. This aspect of the epidemiological process is virtually limitless. The investigator must take special precautions to preclude identifying only the obvious events and factors. The hidden factors often provide the most productive keys to defining the problem completely.

There are some pitfalls to be considered with using epidemiological rates. The most important is in the reporting of the episode. In this case, errors in diagnosis and in establishing the population at risk can lead to the wrong conclusions. Another problem is the complexity of epidemiological nomenclature. Many investigators use terms interchangeably, a practice that hinders valid associations among events. The most common shortcoming is the failure to provide complete background information. For example, the conclusion was reached from survey data that three times as many people own Datsun automobiles as own Ford automobiles. The questions to be asked are: "How many Ford automobiles were available as compared to Datsun automobiles?" "When were these data collected?" "In what geographic area were these automobile owners living?" One could infer from the conclusion that the event occurred in the recent past in the United States, but what if it occurred in Japan in the recent past (Huff, 1954)?

The epidemiological investigation looks for patterns and trends in rates of change within the identified population. This is the analytical phase of the process. Rates are compared and identified factors tested for association, one with another.

Epidemiological patterns, when plotted, enable the investigator to draw cogent inferences as to the nature of the disease episode or event. A bell-shaped (epidemiform) curve of an infectious disease episode indicates a direct transmission of the agent with the shape of the curve being influenced by (1) the incubation period; (2) the period of communicability; (3) the exposure potential; and (4) the surveillance efficiency. A plateau-shaped curve of an infectious disease episode indicates a common vehicle transmission over a span of time. The plateau may be saw-toothed, in which case the exposure would be intermittent. Spiked curves indicate a point-source of the causal agent—either infectious or noninfectious. Combinations of curve types may be present depending upon the nature of and exposure to the causal agents (Russell, 1977).

Evaluation of the data should culminate in the formation of a hypothesis as to the causes of the situation being investigated. The hypothesis is then tested against the collected data or under controlled laboratory conditions.

The foregoing characteristics of the epidemiological process are termed *descriptive epidemiology* or the *retrospective approach* to an event or a disease episode. In this case, the investigator is looking backwards in time in an attempt to establish the causal factors. This approach is relatively quick and inexpensive. It cannot be used as a predictive tool to determine if an event is likely to occur within a given population, however. This latter is the objective of *experimental epidemiology* in which the *prospective approach* to an event or disease episode is taken. In this case, the population at risk is clearly defined—much more so than in the retrospective approach—and followed through a period of time for the occurrence of a pre-established condition. This approach often involves a similar population in which the defined event is not likely to occur.

A third investigational process is the *cross-sectional* (survey) *approach*, which is used to establish the prevalence of one or more factors and their interrelationships. This approach affords the investigator with a quick and inexpensive method to formulate and test hypotheses.

In summary, epidemiology is a way of investigating the occurrences of events within a population during a period of time. The events may be disease processes, births, growth, or any other demonstrable phenomenon. The classic "check-off" list is:

1. Establish that a problem exists.
2. Identify and confirm the nature of the problem and the events.
3. Collect data.
4. Analyze the data, compare rates, and establish trends and patterns.
5. Formulate a hypothesis (or hypotheses) as to the nature of the cause of the problem.
6. Test the hypothesis against the collected data or through controlled experimentation and subsequent prospective studies.
7. Prepare a report detailing the nature of the event, the population at risk, the associated factors, the hypothesis, the results of testing the hypothesis, and conclusions.

3. NATURE OF DISEASES IN FISH POPULATIONS

3.1. Clinical Signs

Diseases, both infectious and noninfectious, are manifested in free-living fish in many ways, depending upon the nature of the causal factors. In a commercial or sport fishery, a disease episode can be detected from the following events:

1. A decline in yield.
2. A change in population structure from the standpoints of male:female ratio, age-class, and size structure.
3. Obvious lesions such as skeletal malformations, fin erosion, tumors, exophthalmia, abscesses, ulcerations, and so forth.
4. A change in catchability.

Further investigation by resource managers may detect other signs of a disease episode in progress. The following events could be noted:

1. Teratological changes.
2. Behavioral changes such as "topping" and alterations in vertical distribution and light sensitivity.
3. A reduced growth rate of individual fish.
4. A change in recruitment in or out of the population in question.

Reports documenting the foregoing clinical signs are numerous. Excellent review articles may be obtained in the texts by Plehn (1924), Almacher (1970), Mawdesley-Thomas (1972), Reichenbach-Klinke (1973), Elkhan and Reichenbach-Klinke (1974), Ribelin and Migaki (1975), Snieszko and Axelrod (1976), Roberts (1978), and Anderson et al. (1979).

Recognition of one or more of the foregoing events begin the first phase of descriptive epidemiology; namely, identification of the problem.

3.2.   Causal Factors

Traditionally, the etiological agents (causal factors) of disease in fish are classified as either infectious (communicable) or noninfectious. The infectious agents of disease are further categorized as bacterial, viral, mycotic, protozoan, metazoan, crustacean, and molluscan. The noninfectious agents of disease are further categorized as biological, chemical, and physical (Reichenbach-Klinke and Elkan, 1965).

3.3.   Factors Influencing Disease

With respect to the infectious diseases, the process begins by establishing the relationship (infection) between an infectious agent and a susceptible host. This relationship may be symbiotic (both host and agent benefit), commensal (either the host or agent benefit without compromising the other), or parasitic (the agent is benefited at the expense of the host). The nonparasitic states, if prolonged, constitute a continuing source of infection (the reservoir) for other fish in the population. However, the nonparasitic states can become one of parasitism if some event, either in the host or in the environment, causes a change in the balance between the agent and the host (Wedemeyer et al., 1976).

During the establishment of the parasitic relationship following infection or the change of a nonparasitic relationship to one of parasitism, there is a period of time—usually a matter of days—in which the agent and the host are struggling for supremacy. This is the incubation period, the period preceding the manifestation of clinical signs of disease. It is during this period that the host is employing its nonspecific defense mechanisms of inflammation and innate immunity. If this is not the first time the host has been exposed to the antigenic nature of the agent, the incubation period is also the time when the host employs its specific antibody defense against the agent (Anderson, 1974).

The length of the incubation period depends upon a number of agent and host factors. Agent factors include infectivity, invasiveness, and virulence. Host factors include physiological status, age, and specific antibody (Klontz, 1979).

If the defensive mechanisms of the host prevail, no disease state ensues. However, if, as in most cases of infectious fish disease, the disease-producing factors of the agent prevail, the host becomes clinically ill (moribund) and usually dies. Transmission of the agent during and after the clinical episode occurs either vertically or horizontally.

Vertical transmission of the infectious agent (pathogen) is defined as passage of the agent from infected adults to their offspring via the egg and/or the sperm. This method of agent transmission is known to occur with at least two systemic bacterial disease agents (*Mycobacterium fortuitum* and *Renibacterium salmoninarum*) and several salmonid viruses (infectious pancreatic necrosis virus, infectious hematopoietic necrosis virus, and Egtvedt virus). There may be many other agents transmitted in this fashion (Anderson et al., 1983).

Horizontal transmission of the agent is defined as the direct or indirect passage of the organism among members of a population. Direct horizontal transmission of the agent among fish is very common in episodes of bacterial, viral, mycotic, and protozoan disease. The major factors influencing direct horizontal transmission are contact conditions (the proximity of one fish to another, which is a function of population density), water temperature and chemical nature, numbers of infectious units, and physiological condition of the susceptible hosts (Anderson et al., 1983).

Indirect horizontal transmission of infectious agents is affected by vehicles and vectors. Vehicles of agent transmission are inanimate objects in or on which infectious agents are transferred from one site to another. An example of this type of transmission would be live fish handling gear that had become contaminated from infected fish and subsequently used to handle healthy but susceptible fish. Vectors are animate entities that transmit the agent either directly (mechanical vectors) or indirectly by functioning to complete the life cycle of the agent (biological vectors). An example of a mechanical vector would be the introduction of contaminated but not infected fish into a susceptible population. An example of a biological vector would be a snail intermediate host for a metazoan that uses the fish as a definitive host.

In most fish populations, there are asymptomatic carriers of infectious agents. These individuals serve as reservoirs of infection. These are the individuals that fish health management specialists attempt to identify both qualitatively and quantitatively within a population (McDaniel, 1979; Hennessen, 1981).

3.4. Diagnostic Methods

The diagnostic process begins with a postmortem examination (necropsy) of clinically affected fish as well as clinically unaffected fish from the subject population. This examination should be a complete, systematic documentation of any and all gross pathological changes (Klontz, 1979).

Many pathological changes, singularly or in concert, enable the prosector to make educated guesses as to the causal agent or agents involved in the disease process. For example, reddened fin bases, serosanguineous peritoneal fluid, and an enlarged spleen most often are associated with an acute to subacute systemic bacterial infection. Whitish, raised, focal areas of necrosis (abscesses) are suggestive of a chronic systemic bacterial infection or a chronic systemic metazoan infection. Pale gills and a pale kidney, indicating an anemia, are suggestive of either chronic blood loss or a failure in the hematopoietic process due, among other things, to exogenous toxins or nutritional imbalances. Bilateral exophthalmia is suggestive of renal dysfunction due to exogenous toxins, systemic bacterial infections, or other damaging agents (Roberts, 1978).

Examination of the gross lesions could be followed by a histopathological examination of tissues from the affected organs. This process further defines the nature of the pathological changes, which then enables the examiner to make more valid associations between effects and possible causes.

In cases where the suspected causative agents are bacterial or viral, several methods are used to confirm or deny their presence. The most common method is staining of body material, followed by in vitro cultivation on enrichment media or selective media in the case of bacteria or by in vivo cultivation on tissue culture in the case of viruses. The cultivation methods require specialized equipment and facilities to accomplish.

In cases where exogenous intoxicants are suspected, fish tissues and body fluids must be specially handled and submitted to a facility having highly sophisticated instrumentation. In addition to the fish samples, samples of water, benthos, and so forth should be submitted. The process of identifying precisely the involvement of a waterborne toxicant is much akin to searching for the proverbial needle in the haystack, as the list of these compounds that can and do affect fish populations is increasing exponentially. The "hit or miss" aspects of the situation are often reduced by the interpretation of the histopathological changes present, especially in the case of neoplasia (tumors). The histopathological changes in fish having been exposed to groups of chemicals are, in many cases, known from experimental studies; thus cause and effect relationships may be deduced (Mawdesley-Thomas, 1972).

Coming more and more into use as diagnostic aids are the physiological and hematological parameters. Physiological parameters that show promise are the serum proteins (albumins and globulins), transaminases, blood glucose, blood cortisol, blood ammonia, and specific antibodies (Hennessen, 1981). Hematological parameters, such as erythrocyte and leukocyte counts, are so sensitive to a variety of stimuli that their utility in other than severe conditions is limited. However, the hematological indices, such as the mean corpuscular volume and the mean corpuscular hemoglobin concentration, are potentially valid methods as changes in them reflect long-term causal factors. Also, the morphological and numerical characteristics of blood cells in the peripheral vascular system and in the hematopoietic centers of the spleen and kidney are often of valuable assistance in assessing cause and effect relationships (Klontz, 1979).

In summary, the diagnostic process is, in importance, second only to the problem identification. As fish health managers become more skilled through training and experience, the level of diagnostic capability will increase. There really is no reason why the majority of the diagnostic techniques currently employed in human and veterinary medicine could not be applied with appropriate modification to fish medicine.

4. APPLICATION OF EPIDEMIOLOGICAL CONCEPTS

Perhaps the best way to illustrate the application of the retrospective and prospective approaches to disease episodes in wild fish is to use an actual case. The case selected was described by A. H. McVicar (1981) for the International Council for the Exploration of the Sea.

The problem, infection of North Sea plaice with *Ichthyophonus hoferi*, was identified through examination of commercial vessel landings of plaice at Aberdeen, Scotland, between May 1976 and March 1978. Examination of 405 landings revealed that the prevalence of infection was 0.45% of all plaice landed. One identified bias in the prevalence assessment was the usual practice of the commercial fishermen rejecting plaice of poor quality while sorting the catches at sea. Subsequent research vessel fishing upgraded the prevalence of *Ichthyophonus* infection in the targeted stocks to 18.7% with one catch of 16 fish having 7 with clinical lesions of infection.

On the basis of the preliminary evidence, a more intensive study of a specified area over a longer period of time (mid-1978 through early 1981) was conducted. The results indicated that, although infection rates differed among individual samples, and there were no seasonal trends, there was a tendency toward increased prevalence of the infection in the target area.

One very significant event described by McVicar was that no plaice smaller than 26.0 cm body length were infected. Also, there was firm evidence that the prevalence of infection increased with increasing host length (McVicar, 1980).

During the laboratory phase of the investigation, experimentally infected plaice were monitored microbiologically, histopathologically, and immunologically. From the acquired data, McVicar was led to conclude that within 36 days following infection there was a detectable immune response and that all infected fish died within 62 days of infection regardless of the immune status. From these conclusions, he was able to calculate an annual mortality rate within an affected plaice population from the prevalence rate of infection. The estimated annual mortality rate assumed that there was a constant daily infection rate and that all fish responded immunologically.

The foregoing description of the *Ichthyophonus* episodes in commercially harvestable plaice off the north coast of Scotland is an excellent example of the descriptive phase of an epidemiological study. The problem was identified in the quality of the commercial catches. The prevalence rates and histopathological characteristics of specific infection were determined in several populations. And a report was written and disseminated.

Prospective studies of naturally occurring *Ichthyophonus* infection in North Sea plaice are currently in progress. The studies are concentrating on extending the analyses to other populations in areas adjacent to the identified population at risk. In addition, catch composition data are being examined for indications of infection-induced changes.

5. SUMMARY

Episodes of infectious and noninfectious disease in populations of fish are an inescapable factor in fishery resource management. Population structures ebb and flow under their influence (Anderson et al., 1979). It will be only through increased awareness of the causal factors and their interactions among themselves and with susceptible hosts that resource managers may set attainable goals for population health.

One might ask, "What is a healthy population of fish? Is it a population that is free of disease?" The answer to the first question should come from the management sector because the response should be framed around population dynamics, which would include age composition, size composition, stock recruitment, harvest, and stability. The answer to the second question should come from the biomedical sector. To have a free-living population of animals—fish, mammals, birds, or whatever—free of disease is a nice dream but certainly an unrealistic goal at this point in time. Far too little is known about disease per se. In the case of fish, for them to be free of disease is questionable because of their innate irrevocable orientation to their environment. Thus if the environment changes physically, chemically, or biologically, changes are induced in the fish. These changes are detectable in many cases and often lead to the disease state if the fish is unable to adapt to the change.

Application of the principles and techniques of the art and science of epidemiology enable the biomedical researcher to detect these changes, evaluate their effects under controlled conditions, and then, in concert with resource managers, test the laboratory findings under field conditions. The major limiting factor in this entire process is unimaginative thinking. Breakthroughs in problem solving occur out of the interactions of dedicated, imaginative groups comprised of many related disciplines. The fishery resource management discipline has for years tried to ignore the scientific potential of nonfishery scientists from the fields of physiology, pharmacology, biochemistry, and medicine. This era is coming to a close as evidenced by the discipline composition of the group involved in preparing this report.

REFERENCES

Amlacher, E. (1970). *Textbook of fish diseases*. T.F.H. Publications, Neptune, NJ.

Anderson, D. P. (1974). Fish immunology. In: S. F. Snieszko and H. R. Axelrod (Eds.), *Diseases of fishes: Book 4*. T.F.H. Publications, Neptune, NJ.

Anderson, D. P., Dorson, M. M., and Dubourget, P. (1983). Antigens of fish pathogens: Development and production of vaccines and serodiagnostics. Association Corporative des Etudiants en Médecin de Lyon, Lyon, France.

Anderson, R. M., Turner, B. D., and Taylor, L. R. (Eds.) (1979). *Population dynamics*. Blackwell Scientific Publications, Oxford, England.

Austin, D. F. and Werner, S. B. (1974). *Epidemiology for the health sciences*. Charles C. Thomas, Springfield, IL, p. 4.

Elkan, E. and Reichenbach-Klinke, H. H. (edited by M. Landolt) (1974). *Color atlas of the diseases of fishes, amphibians, and reptiles*. T.F.H. Publications, Neptune, NJ.

Hennessen, W. (Acting Ed.) (1981). *Developments in biological standardization, Vol. 49: Fish biologics: Serodiagnostics and Vaccines*. S. Karger, Basel, Switzerland.

Hippocrates (377 B.C.). On airs, waters, and places. In: *Hippocratic writings* (translated by F. Adams), Great Books of the Western World, Vol 10. Univ. of Chicago, Chicago, IL, pp. 9–19.

Huff, D. (1954). *How to lie with statistics*. W. W. Norton & Co., New York, pp. 53–60.

Klontz, G. W. (1979). *Fish health management: Vol. II*. Office of Continuing Education, Univ. of Idaho, Moscow, ID.

Mawdesley-Thomas, L. E. (Ed.) (1972). *Diseases of fish*. Academic Press, New York.

McDaniel, D. (1979). Procedures for the detection and identification of certain fish pathogens. Am. Fish. Soc., Fish Health Section, Washington, DC.

McVicar, A. H. (1980). The effects of *Ichthyophonus* infection in haddock *Melanogrammus aeglefinus* and plaice *Pleuronectes platessa* in Scottish waters. ICES special meeting on diseases of commercially important marine fish and shellfish, Copenhagen, 1980/12, 12 pp.

McVicar, A. H. (1981). An assessment of *Ichthyophonus* disease as a component of natural mortality in plaice populations in Scottish waters. Dimersal Fish Committee, Int. Council for the Exploration of the Sea, 8 pp.

Plehn, M. (1924). *Manual of fish diseases.* Erwin Nagele, Ltd., Stuttgart, Germany.

Reichenbach-Klinke, H. H. (1973). *Fish pathology.* T.F.H. Publications, Neptune, NJ.

Reichenbach-Klinke, H. H. and Elkan, E. (1965). *The principal diseases of lower vertebrates: Diseases of fishes.* T.F.H. Publications, Neptune, NJ.

Ribeline, W. E. and Migaki, G. (Eds.). (1975). *The pathology of fishes.* Univ. of Wisconsin Press, Madison, WI.

Roberts, R. J. (1978). *Fish pathology.* Bailliere Tindall, London.

Russell, L. H. (1977). *A syllabus: The principles of epidemiology.* Texas A&M Univ., College Station, TX, pp. 73–83.

Schwabe, C. W. (1969). *Veterinary medicine and human health,* 2nd ed. Williams and Wilkins Co., Baltimore, MD, p. 6.

Snieszko, S. F. and Axelrod, H. R. (Eds.) (1976). *Diseases of fishes. Books 1–6.* T.F.H. Publications, Neptune, NJ.

Wedemeyer, G. A., Meyer, F. P., and Smith, L. (1976). Environmental stress and fish diseases. In: S. F. Snieszko and H. R. Axelrod (Eds.), *Diseases of fishes: Book 5.* T.F.H. Publications, Neptune, NJ.

3

USE OF HISTOPATHOLOGY IN AQUATIC TOXICOLOGY: A CRITIQUE

Rodney D. Johnson and Harold L. Bergman

Fish Physiology and Toxicology Laboratory
Department of Zoology and Physiology
University of Wyoming
Laramie, Wyoming

1. INTRODUCTION

Histopathology has been accepted as an important tool in biomedical pathology for many years. Ever since Virchow published his historic treatise *Cellular Pathology*, histopathology—or more broadly, structural pathology—has been a cornerstone in the larger field of biomedical pathology (Rather, 1971). In addition to this important role in biomedical pathology, histopathology also is used extensively in biomedical toxicology. In fact, histopathology is accepted as one of the most important determinants for establishing acceptable concentrations of environmental contaminants, drugs, food additives, cosmetic additives, and other chemicals that contact people.

In contrast, the value of histopathology in aquatic toxicology has yet to be fully established. At least four major reasons help explain this difference between biomedical and aquatic toxicology. First, the focus of biomedical applications of histopathology is narrow, inasmuch as humans and a few human surrogates are the main species considered. In contrast, laboratory and field histopathology studies of aquatic biota encompass several phyla and numerous species. Second, the cumulative effort devoted to histopathology in biomedical disciplines has been immense compared to the total effort expended thus far in applying histopathology to problems in aquatic biology. In particular, there are a large number of trained, certified biomedical pathologists, while there are very few aquatic biologists properly trained in histotechnique and pathobiology. Third, because of the narrower focus and substantially greater effort, the histological baseline is much greater in biomedical toxicology than in the rather young field of aquatic toxicology. Because of the wealth of baseline information available, biomedical histopathologists are better able to distinguish between contaminant-caused lesions and normal structural variation in cells and tissues. Fourth, and possibly most important, there is a substantial difference between the significance accorded a tissue lesion observed in humans and a lesion observed in fish. Any lesion observed in human or surrogate species is cause for concern and such lesions often provide sufficient basis for limiting human exposure to a chemical. But for fish and other aquatic organisms "significant adverse effect" is acknowledged only after demonstration of adverse effects at the population level,

Dr. Johnson's present address: Physiology and Biophysics Department, University of Alabama in Birmingham, Birmingham, Alabama 35294.

such as reduced growth or reproduction. And, in fish, histological lesions cannot necessarily be used to infer a population-level effect because few studies have linked effects at the tissue or cellular level to effects at the population level.

These differences between the uses of histopathology in biomedical and aquatic toxicology would not be important except that histopathologists studying aquatic organisms have usually designed and executed research as if they were biomedical histopathologists. And in our opinion biomedical approaches and techniques usually are not appropriate for the objectives of aquatic toxicology, which are to estimate "safe" concentrations and to detect adverse effects of toxicants in the environment.

To effectively address the objectives of aquatic toxicology, we believe that histopathology studies will require changes in approach and changes in technique. In this paper, we evaluate current approaches and techniques for the use of histopathology in toxicology, and we suggest alternative approaches and techniques that would increase the utility of histopathology in studies of contaminant effects on fish and other aquatic animals.

2. APPLICATION TO LABORATORY AND FIELD STUDIES

In the past 10–15 years a number of reports have been published that use histopathology to help determine the effects of environmental contaminants on fish and other aquatic animals. But the approaches used, whether for laboratory or field studies, frequently are limited in scope or flawed in design. Consequently, these studies often lead to results that can only be applied narrowly to actual contaminant problems in the field or, sometimes, they lead to conclusions that are clearly invalid.

Appropriate laboratory and field approaches for the use of histopathology in environmental contaminant studies should be interrelated. For instance, expected contaminant concentrations in the field should help define exposure concentrations to be used in the laboratory, and cellular or tissue lesions observed in the laboratory should help define the parameters used in field monitoring studies. But for clarity, we discuss appropriate laboratory and field approaches separately.

2.1. Laboratory Approaches

Histopathology techniques can be used in laboratory contaminant studies to pursue at least two interrelated problems in aquatic toxicology, identification of toxic mechanisms and prediction of "safe" contaminant concentrations. The primary objective of studies designed to identify mechanisms of toxicity is to describe the mode of action of a toxicant on an organism by identifying the structures (e.g., target organs, cells, or organelles) and the functions (e.g., respiration, reproduction, ion transport) that the toxicant affects directly. The primary objective of studies designed to predict "safe" contaminant concentrations is to determine in the laboratory, using measurements taken from individual organisms (e.g., growth, fecundity, swimming stamina, histological lesions), the concentrations of toxicants that will not adversely affect populations or communities in the environment.

The ways in which mechanistic and predictive toxicology are, or should be, interrelated in aquatic toxicology were explained by Warren (1971), using the ideas of Bartholomew (1964) about the relationships among different levels of biological organization. Bartholomew observed that, in the sequence of biological organization from subcellular and cellular through populations and communities, each level of organization finds its explanation of mechanism in the levels below and its significance in the levels above. Warren then pointed out that, "From a strictly biological point of view, as well as a practical point of view, it is the population and not the individual organism that is important." So then, histological techniques, which are applied at the suborganismal levels of organization, can be used directly to study toxic mechanisms, but results must be correlated with responses at higher levels of biological organization to ascertain the ecological significance of any observed structural lesions.

2.1.1. *Contaminant Concentration*

The most common errors in approach that we note in published laboratory histopathological studies, whether the studies are directed at mechanistic or predictive toxicology problems, or both, involve inappropriate contaminant exposure protocols or inadequate correlation with other effects. Contaminant exposure protocols often use very high concentrations and short exposure durations. These kinds of studies are appropriate for determining the mechanisms of acute response or for predicting the specific structural lesions expected after brief exposure to high contaminant concentrations as might occur following a chemical spill. But investigators sometimes make inferences about chronic toxicity responses from these acute exposures, which is clearly in error (Klaassen and Doull, 1980). The physiological mechanism of effect and the structural lesions observed may be quite different in short-term, high-concentration exposures than in long-term, low-concentration exposures, which are more typical of most environmental contamination problems. In a critical review of structural pathology studies in marine organisms exposed to petroleum, Malins (1982) strongly criticized laboratory histopathological studies because they often used exposure concentrations that were several orders of magnitude higher than expected in the environment.

We suspect that investigators' use of unrealistically high exposure concentrations may often stem from their desire to observe at least some structural response to the contaminant. But the histological techniques used are often too insensitive to allow detection of lesions in organisms exposed chronically to contaminant concentrations more likely to occur in the environment. So rather than evaluating the tissue response using more sensitive methods, the exposure concentrations are increased so that tissue damage is extensive enough to be observed with insensitive histological technique.

2.1.2. *Interpretation*

The second common error in experimental approach, inadequate correlation of structural pathologies with other effects, also is found in both mechanistic and predictive toxicology studies. In mechanistic studies, investigators often report specific structural lesions after toxicant exposure but do not include the functional consequences of the exposure. Hence, structural responses to toxicants cannot be correlated with physiological mechanisms of toxicity. Woodward et al. (1983) demonstrated an experimental approach that correlated gill lesions to swimming stamina in trout that were chronically exposed to crude oil. Their results clearly show that fish with gill hyperplasia also had less swimming stamina, thus supporting the hypothesis that at least one of the chronic modes of toxic action of crude oil in trout may be impaired gill function, perhaps respiration.

As in mechanistic toxicity studies, predictive studies to determine "safe" concentrations of contaminants should attempt to correlate observed structural lesions with other measured contaminant effects. Whereas biochemical and physiological function must be correlated with observed lesions to establish mechanism, population-level responses must be correlated with structural lesions to determine ecological significance in predictive toxicology studies. All too often, however, the ecological significance of reported structural lesions cannot be ascertained because other contaminant effects accorded greater ecological importance are not even measured. Malins (1982) specifically criticizes most published histopathological studies on the effects of petroleum on aquatic biota because the studies are of little value in determining the "significance of the observed morphological changes." As Malins also points out, identifying histological lesions does advance knowledge about the effects of contaminants. But the value of observations about structural lesions would be substantially increased if only the ecological significance of these effects were established.

2.1.3. Predictive Value

Obviously, ecologically significant effects of contaminants such as those at the population level are difficult to measure in the laboratory. Growth and various measures of reproductive success in full life-cycle toxicity tests with fish (e.g., McKim and Benoit, 1971), as well as egg hatchability, fry survival, and fry growth in early life-stage fish tests (McKim, 1977), are commonly accepted as reliable predictors of contaminant concentrations that would be safe for populations of fish in the field (Lloyd, 1980). However, full life-cycle and early life-stage toxicity studies with fish are costly and require long exposures. In our opinion the potential value of histological measurements in predictive toxicology would be in predicting such parameters as long-term growth or reproduction responses based on structural lesions observed early in the exposure period. It might be possible then to shorten the time and minimize the costs required to predict safe concentrations of environmental contaminants. But this potential value will first have to be established through a series of correlated studies that concurrently measure both structural lesions and the accepted measures that establish ecological significance.

Woodward et al. (1983), in the same study referred to previously about the effects of crude oil on trout, correlated growth with the histopathological response in gills and liver. They found that growth and gill structural pathology were equally sensitive to the toxicant exposure. Johnson (1983) correlated a series of gill structural lesions to traditional early life-stage effect measures in a study using rainbow trout.

He found that several quantified parameters, white blood cell infiltration, chloride cell and mucous cell proliferation, mitotic index and respiratory diffusion distance, were all more sensitive to toxicant exposure than either growth or survival. If further studies confirm that histological methods can predict "safe" contaminant concentrations reliably and cheaply, then predictive studies might be performed using histological methods alone.

2.2. Field Approaches

In our view, the potential uses of histological techniques in field contaminant investigations fall into two interrelated and somewhat overlapping categories, diagnosis and monitoring. In a way, diagnostic and monitoring studies in the field are analogous to the mechanistic and predictive toxicity studies discussed above under laboratory approaches.

2.2.1. Diagnosis

Ideally, the main objective of what we call diagnostic toxicology studies would be to link contaminant levels in the environment with significant adverse population-level effects through identification of contaminant-specific structural and functional lesions in organisms from the affected populations in the field. To meet this objective it would be necessary to have a known or suspected population effect potentially attributable to an environmental contaminant(s). Also it would be necessary to have linked specific structural or functional lesions with exposure to or accumulation of that contaminant(s) in the laboratory. Thus, diagnostic studies would be pursued "after the fact" to identify the cause of a known or suspected population health problem.

2.2.2. Monitoring

The objective of monitoring studies, on the other hand, is to use repeated measures of selected parameters and pollutant concentrations over a period of time with the implicit understanding that the observations would form a warning system to determine unnatural, possibly harmful, changes in the environment (Lassig and Lahdes, 1980). To be effective a monitoring program should not only signal impending problems early, but it should also suggest probable causes of the problem (Sindermann et al., 1980). Implied in this definition of monitoring is that the parameters selected for biological measurement certainly are not population-level parameters. Instead, they are selected parameters measured in lower levels of biological organization (e.g., gross anatomy, histology, biochemistry, growth, physiology), which will predict effects at higher levels (e.g., population, community). Thus field monitoring studies should be predictive and, in this sense, would parallel the predictive toxicology tests discussed above under laboratory approaches. And in contrast to "after the fact" diagnostic studies we defined above, monitoring studies could be characterized as "before the fact" in that they should predict problems at the population level before they develop. Of course, the definitions we have offered or cited for diagnostic and monitoring studies might be considered ideal, and thus the objectives unattainable. Although we believe that the objectives of such field studies can be achieved some day if properly approached, using the best histological techniques available, we would agree that a considerable amount of research is needed before histopathology can be accepted as a routine monitoring tool.

2.2.3. Field Problems

The problems we see in published field studies, whether they are diagnostic or monitoring, are not so often a result of poor study design as they are problems related to the immense scope of the study objectives and the inability to control variables. Because variables cannot be controlled in field studies, investigators must rely on an epidemiological approach to most of their research questions. But very few comprehensive, coordinated research efforts are in effect now with the breadth necessary to answer some of the complex questions being asked. Instead, many field studies tend to be disorganized and somewhat haphazard. For instance, studies are rare that combine fish population health assessment, contaminant monitoring, and systematic measures of structural or functional parameters at and below the organism level. And studies that do combine such measurements are often not based on sound hypotheses from previous laboratory studies or are hindered by inadequate statistical sampling designs.

Monitoring studies have been discussed and implemented more extensively for marine environments than for freshwater environments. Participants in a workship on "Biological Effects of Marine Pollution and the Problems of Monitoring" (McIntyre and Pearce, 1980) critically reviewed monitoring strategies and parameters. Although many of the specific recommendations from this workshop apply only to marine problems, many recommendations would be of value in designing a contaminant effects monitoring program in freshwater environments as well. A number of extensive recommendations were presented on virtually all aspects of biological effects monitoring related to marine contaminants.

2.3. Selection of Organisms

Selection of organisms for monitoring studies was discussed by several authors in the workshop proceedings (Bang, 1980; Phelps and Galloway, 1980; Uthe et al., 1980) and, collectively, they presented the following characteristics as criteria to consider during species selection:

1. Common and cosmopolitan species.
2. Sedentary or, at least, not migratory.
3. Well-understood physiologically, biochemically, and genetically.
4. Well-known life history characteristics.
5. Well-known laboratory rearing and reproduction requirements.
6. Importance to the ecosystem.
7. Economic importance.

8. Different feeding habitats and mechanisms (if several species are included in the same monitoring program).

An obvious criterion that we would add to this list is that selected species in the biological effects monitoring program would have to be included in parallel population health assessment and contaminant monitoring programs.

2.4. Selection of Parameters

Recommendations regarding monitoring parameters in the workshop proceedings include biochemical, physiological, genetic, behavioral, and structural measures of biological effects. A panel report also specifically discussed the use of pathobiology in monitoring (Sindermann et al., 1980). The panel recommended the use of several disease syndromes such as ulcers, fin erosion, skeletal anomalies, and neoplasms as monitoring parameters, and furthermore, they concluded that these disease syndromes are available for immediate use in monitoring programs. They also suggest that histological parameters like gill and liver pathology, while potentially useful, are not ready as yet for monitoring biological effects of contaminants. We agree that histological parameters cannot yet be used for routine monitoring, but suggest that structural pathology is ready for immediate use in field diagnostic studies. In fact, although not unequivocally related to contaminants, a recent study by Moccia et al. (1981) on structural anomalies in thyroid tissues of Great Lakes salmon and a study by Ruby (1981) on gonadal constrictions in male lake trout clearly demonstrate the utility of structural pathology examinations in diagnosing problems in the field. Furthermore, we also disagree with the panel in their conclusion that various disease syndromes are useful now in monitoring programs to detect contaminant effects.

2.4.1. Disease Syndromes

Our criticism of disease syndrome frequency as a monitoring tool in fish or other aquatic organisms is twofold. First, many of the proposed disease syndromes occur with a low frequency. Thus detecting significant changes in the frequency of these rare events becomes a statistical problem requiring very large sample sizes, unless the proportion of diseased animals changes drastically. For example, Dethlefsen (1980) attempted to correlate changes in disease frequency (skeletal deformities, finrot, ulcers, lymphocystis, epidermal papillomas, stomato papillomas, pseudo-branchial tumors, and hemorrhages) with proximity to contaminant dumping sites in the German Bight. But even with huge samples (i.e., 20,000 cod, 35,000 dab, and 40,000 whiting) they were unable to unequivocally correlate proximity to contaminant dumping with disease incidence. Moreover some of their results suggested that pollution was not the only factor influencing the occurrence and abundance of fish diseases. This observation supports our second contention, that disease frequency usually cannot be directly related to contaminants. This is because contaminants as well as many other "natural" variables can impinge simultaneously on the disease-causing organism and the disease host, causing changes in the disease frequency. Even without contaminant effects the factors affecting changes in disease frequency are usually unknown. Thus it would be extremely difficult to link disease prevalence unambiguously to contaminant effects. This appears to be especially true with freshwater fish diseases. Lloyd (1980) states that, "The concept that stressed fish are more prone to disease is poorly founded, certainly in the freshwater fish field, in that major outbreaks of disease (such as perch ulcer disease and UDN of salmon) do not appear to be correlated with chemical pollution but are associated with natural factors."

2.4.2. Histological Parameters

In contrast, our experience has demonstrated that at least some histological gill lesions occur in all organisms exposed at contaminant concentrations above some threshold exposure (Johnson, 1983). This would suggest that it might be possible to use smaller sample sizes than appear to be required in monitoring for incidence of disease syndromes. Also, observed structural lesions might be easier to link directly to the presence of contaminants, because complex disease-host-contaminant interactions would be avoided and because it would at least be possible to use laboratory studies to determine whether contaminant exposure in fact causes the structural lesions observed in the field.

3. HISTOLOGICAL TECHNIQUES

The histological techniques being used currently in aquatic toxicology studies can usually be criticized in one or more of the following ways. First, tissues are usually embedded in paraffin even though plastic embedded tissues are far superior, especially with respect to structural fidelity and resolution attainable. Second, tissue heterogeneity and control of section orientation often are not considered. Thus comparisons among samples are often compromised and, in some cases, may even be invalid. Third, the evaluation of tissue and cellular responses to contaminant exposure is often too subjective, since results are rarely quantified. When these technical deficiencies are combined with poor laboratory or field approach strategies, the results obtained are often of little value.

Few of the recently published histopathological studies in aquatic toxicology have all of the technical flaws described above. In fact, some use nearly perfect histological technique, but suffer from poor experimental design. However, because many published histopathological studies did not use the best available techniques, structural lesions, if detected at all, were insensitive to low-level contaminant exposure. This has sometimes led to the conclusion that histological methods are not useful in studying aquatic toxicology problems.

To be useful in aquatic toxicology, histological methods must be sensitive enough to detect lesions in organisms exposed to toxicants at or below the chronic no observed effects concentration (NOEC) derived from traditional laboratory toxicity testing. In our experience, responses observed in fish tissues after chronic contaminant exposures are both subtle and positively correlated with exposure level. Thus to be useful the histological techniques selected must maximize resolution, minimize artifact, provide tissue sections that can be validly compared, and allow quantification of the structural response. Brief descriptions of optimal histological techniques that would best meet these criteria are presented below.

3.1. Tissue Fixation and Embedding

There are several options for fixing and embedding tissues for histological examination. The traditional procedure probably most familiar to aquatic biologists and toxicologists uses formalin fixation followed by paraffin embedment. The main alternative to paraffin, buffered fixation followed by plastic embedment, has been developed over the past 10–15 years for electron microscopy and high-resolution light microscopy. Based on our experience with both paraffin and plastic embedments, we believe that the modern buffered fixatives and plastic embedments provide results that are so superior to paraffin that paraffin should not be used at all in histopathological studies.

Many other investigators agree with this position. Lindner and Richards (1978) and Bennett et al. (1976) critically examined the relative merits of embedding tissues in paraffin and several types of plastic embedments, including epoxy plastics such as Araldite 502 and Epon 812 as well as the water soluble plastic, glycol methacrylate, sold under the trade name JB-4 (Polysciences Inc., Warrington, NJ). While they conclude that there are no great disadvantages to using plastic embedments, there are some substantial advantages. The two most important advantages are increased fidelity (principally due to fixation improvements) and increased resolution of specimen detail (due to section thinness). Moreover, Lindner and Richards also suggest that there are more staining techniques available for plastic-embedded material than there are for paraffin-embedded tissues. Butler (1979) concluded that inferior methods, which include formalin fixation and paraffin embedment, should only be used in routine pathology where the resolution of fine cytological detail is generally unnecessary. Burns and Bretschneider (1981) state that, "The cytological detail seen in 1-μm thick sections of plastic embedded material, viewed under the light microscope, is far superior to that which is available in paraffin embedded sections," and that, "Any tissue submitted for light microscopic examination will show optimal morphology if it is embedded in plastic." In their comparison of plastic and paraffin embedding procedures, Bennett et al. (1976) are even more emphatic in concluding that there are no compelling reasons to use paraffin embedments, even for routine histology.

3.1.1. Tissue Size

When deciding which fixation and embedment protocol to use, one must keep the following points in mind:

1. Smaller tissue samples fix and embed with less artifact than large tissue samples. Thus use tissue samples that are as small as is practical.
2. Tissue size also dictates the type of embedment plastic that can be used, which in turn dictates observation options available. Large tissues or tissue pieces (i.e., 10–20 mm) must be embedded in JB-4, which can be used only for light

Table 1.   **Comparison of Different Plastic Embedding Techniques**

Criteria	Embedding Options		
	JB-4[a]	Epoxy—LM[b]	Epoxy—EM[b]
Maximum specimen size[c]	10–20 mm	5 mm	1 mm
Fixation	Buffered aldehydes	Primary fixation, buffered aldehydes; secondary fixation, buffered osmium	Primary fixation, buffered glutaraldehyde; secondary fixation, buffered osmium
Sectioning	Modified rotary[d] or ultramicrotome, thickness 1–2 μm	Modified rotary[d] or ultramicrotome, thickness 0.5–1.5 μm	Ultramicrotome, thickness 50 nm–1.5 μm
Staining	Many available, some very specific	Few available, usually nonspecific	Few available, usually nonspecific
Microscopy	LM only	LM and poor EM	LM and EM

[a] Water soluble plastic, glycol methacrylate, sold under the trade name JB-4 by Polysciences Inc., Warrington, NJ.

[b] Epoxy—LM, epoxy embedment protocols appropriate for light microscopy; Epoxy—EM, epoxy embedment protocols appropriate for electron microscopy.

[c] This is the maximum distance for effective penetration of fixatives and unpolymerized embedment plastic. The distance reported is the maximum distance allowed for the smallest dimension of the specimen. For example, large flat sheets of tissue 1 mm or less thick process nicely in Epoxy-EM procedures, but large flat sheets, if thicker than 1 mm, do not.

[d] Butler (1979).

microscopy (LM) observation. Smaller tissues (i.e., 5 mm) can be embedded in JB-4 or epoxy but will still be suitable only for LM observation. Very small tissues (i.e., 1 mm) can be embedded in JB-4 or epoxy but, if doubly fixed and embedded in epoxy, can be used for both electron microscopy (EM) and LM observation.

3.   Most specific staining reactions must be done with JB-4 embedded tissues, epoxy embedments being unacceptable.

These and other tradeoffs among the alternative plastic embedding procedures are summarized in Table 1.

3.1.2.   Tissue Fixation

As mentioned earlier, the embedment determines to some extent the appropriate fixation procedures. The fixatives used with good success are buffered aldehydes such as formaldehyde and glutaraldehyde, and osmium tetroxide. Which fixatives to use and the protocols necessary for each are determined by the fidelity required and the tissue size. But generally, large specimens for LM are fixed in buffered formalin because it penetrates the tissues rapidly. Smaller LM specimens are fixed in either buffered formalin or buffered glutaraldehyde. Glutaraldehyde generally provides better structural fidelity than formalin but penetrates tissues more slowly than formalin. Thus the fixative to use is dictated by tissue size (see Table 1). The smallest specimens, which allow either EM or LM observation, should be doubly fixed. The first, or primary, fixation is in buffered glutaraldehyde and the secondary fixation is in buffered osmium tetroxide. These fixation protocols with small specimens maximize fidelity and resolution but may compromise the study by reducing the maximum allowable tissue size.

3.1.3. Tissue Embedding

Several aquatic toxicology studies have used plastic embedments successfully. Hughes et al. (1979) used epoxy plastic embedments for studying the effects of heavy metals on rainbow trout gills. Investigations concerning the effects of petroleum hydrocarbons on several different fish tissues (Hawkes, 1977) and on fish gills (Engelhardt et al., 1981) were both done with epoxy plastic embedments. Hinton and Walker (1980) used epoxy plastic embedments for some of their histological work in fish exposed to mine water. The structural effects of coal particulate-bound phenathrene on fathead minnows was studied by Gerhart et al. (1981), where whole fish embedded in JB-4 were used for LM and pieces of gut tissue were embedded in epoxy plastic for EM study.

3.2. Tissue Orientation and Sectioning

Systematic orientation of the tissue sectioning plane helps ensure that comparisons among sections of a tissue from different organisms are valid. While this is especially important in highly heterogeneous and anisotropic tissues like gills, few tissues are truly homogeneous or isotropic. Even liver tissues are apparently not as homogeneous as was once believed (see Pang and Terrell, 1981). Thus we feel it is wise to control section orientation in all tissues.

The following criteria should be considered when deciding on section plane orientation:

1. The appropriate sectioning plane should be located easily and reproducibly in each embedded tissue sample.
2. The sectioning plan should traverse sensitive tissue regions in a manner that optimizes lesion recognition and observation.
3. Each section should contain a sample of all the cell types and structures contained in a heterogeneous tissue.

Hughes et al. (1979) and Johnson (1983) demonstrate tissue orientation techniques in histological studies with fish gills.

The relative difficulty of sectioning plastic-embedded versus paraffin-embedded tissues may be a matter of some concern to investigators who have only worked with paraffin. However, we believe that this concern is based on predjudices from personal experience rather than on fact. With proper equipment (see Table 1) any investigator adept at paraffin histotechnique could learn plastic techniques either from the literature or from a short visit to a lab using plastics. Investigators completely unfamiliar with histotechnique would perhaps find it best to take a course in electron microscopy to learn plastic embedment and sectioning techniques that can be used for both LM and EM.

3.3. Quantification

The quantification of tissue structure increases the power of histopathology in at least three ways. First, detection of subtle responses to contaminant exposure is improved because data from many successive microscope fields can be accumulated. Second, the accumulated data can be analyzed statistically to determine threshold exposure concentrations and to compare the relative sensitivities of various structural responses. These structural responses may then be compared with ecologically significant response parameters including the NOEC from traditional toxicity tests. Third, quantified responses can be reported more accurately and with less subjectivity, thereby facilitating the accumulation of a data base that would allow objective comparison of structural responses among studies and investigators.

3.3.1. Computer Image Analysis

There are basically two quantification methods being used currently. The first method, computer-assisted image analysis, requires relatively expensive, specialized equipment and quantifies structural characteristics of cells and tissues using computer processing of digitized images (see Winkler et al., 1981). The images can be processed directly from the light or electron microscope, from a video monitor, or from photographs. These methods are being applied very successfully to routine clinical problems like differential blood cell counts. But for use of these methods in diverse research applications where many different types of cell or tissue structures might often be of interest, the investigator would face problems such as devising appropriate sampling methods and programming the computer to process the different types of images encountered. From our point of view, current technology for computer-assisted image analysis might be applied successfully to structural pathology evaluations only when the measurements to be made were well defined and, even, routine. For instance, the techniques might be useful in routine field monitoring programs after selecting the species and structures to be monitored and after having sufficient experience with the parameters to be monitored.

3.3.2. Stereology

Stereology, the second method, is a statistical sampling method for quantifying tissue and cellular morphology. Simply, a grid is superimposed on an image of the tissue of interest and at grid intersections appropriate sampling data are recorded (e.g., cell type, tissue structure, or perhaps, distance to some other structure of interest). By doing this a number of times on several tissue sections, many different morphological features can be quantified. A few examples are cellular or tissue surface area, cell volume, relative proportions of cell types in a tissue, and nuclear to cytoplasmic volume ratios. An excellent book authored by Weibel (1979), which we recommend to interested investigators, carefully details stereological procedures and demonstrates the technique using several different tissues.

Several histopathological studies have been published recently that demonstrate stereological techniques with tissues from fish exposed to environmental contaminants. Hughes et al. (1979) quantified detailed morphological responses (e.g., mean thickness of the water-blood barrier) in gills from rainbow trout exposed to several heavy metals. These techniques allowed them to quantify the effect of metals on diffusing capacity and to establish the relative importance of epithelial cell hypertrophy and epithelial separation on changes in diffusion distance. Zuchelkowski et al. (1981) quantified responses to acid stress in bullhead skin and showed that the observed increase in mucus was "due to production of more mucous cells, with no change in volume of individual cells (i.e., hyperplasia without hypertrophy)." Several structural gill parameters were measured by Johnson (1983) in contaminant-exposed rainbow trout. While the methods used were not strictly stereological, mucous cell numbers, chloride cell numbers, white blood cell infiltration, mitotic index, and respiratory diffusion distance were all shown to respond to lower contaminant concentrations than did either fish fry growth or fry survival.

4. SUMMARY

The value of histopathology in aquatic toxicology has yet to be fully established. Although numerous reports have been published recently that use histopathology as a tool to determine effects of contaminants on aquatic biota, we are unaware of any studies that combine sensitive, high-resolution histological techniques with a comprehensive experimental approach. Thus the reported results are often of little practical value.

In this paper we have discussed the application of histopathology in both laboratory- and field-oriented aquatic toxicology studies. The major objectives of aquatic toxicology studies in the laboratory are to identify mechanisms of toxicity and to predict "safe" contaminant concentrations in the environment. We have identified what we believe to be two common problems in laboratory approaches as applied by many investigators to either mechanistic or predictive toxicology problems. First, many studies use contaminant exposure concentrations that are too high to be applicable to most contaminant problems in the aquatic environment. Second, too few studies correlate structural pathologies with other contaminant effects, such as functional responses that establish toxic mechanisms, or accepted predictors of population-level responses that establish ecological significance.

Potential applications of histopathology to aquatic contaminant problems in the field can also be divided into two categories, diagnosis, where the objective is to identify the cause of observed adverse population effects, and monitoring, where the objective is to warn of impending contaminant-caused environmental degradation.

In contrast to the increasing use of histopathology in laboratory studies, very few field contaminant studies have used histopathological techniques; most that have are marine studies, usually with shellfish.

Studies that have been performed or contemplated all face a set of formidable problems. First, to establish cause-effect relationships, contaminant levels, structural pathologies, and population health must be evaluated in the field simultaneously. Although such correlated studies would be difficult and costly, we believe that a number of carefully conducted studies of this kind will be necessary before histopathology can be accepted as a routine field monitoring tool for realiably predicting ultimate population decline. Second, immense sample sizes are required for adequate statistical designs in many field studies, especially for effective population stock assessment or for detection of secondary contaminant effects such as disease incidence. But the requirements of good statistical design are often too costly to be met; thus less than adequate studies are performed. As a consequence, data from these studies are difficult if not impossible to interpret. Third, without knowing what to look for, histopathologists cannot effectively address field problems. Haphazard "fishing expeditions" in search of histological lesions are simply inadequate. Laboratory studies should provide the basis for testable hypotheses about specific structural lesions expected in organisms from the field. Fourth, parameters must be chosen that can be readily studied in the laboratory. Such laboratory studies are usually going to be necessary to confirm suspected cause-effect relationships in the field. This may be difficult in studies of stress-related disease syndromes because disease responses are a secondary contaminant effect.

Because of this formidable set of problems, we believe that a considerable amount of research will be necessary before structural pathology, or for that matter disease incidence or functional responses in aquatic organisms, can be reliably used as a predictor of population decline. The type of research necessary would be carefully conducted field demonstrations that avoid the problems listed above. Moreover, we believe that the use of structural pathology to diagnose suspected contaminant-caused population health problems will be much easier to demonstrate than the use of structural pathologies in routine field monitoring.

Histological techniques used in either laboratory or field investigations are rarely as sensitive as they could be. Consequently, studies that would be expected to reveal histopathological responses to contaminant exposure often do not. The histological techniques used in these studies are often inferior in one or more of the following ways. First, fixation protocols are not optimized and the tissues are embedded in paraffin, which is inferior to available plastics. Second, because tissue orientation during sectioning is often not controlled, comparisons between samples are sometimes not valid. And third, histological responses are usually not quantified. As a result, evaluation and reporting of contaminant effects on tissue structure are too subjective.

5. Recommendations

To improve the utility of histopathology in laboratory studies we would recommend the following:

1. Reasonable exposure concentrations should be used so that the experimental results can be applied to actual contaminant problems in the field.
2. Histopathological responses should be correlated to responses measured at several other levels of biological organization to help ascertain mechanisms of toxicity as well as the ecological significance of observations.

To pursue the monumental problems of correlating environmental contaminant levels to important population-level effects in the field using histopathology, as well as other methods, we suggest the following:

1. Contaminant levels, structural parameters, and population health must be measured simultaneously to understand contaminant problems in the field; this will be necessary at least until such time as histopathology is more widely accepted as a reliable predictor of ecologically significant effects at the population level.
2. All aspects of field research must use adequate statistical designs.
3. Organism response parameters measured in the field should be based on previous laboratory study.
4. It should be possible to study chosen response parameters in the laboratory so as to confirm field observations.

To detect the subtle structural responses produced by low-level contaminant exposures will require very sensitive histological techniques. To maximize the sensitivity of histology we recommend the following:

1. Carefully selected fixation protocols and plastic embedment should be used to maximize preparation fidelity and resolution.
2. Orientation of the tissue sectioning plane should be carefully selected and controlled.
3. Stereological methods should be used to quantify tissue structure, both normal and pathological, to help establish an extensive data base to facilitate further comparisons.

ACKNOWLEDGMENTS

Preparation of this article was supported in part by the U.S. Environmental Protection Agency under Cooperative Agreement Number CR808671 between the Environmental Research Laboratory—Duluth and the University of Wyoming. We appreciate the valuable insights on the application of histopathology to aquatic toxicology problems gained through discussions with James McKim, Alan Maki, Don Mount, Mike Mac, and Joe Meyer.

REFERENCES

Bang, F. B. (1980). Monitoring pathological changes as they occur in estuaries and in the ocean in order to measure pollution (with special reference to invertebrates). In: A. D. McIntyre and J.

B. Pearce (Eds.), *Biological effects of marine pollution and the problems of monitoring.* Conseil Int. pour l'Exploration de la Mer, Copenhagen, Denmark, pp. 118–124.

Bartholomew, G. A. (1964). The roles of physiology and behavior in the maintenance of homeostasis in the desert environment. In: *Homeostasis and feedback mechanisms, Symposia of the Society for Experimental Biology,* Vol. 18. Academic Press, New York, pp. 7–21.

Bennett, H. S., Wyrick, A. D., Lee, S. W., and McNeil, J. H. (1976). Science and art in preparing tissues embedded in plastic for light microscopy, with special reference to glycol methacrylate, glass knives and simple stains. *Stain Tech.* **51**, 71–97.

Burns, W. A. and Bretschneider, A. (1981). Thin is in: Plastic embedding of tissue for light microscopy. Amer. Soc. of Clinical Pathologists, Chicago, IL, 58 pp.

Butler, J. K. (1979). Methods for improved light microscope microtomy. *Stain Tech.* **54**, 53–69.

Dethlefsen, V. (1980). Observations of fish diseases in the German Bight and their possible relation to pollution. In: A. D. McIntyre and J. B. Pearce (Eds.), *Biological effects of marine pollution and the problems of monitoring.* Conseil Int. pour l'Exploration de la Mer, Copenhagen, Denmark, pp. 110–117.

Engelhardt, F. R., Wong, M. P., and Duey, M. E. (1981). Hydromineral balance and gill morphology in rainbow trout (*Salmo gairdneri*) acclimated to fresh and sea water, as affected by petroleum exposure. *Aquat. Tox.* **1**, 175–186.

Gerhart, E. H., Liukkonen, R. J., Carlson, R. M., Stokes, G. N., Lukasewycz, M., and Oyler, A. R. (1981). Histological effects and bioaccumulation potential of coal particulate-bound phenanthrene in the fathead minnow (*Pimephales promelas*). *Environ. Pollut. (Ser. A)* **25**, 165–180.

Hawkes, J. W. (1977). The effects of petroleum hydrocarbon exposure on the structure of fish tissues. In: D. A. Wolfe (Ed.), *Fate and effects of petroleum hydrocarbons in marine ecosystems and organisms.* Pergamon Press, New York, pp. 115–128.

Hinton, D. E. and Walker, E. R. (1980). Cellular bioassay of fish as a sensitive index of water quality related to mine water. U.S. Dep. of Interior No. A-037-WVA. Washington, DC.

Hughes, G. M., Perry, S. F., and Brown, V. M. (1979). A morphometric study of effects of nickel, chromium and cadmium on the secondary lamellae of rainbow trout gills. *Water Res.* **13**, 665–679.

Johnson, R. D. (1983). The use of quantitative histopathology in aquatic toxicology: Correlations between gill structure and routine toxicity measures in rainbow trout. II. Changes in rainbow trout gill structure as affected by chronic exposure to oil-shale retort water. Ph.D. thesis, Univ. of Wyoming, Laramie, WY.

Klaassen, C. D. and Doull, J. (1980). Evaluation of safety: Toxicologic evaluation. In: J. Doull, C. D. Klaassen, and M. O. Amdur (Eds.), *Casarett and Doull's toxicology: The basic science of poisons.* Macmillan Publishing Company, New York, pp. 11–27.

Lassig, J. and Lahdes, E. (1980). A review of biological monitoring and effects studies in the Baltic Sea with special reference to research in Finland. In: A. D. McIntyre and J. B. Pearce (Eds.), *Biological effects of marine pollution and the problems of monitoring.* Conseil Int. pour l'Exploration de la Mer, Copenhagen, Denmark, pp. 212–227.

Lindner, M. and Richards, P. (1978). Long-edged glass knives ('Ralph knives')—Their use and the prospects for histology. *Sci. Tools* **25**, 61–67.

Lloyd, R. (1980). Toxicity testing with aquatic organisms: A framework for hazard assessment and pollution control. In: A. D. McIntyre and J. B. Pearce (Eds.), *Biological effects of marine pollution and the problems of monitoring.* Conseil Int. pour l'Exploration de la Mer, Copenhagen, Denmark, pp. 339–341.

Malins, D. C. (1982). Alterations in the cellular and subcellular structure of marine teleosts and invertebrates exposed to petroleum in the laboratory and field: A critical review. *Can. J. Fish. Aquat. Sci.* **39**, 877–889.

McIntyre, A. D. and Pearce, J. B. (Eds.) (1980). *Biological effects of marine pollution and the problems of monitoring.* Conseil Int. pour l'Exploration de la Mer, Copenhagen, Denmark, 346 pp.

McKim, J. M. (1977). Evaluation of tests with early life-stages of fish for predicting long-term toxicity. *J. Fish. Res. Board Can.* **34**, 1148–1154.

McKim. J. M. and Benoit, D. A. (1971). Effects of long-term exposures to copper on survival, growth and reproduction of brook trout (*Salvelinus fontinalis*). *J. Fish. Res. Board Can.* **28**, 655–662.

Moccia, R. D., Leatherland, J. F. and Sonstegard, R. A. (1981). Quantitative interlake comparison of thyroid pathology in Great Lakes coho (*Oncorhynchus kisutch*) and chinook (*Oncorhynchus tschawytscha*) salmon. *Cancer Res.* **41**, 2200–2210.

Pang, K. S. and Terrell, J. A. (1981). Retrograde perfusion to probe the heterogeneous distribution of hepatic drug metabolizing enzymes in rats. *J. Pharmacol. Exp. Ther.* **216**, 339–346.

Phelps, D. K. and Galloway, W. B. (1980). A report on the coastal environmental assessment stations (CEAS) program. In: A. D. McIntyre and J. B. Pearce (Eds.), *Biological effects of marine pollution and the problems of monitoring.* Conseil Int. pour l'Exploration de la Mer, Copenhagen, Denmark, pp. 76–81.

Rather, L. J. (1971). The place of Virchow's "Cellular pathology" in medical thought. In: *Cellular pathology*, Dover Publications, New York, pp. v–xxvii.

Ruby, S. M. (1981). Use of histological techniques in assessing reproductive potential among fish populations. Presented at Great Lakes Fishery Commission workshop on fish health, Geneva Park, Ont. Nov. 16, 1981.

Sindermann, C. J., Bang, F. B., Christensen, N. O., Dethlefsen, V., Harshbarger, J. C., Mitchell, J. R., and Mulcahy, M. F. (1980). The role and value of pathobiology in pollution effects monitoring programs. In: A. D. McIntyre and J. B. Pearce (Eds.), *Biological effects of marine pollution and the problems of monitoring.* Conseil Int. pour l'Exploration de la Mer, Copenhagen, Denmark, pp. 135–151.

Uthe, J. F., Freeman, H. C., Mournib, S., and Lockhart, W. L. (1980). Selection of biochemical techniques for detection of environmentally induced sublethal effects in organisms. In: A. D. McIntyre and J. B. Pearce (Eds.), *Biological effects of marine pollution and the problems of monitoring.* Conseil Int. pour l'Exploration de la Mer, Copenhagen, Denmark, pp. 39–47.

Warren, C. E. (1971). *Biology and water pollution control.* Saunders, Philadelphia, PA, 434 pp.

Weibel, E. R. (1979). *Stereological methods, Vol. 1: Practical methods for biological morphometry.* Academic Press, New York, 415 pp.

Winkler, D. G., Baky, A. A., Hunter, N. R., Greenberg, S. D., Rogers, T. D., Spjut, H. J., and Estrada, R. (1981). Image analysis of atypical bronchial epithelial cells—A quantitative and qualitative examination of squamous cell carcinogenesis. *Anal. Quant. Cytol.* **3**, 295–298.

Woodward, D. F., Riley, R. G., and Smith, C. E. (1983). Accumulation, sublethal effects, and safe concentration of a refined oil as evaluated with cutthroat trout. *Arch. Environ. Contam. Toxicol.* **12**, 455–464.

Zuchelkowski, E. M., Lantz, R. C., and Hinton, D. E. (1981). Effects of acid-stress on epidermal mucous cells of the brown bullhead *Ictalurus nebulosus* (LeSeur): A morphometric study. *Anat. Rec.* **200**, 33–39.

4

BIOCHEMICAL INDICATORS OF STRESS IN FISHES: AN OVERVIEW

Dora R. May Passino

Great Lakes Fishery Laboratory
U.S. Fish and Wildlife Service
Ann Arbor, Michigan 48105

---

1. GENERAL REVIEW

1.1. Enzymes

During the past two decades, water quality biologists have been searching for sensitive indicators of sublethal effects of contaminants on fish in an attempt to understand the mode of action of toxicants and to develop a basis for corrective action in cleaning up water bodies before the health of aquatic populations is seriously threatened. Enzymes are attractive as indicators because they are more easily quantified than are other indicators, such as changes in behavior. Useful precedents have been set in clinical medicine in the successful diagnosis of disease and evaluation of exposure to industrial chemicals or drugs by analyses of such variables as enzymes, blood chemistry, and liver function. In the fishery field, however, enzyme research is in its infancy: only two fish enzymes have yet been proven to be useful indicators of toxic effects of specific contaminants or classes of contaminants.

1.1.1. Acetylcholinesterase

The first of these enzymes is acetylcholinesterase, which modulates the amounts of the neurotransmitter acetylcholine in the nervous system. The mode of action of malathion, parathion, and other widely used organophosphate insecticides is inhibition of this enzyme. Measurement of acetylcholinesterase in the brains of distressed fish collected in the field has been successfully used to diagnose organophosphate poisoning of fish in natural waters (Coppage et al., 1975).

1.1.2. Delta-Amino Levulinic Acid Dehydratase

The second indicator enzyme is δ-amino levulinic acid dehydratase (ALA-D), which is in the pathway of heme synthesis. Fish exposed chronically to lead suffer from anemia, which is probably a result of the disturbance of hemoglobin synthesis by the blocking effect of lead on ALA-D. Inasmuch as Hodson et al. (1977) reported that near-lethal exposures of rainbow trout (*Salmo gairdneri*) to cadmium, copper, zinc, and mercury did not significantly inhibit erythrocyte ALA-D, the enzyme appears to be specific for lead. In addition, the concentrations of lead that inhibit ALA-D activity during short-term exposures of fish are similar to concentrations causing other sublethal symptoms such as black tail, spinal curvature, and erosion of the caudal area of rainbow trout after many months of exposure. Based on laboratory tests, ALA-D fits the criteria for a short-term enzyme indicator of long-term sublethal effects of a contaminant on fish. Hodson et al. (1984) reported that ALA-D showed promise as an indicator of field exposure to lead, but further work was necessary before routine application of this enzyme would be possible. The

Columbia National Fishery Research Laboratory (U.S. Fish and Wildlife Service) has been measuring blood ALA-D in fish exposed environmentally to lead released from mine tailings in Missouri. Their work should provide additional data on the usefulness of this enzyme in the field.

1.1.3. Adenosine Triphosphatase

Adenosine triphosphatase (ATPase), a widely distributed enzyme involved in the energy-requiring active transport of electrolytes across membranes, shows promise as an indicator. The pesticide DDT inhibits mitochondrial ATPase from several fish tissues. However, different types of ATPases from some tissues may be activated in response to toxicants. The ATPases are not specific indicators of DDT exposure— inhibition by toxaphene, PCBs, and pyrethrins has also been reported. Lock et al. (1981) concluded that change in permeability characteristics of gills for water, rather than inhibition of the Na^+-K^+-ATPase of the gills, was the primary reason for mercury-induced effects on osmoregulation in rainbow trout. Mercury inhibited Na^+-K^+-ATPase only at lethal concentrations.

1.1.4. Leucine Aminonaphthylamidase

Since stress can cause wasting disorders, Bouck (1980) suggested measuring leucine aminonaphthylamidase (LAN) to assess fish health. This proteolytic enzyme is found mainly in the lysosomes of cells and is activated by and assists in the autolytic process of cell death. He provided baseline data on LAN in several tissues of rainbow trout and proposed relating plasma LAN to the amount of liver destruction in vitro that would result in an equivalent amount of LAN.

1.1.5. Mixed Function Oxidase Enzymes

Another approach to using enzymes as indicators of toxicant exposure is the measurement of enzymes that metabolize or detoxify contaminants (Gingerich and Weber, 1979). Generally, when animals are exposed to toxicants, the detoxifying enzymes are induced—that is, more enzyme is synthesized (Melancon et al., 1981; Addison, 1984). A good example of how detoxifying enzymes can be used to monitor specific pollutants was given by Walton et al. (1978), who found that hepatic aryl hydrocarbon hydroxylase (AHH) in the cunner (*Tautogolabrus adspersus*), a marine fish living off New England, was a sensitive indicator of petroleum contamination. Oil concentrations of 1–2 mg/l caused twofold to sixfold induction of AHH above normal levels. Observations indicated that elevated AHH followed oil contamination within 1 or 2 days. Because induction of AHH decayed in less than 7 days after exposure ceased, AHH reflected the current state of oil contamination. Feeding of crude oil or the tissue of mussels contaminated with oil at a concentration of 500 mg/kg caused up to fivefold induction. In vitro tests with trout showed that AHH was not induced by polychlorinated biphenyls (PCBs) nor by representative organochlorine, organophosphate, or carbamate pesticides.

Stegeman (1978) reported elevated levels of benzo[a]pyrene hydroxylase, aminopyrine demethylase, aldrin epoxidase, and cytochrome P-450 of killifish (*Fundulus heteroclitus*) from a marsh that was the site of an oil spill 8 years previously. He postulated that the elevated enzyme levels were a response to the substantial amounts of aromatic hydrocarbons still remaining in the marsh. Vodicnik et al. (1981) reported that 2,3,7,8-tetrachlorodibenzo-p-dioxin (TCDD), a highly toxic environmental contaminant, induced cytochrome P-448-type hepatic monooxygenase activity in rainbow trout, but demonstration of the induction of these detoxifying enzymes in field-collected fish exposed to TCDD has not been attempted.

2. RESEARCH AT GREAT LAKES FISHERY LABORATORY

2.1. Allantoinase

My colleagues and I at the Great Lakes Fishery Laboratory did some in situ enzyme work a few years ago. After completing in vitro experiments to determine effects of heavy metals, PCBs, and DDT on the uricolytic enzyme allantoinase in lake trout (*Salvelinus namaycush*) liver (Passino and Cotant, 1979) and in vivo experiments with PCBs, we collected lake trout from two areas—one near Isle Royale, a relatively clean area in northern Lake Superior, and one near Saugatuck, Michigan, a contaminated area influenced by industrialization, typical of southeastern Lake Michigan. We found significantly ($P < 0.01$) greater enzyme activity in small than in large lake trout from Lake Michigan (see Table 1). Enzyme activity was also higher, but not significantly so, in the smaller fish from Lake Superior and from stocks in the laboratory.

Although the interpretation of data from field samples is complicated by many unknown variables operating on enzyme systems in fish, we demonstrated an inverse correlation between enzyme activity and length of lake trout. This clear trend in activity may have been caused in part by higher levels of contaminants in older fish—especially in Lake Michigan (see Table 2). This example illustrates the difficulty in establishing a causal relationship between residue levels and observed enzyme activities in fish collected in the field. The only exceptions may arise after clear-cut demonstration in the laboratory of specificity of a particular contaminant or class of contaminants for a particular enzyme.

2.2. Subcellular Contaminant Distribution

We also measured the subcellular distribution of chemical contaminants in lake trout liver (Passino and Kramer, 1982). Such determination holds promise as a means for predicting which metabolic pathways are most likely to be affected, because each subcellular fraction or organelle is specialized for particular functions. Although we investigated mercury, PCBs, and DDT plus DDE (a product of DDT metabolism), only the data on mercury are relatively free of problems. These data were obtained by collecting a large live lake trout in Lake Michigan near Saugatuck, bringing it live to the laboratory, killing it, and immediately removing the liver and separating the subcellular fractions by differential centrifugation. One set of

Table 1. Allantoinase Activity of Individual Samples of Frozen Livers from Wild Lake Trout Collected in Lake Michigan off Saugatuck and Lake Superior off Isle Royale in 1974, and from Laboratory Stocks of Lake Trout (± SE in Parentheses)[a]

Source, and Number of Fish	Total Length (mm)		Allantoinase Specific Activity (Unit/mg Protein)
	Group	Mean	
Lake Michigan			
19	>500	707	0.0701[b]
		(14)	(0.00755)
18	<500	310	0.140[b]
		(26)	(0.0109)
Lake Superior			
18	>500	592	0.0712
		(10)	(0.0103)
32	<500	350	0.0749
		(11)	(0.00869)
Laboratory[c]			
3	>500	564	0.109
		(6)	(0.0127)
30	<500	290	0.197
		(14)	(0.0125)

[a] From Passino and Cotant, 1979.

[b] Difference between means of allantoinase activity within fish length is highly significant ($P < 0.01$) by analysis of variance.

[c] Fish maintained at Great Lakes Fishery Laboratory.

subcellular fractions was examined by electron microscopy to quantify the organelles present and the other set was analyzed for total mercury by a combustion amalgamation technique. (In addition, we had earlier obtained analogous data from a frozen liver.)

Mercury was primarily (80%) in the submicroscopic material (Fig. 1)—that is, in material so small it could not be seen at $7700 \times$ magnification. Of the remainder, 5% was in the mitochondria, 5% in the rough endoplasmic reticulum, and 1–2% in some of the other organelles.

What does this distribution mean in terms of fish metabolism and the health of the fish? In other vertebrates, such as the rat, a metal-binding protein, metallothionein, has been found in liver and kidney. This protein serves as a protective mechanism in animals, because the protein-bound metal is not available to interfere with metabolic processes. Therefore, I suggest that lake trout may be able to protect themselves to a large extent from the toxic effects of mercury by concentrating mercury in the submicroscopic material of the liver, where the mercury is probably bound to a metallothionein-like protein. However, since other recent investigations on rainbow trout (Olson et al., 1978) and goldfish (*Carassius auratus*) (Marafante, 1976) showed

Table 2. Contaminant Analysis of Composite Samples of Livers and Individual Samples of Fillets of Lake Trout Collected off Saugatuck in Lake Michigan and off Isle Royale in Lake Superior in 1974[a]

Source and Number of Fish	Total Length (mm)		Livers			Fillets, Total Hg (μg/g)[b,d]
	Group	Mean ± SE	PCBs (μg/g)[b]	DDT + DDE (μg/g)[b]	Total Hg (μg/g)[b,c]	
Lake Michigan						
10	>500	717 ± 23	21.4	4.68	0.939	0.594
10	<500	210 ± 14	<0.6	0.141	0.479	0.0552
Lake Superior						
10	>500	586 ± 14	4.55	1.86	0.690	0.581
8	<500	284 ± 4	<0.6	0.206	0.105	0.152

[a] From Passino, 1981.
[b] μg/g wet weight.
[c] Average value of three (fish <500 mm long) or four (>500 mm) analyses per composite sample.
[d] Average value of two analyses per fillet for specified number of fish.

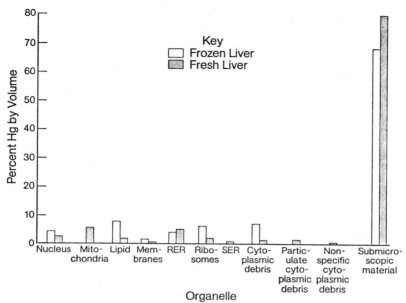

Figure 1. Percent mercury by volume in organelles of lake trout liver. RER = rough endoplasmic reticulum; SER = smooth endoplasmic reticulum. (Data from Passino and Kramer, 1982.)

that no more than 50% of mercury was bound to metallothionein-like proteins, some interference of mercury with metabolic pathways in the cytosol is possible. These pathways include fatty acid synthesis, glycolysis, many reactions in gluconeogenesis, the hexose monophosphate shunt, and activation of amino acids for protein synthesis.

Similar results were obtained for the distribution of PCBs and DDT plus DDE in frozen liver. Although 96–98% of these organochlorines were found in the soluble fraction, the mitochondria were ruptured during the freezing and thawing process; consequently, confirming work needs to be done with fresh liver. The logical follow-up research would be to determine if the levels of contaminants present are inhibitory to key enzymes in these fractions, especially those in the cytosol. We did not do the follow-up research because it would have entailed a mode-of-action study at the subcellular level. Inasmuch as fishery and water quality managers need to have technical information that is directly useful in their decision-making process, we are now conducting whole-animal studies, with emphasis on effects on mortality, growth, and reproduction.

2.3. Reproduction

As part of the continuing research on the reproductive failure of lake trout in Lake Michigan, the Great Lakes Fishery Laboratory has been involved in two evaluations of the toxicity of major organochlorine contaminants to lake trout eggs and fry. In one study (Willford et al., 1981), 20,000 fry from Lake Michigan lake trout were exposed to simulated Lake Michigan levels of PCBs (Aroclor 1254) and DDE, and to levels 5 and 25 × higher, in food and water. Ambient levels were 10 ng/l PCBs and 1.0 ng/l DDE in water and 1.0 μg/g PCBs and 0.1 μg/g DDE in food. Cumulative mortality of the fry exposed to simulated Lake Michigan levels of PCBs and DDE for 6 months was 40.7%—nearly twice that of unexposed (control) fry— and mortality at the highest exposure level was 46.5%. Willford et al. (1981) concluded that, although several factors have undoubtedly contributed to the re- productive failure of lake trout, the levels of PCBs and DDE in Lake Michigan during the early and mid-1970s were high enough to significantly reduce survival of fry in the lake and thereby impede restoration of the lake trout population to self-sustainability.

2.4. Growth

During the course of this study, two interesting trends were observed that warrant further investigation. One trend was toward a slight increase in growth rate of the exposed fry. A second trend was toward an alteration in glycolysis in exposed fry that had been exercised to exhaustion in a swimming apparatus. Further work needs to be done to measure oxygen consumption, muscle lactate, and muscle glycogen in fry before and after swimming to exhaustion to determine if their stamina is significantly affected by exposure to organochlorine contaminants.

In the second study, eggs and fry from Lake Michigan lake trout were exposed to simulated Lake Michigan levels of organochlorines. Nominal concentrations were 1 ng/l PCBs (Aroclor 1254), 0.2 ng/l DDE, 1 ng/l dieldrin, and 0.9 ng/l technical grade chlordane, and multiples (10×) of these concentrations. No differences in mortality were observed between exposed and unexposed eggs and fry for 180 days (to the "swim-up" stage). Measurements of lengths taken from the time of hatch almost until swim-up (age about 60 days) showed a trend, significant at the 10% level, for increased growth of fry exposed to the lower levels of organochlorines, by time of swim-up.

3. OTHER INDICATORS OF CONTAMINANT EFFECTS

3.1. Instantaneous Growth

A further biochemical indicator of instantaneous growth consisted of the measurement of RNA–DNA ratios. A high RNA–DNA ratio is correlated with rapid protein synthesis and weight gain. At swim-up, the RNA–DNA ratio of the control fry (3.85 ± 0.11 [± SE]) was significantly lower ($P < 0.10$) than that of the fry receiving higher levels of the organochlorines (4.38 ± 0.04). The RNA–DNA ratio for the low treatment fry was intermediate (4.08 ± 0.15). Thus the trend toward increased growth of fry exposed to organochlorines was supported by the RNA– DNA data.

Kearns and Atchison (1979) demonstrated that field-collected yellow perch (*Perca flavescens*) showing different growth rates, as measured by changes in weight or length, had parallel changes in the RNA–DNA ratios. In their study, the growth rate of yellow perch exposed environmentally to cadmium and zinc from an electroplating plant was slower than that of controls in a cleaner area upstream from the plant. However, contaminated fish grew fastest in late summer, and uncontaminated fish in midsummer. The authors cautioned against comparing growth rates in two populations at a single sampling period.

Measurement of RNA–DNA ratios shows promise as an in situ indicator of contaminant effects.

3.2. Bioselective Membrane Electrode Probe

A new tool that shows potential as a means of sensing contaminant effects is the bioselective membrane electrode probe (Rechnitz, 1981). The first bioselective membrane electrode made with intact animal slices (Anonymous, 1978) was rather crude (Fig. 2). This electrode required both beef liver tissue and isolated urease enzyme to mediate the conversion of the amino acid to be measured, arginine, to the electroreactive product, ammonia. In the more advanced design of the glutamine-selective electrode (Rechnitz et al., 1979), the need for an auxiliary enzyme was eliminated because the complete biocatalytic functions were carried out by a slice

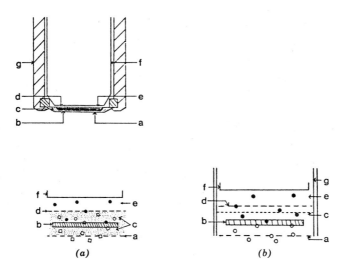

Figure 2. (*a*) Schematic of liver tissue-enzyme electrode for arginine; a, dialysis membrane; b, bovine liver tissue slice; c, urease enzyme suspension; d, gas-permeable membrane; e, internal electrolyte; f, combination pH electrode; g, plastic electrode body; □, arginine substrate; ○, urea intermediate; ●, ammonia product. (Components d–g constitute the ammonia gas-sensing electrode.) (*b*) Expanded view of porcine kidney tissue sensor for glutamine; a, nylon mesh or dialysis membrane; b, kidney tissue slice; c, dialysis membrane; d, gas-permeable membrane; e, internal electrolyte; f, combination pH electrode; g, plastic electrode body; ○, glutamine substrate; ●, ammonia product. (Rechnitz, 1981).

of porcine kidney (Fig. 2). Meyerhoff (1981) predicted that reliable electrodes could be developed for bioassays of heavy metals and organic contaminants. If fish tissue such as liver were used as the biocatalyst, then these electrodes could be used to screen contaminants or effluents to determine substances that would stress fish at the biochemical level.

3.3. Reproductive Hormones

Reproductive hormones are biochemical indicators that could be examined when investigating reproductive failure of fish populations. The importance of reproductive hormones in enabling fish to spawn and produce normal young is well-known. Hence, alterations in reproductive hormone levels by exposure to contaminants is a serious threat to the productivity of populations. Sangalang and Freeman (1974) monitored testosterone and 11-ketotestosterone in brook trout (*Salvelinus fontinalis*) during low-level (1 μg/l) exposure to cadmium. The concentrations of these two androgens were different in the fish exposed to cadmium, suggesting impairment in the clearance and use of testosterone and 11-ketotestosterone. The impaired clearance appeared to be related to the abnormal vasculature and hemorrhagic necrosis observed in the testes of some of the cadmium-treated fish. These results are significant because brook trout spawn in rivers and streams that are prone to industrial pollution. Further studies of fish reproductive hormones are needed, especially field verification by the measurement of reproductive hormone levels in environmentally exposed fish.

3.4. Bone Composition

Investigators at the Columbia National Fishery Research Laboratory, who used bone composition and strength as sublethal indicators of contaminant effects, found that toxaphene exposure of fathead minnows (*Pimephales promelas*) led to decreased collagen concentration, altered amino acid composition, increased calcium concentration, and decreased bone strength of the vertebrae (Mehrle and Mayer, 1975, 1981). These biochemical manifestations of the "broken back" syndrome have been recently observed in striped bass (*Morone saxatilis*) collected in tributaries of Chesapeake Bay. Collagen was lowered and bone strength decreased in striped bass with high body burdens of organochlorines and metals (P. M. Mehrle, personal communication).

4. PRIMARY AND SECONDARY STRESS INDICATORS

4.1. Primary

Other types of biochemical indicators are those associated with the generalized response of fish to stress, contaminants being one type of stress (Pickering, 1981).

Response to stress involves both the primary effects (hormonal changes) and secondary effects (Fig. 3). Steroid hormones (epinephrine, norepinephrine) that have been investigated relative to contaminants are part of the adrenal-pituitary response of vertebrates to stress. In an earlier (1977) study on lake trout fry (data unpublished) at the Great Lakes Fishery Laboratory, cortisol was measured by radioimmunoassay. No differences were found between exposed and unexposed fry. However, the variability of cortisol within treatments was high. Because fry were too small for sampling blood, the cortisol had to be extracted with methylene chloride from homogenized whole fry. Although cortisol may have been elevated during the initial exposure of the fry, the fish may have become adapted to the low-level chronic exposure. Swift (1981) reported maximum plasma cortisol values within the first few hours of exposure of rainbow trout to phenol, followed by a gradual return to normal values during the remainder of the exposure time.

Schreck and Lorz (1978) found that exposure to copper resulted in a marked, dose-dependent serum cortisol elevation in coho salmon (*Oncorhynchus kisutch*). However, treatment with cadmium did not elicit a cortisol elevation, even in moribund fish. They speculated that some, but not all, stresses produce elevations of cortisol, and that the concept of General Adaptation Syndrome for mammals may not completely apply to fish. In addition, cortisol may not be useful in detecting effects of continuous exposure to sublethal contaminants, since cortisol is elevated primarily during the initial period of exposure to the stressor and may later return to normal levels.

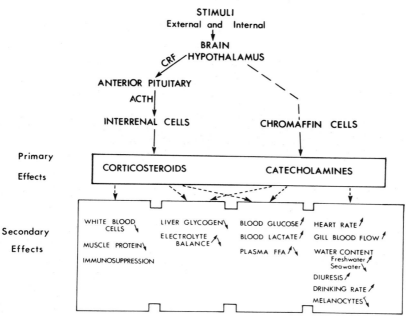

Figure 3. The interrelationship between primary and secondary effects of stress in fish. (From Mazeaud et al., 1977.)

4.2. Secondary

Secondary response to stress includes changes in blood glucose and lactate, plasma free fatty acids, electrolyte balance, liver glycogen, immunosuppression, and so on (Mazeaud et al., 1977). The clinical approach to blood chemistry is successfully applied to human beings because normal baseline values for such properties as electrolytes, glucose, and hematocrit have been established. For fish, however, such baseline values have yet to be established, especially for wild populations, although some progress is being made. Since homeostasis is not so finely tuned in poikilothermic fish as in homothermous animals, the range for normal values is much broader for fish. Consequently, deviations from normal must be greater to be statistically significant. In addition, the physiology of fish is profoundly influenced by factors such as season, spawning condition, and migrations.

5. RESEARCH NEEDS

Biochemical indicators are helpful to managers of fishery resources and water quality, especially those charged with establishing water quality criteria, only if they can be shown to influence higher levels of organization, ultimately, the population level. An observed change in a biochemical indicator must be recognizable as being caused only by contaminants and not by natural or endogenous agents. Background data on normal values and expected variability are required (Sastry and Miller, 1981), and the biochemical indicators should not be altered by the stress caused in capturing the fish. Hence, blood chemistry indicators such as glucose, electrolytes, and hemoglobin are not very useful because of the difficulty of establishing normal values and the changes in these indicators that occur during the capture of the fish.

Enzymes that are specific for particular contaminants appear to be the most promising, together with reproductive hormones that show a significant response to contaminants. More research is needed to find enzymes and hormones that show specificity toward particular contaminants. Most likely, no one biochemical indicator will suffice to diagnose the health of fish, but rather a combination of indicators will be needed. Practitioners of clinical medicine normally measure an array of physiological and biochemical indicators before diagnosing disease. Methods for clinical indicators useful in assessing the health of fish populations have been published by Wedemeyer and Yasutake (1977).

Research is needed to integrate laboratory and field studies and to develop correlations between adverse effects in laboratory animals and adverse effects in wild populations. Laboratory studies provide high precision for predicting effects of individual contaminants under controlled conditions, but these studies fail to duplicate the many interacting environmental variables. Field studies address real environmental problems but fail to establish cause-and-effect relationships and are usually not sensitive enough to detect adverse effects before they reach crisis proportions. Through discussion of both biochemical and physiological indicators and population and community indicators in this book, readers should strive to achieve an integrated approach, using multidisciplinary field and laboratory studies to interpret and understand the response of fish to contaminants.

REFERENCES

Addison, R. F. (1984). Hepatic mixed function oxidase (MFO) induction in fish as a possible biological monitoring system. In this volume, Chapter 5.

Anonymous (1978). Biochemical electrode uses tissue slices. *Chem. Eng. News* **56**(41), 16.

Bouck, G. R. (1980). Concentrations of leucine aminonaphthylamidase (LAN) and soluble protein in tissues of rainbow trout, *Salmo gairdneri*. *Can. J. Fish. Aquat. Sci.* **37**, 116–120.

Coppage, D. L., Matthews, E., Cook, G. H., and Knight, J. (1975). Brain acetylcholinesterase inhibition in fish as a diagnosis of environmental poisoning by malathion, *O,O*-dimethyl *S*-(1,2-dicarbethoxyethyl) phosphorodithioate. *Pestic. Biochem. Physiol.* **5**, 536–542.

Gingerich, W. H. and Weber, L. J. (1979). Assessment of clinical procedures to evaluate liver intoxication in fish. Environmental Protection Agency, Duluth, MN, EPA-600/3-79-088.

Hodson, P. V., Blunt, B. R., Spry, D. J., and Austen, K. (1977). Evaluation of erythrocyte δ-amino levulinic acid dehydratase activity as a short-term indicator in fish of a harmful exposure to lead. *J. Fish. Res. Board Can.* **34**, 501–508.

Hodson, P. V., Blunt, B. R., and Whittle, D. M. (1984). Monitoring lead exposure of fish. In this volume, Chapter 8.

Kearns, P. K. and Atchison, G. J. (1979). Effects of trace metals on growth of yellow perch (*Perca flavescens*) as measured by RNA-DNA ratios. *Environ. Biol. Fish.* **4**, 383–387.

Leninger, A. L. (1975). *Biochemistry*, 2nd ed. Worth, New York.

Lock, R. A. C., Cruijsen, P. M. J. M., and VanOverbeeke, A. P. (1981). Effects of mercuric chloride and methylmercuric chloride on the osmoregulatory functions of the gills in rainbow trout, *Salmo gairdneri* Richardson. *Comp. Biochem. Physiol.* **68C**, 151–159.

Marafante, E. (1976). Binding of mercury and zinc to cadmium-binding protein in liver and kidney of goldfish (*Carassius auratus* L.). *Experientia* **32**, 149–150.

Mazeaud, M. M., Mazeaud, F., and Donaldson, E. M. (1977). Primary and secondary effects of stress in fish: Some new data with a general review. *Trans. Amer. Fish. Soc.* **106**, 201–212.

Mehrle, P. M. and Mayer, F. L., Jr. (1975). Toxaphene effects on growth and bone composition of fathead minnows, *Pimephales promelas*. *J. Fish. Res. Board Can.* **32**, 593–598.

Mehrle, P. M. and Mayer, F. L., Jr. (1981). Clinical tests in aquatic toxicology. Presented at the Fish health workshop, Great Lakes Fishery Commission, Ann Arbor, MI.

Melancon, M. J., Elcombe, C. R., Vodicnik, M. J., and Lech, J. J. (1981). Induction of cytochromes P450 and mixed function oxidase activity by polychlorinated biphenyls and β-naphthoflavone in carp (*Cyprinus carpio*). *Comp. Biochem. Physiol.* **69C**, 219–226.

Meyerhoff, M. E. (1981). Potentiometric biosensors based on heterogeneous catalytic systems. In: *Proceedings of the 2nd interagency workshop on in-situ water-quality sensing: Biological sensors*. Nat. Oceanic and Atmospheric Admin., Rockville, MD, pp. 167–172.

Olson, K. R., Squibb, K. S., and Cousins, R. J. (1978). Tissue uptake, subcellular distribution, and metabolism of $^{14}CH_3HgCl$ and $CH_3^{203}HgCl$ by rainbow trout, *Salmo gairdneri*. *J. Fish. Res. Board Can.* **35**, 381–390.

Passino, D. R. M. (1981). Enzymes and other indicators of toxicant effects in fishes. In: *Proceedings of the 2nd interagency workshop on in-situ water-quality sensing: Biological sensors*. Nat. Oceanic and Atmospheric Admin., Rockville, MD, pp. 19–43.

Passino, D. R. M. and Cotant, C. A. (1979). Allantoinase in lake trout (*Salvelinus namaycush*): In vitro effects of PCBs, DDT, and metals. *Comp. Biochem. Physiol.* **62C**, 71–75.

Passino, D. R. M. and Kramer, J. M. (1982). Subcellular distribution of mercury in liver in lake trout (*Salvelinus namaycush*). *Experientia*, **38**, 689–690.

Pickering, A. D. (Ed.) (1981). *Stress and fish*. Academic Press, New York, 367 pp.

Rechnitz, G. A. (1981). Bioselective membrane electrode probes. *Science* **214**, 287–291.

Rechnitz, G. A., Arnold, M. A., and Meyerhoff, M. E. (1979). Bio-selective membrane electrode using tissue slices. *Nature (London)* **278**, 466–467.

Sangalang, G. B. and Freeman, H. C. (1974). Effects of sublethal cadmium on maturation and testosterone and 11-ketotestosterone production in vivo in brook trout. *Biol. Reprod.* **11**, 429–435.

Sastry, A. N. and Miller, D. C. (1981). Application of biochemical and physiological responses to water quality monitoring. In: F. J. Vernberg, A. Calabrese, F. P. Thurberg, and W. B. Vernberg (Eds.), *Biological monitoring of marine pollutants*. Academic Press, New York, pp. 265–294.

Schreck, C. B. and Lorz, H. W. (1978). Stress response of coho salmon (*Onchorhynchus kisutch*) elicited by cadmium and copper and potential use of cortisol as an indicator of stress. *J. Fish. Res. Board Can.* **35**, 1124–1129.

Stegeman, J. J. (1978). Influence of environmental contamination on cytochrome P-450 mixed function oxygenase in fish: Implications for recovery in the Wild Harbor Marsh. *J. Fish. Res. Board Can.* **35**, 668–674.

Swift, D. J. (1981). Changes in selected blood component concentrations of rainbow trout, *Salmo gairdneri* Richardson, exposed to hypoxia or sublethal concentrations of phenol or ammonia. *J. Fish Biol.* **19**, 45–61.

Vodicnik, M. J., Elcombe, C. R., and Lech, J. J. (1981). The effects of various types of inducing agents on hepatic microsomal monooxygenase activity in rainbow trout. *Toxicol. Appl. Pharmacol.* **59**, 364–374.

Walton, D. G., Penrose, W. R., and Green, J. M. (1978). The petroleum-inducible mixed-function oxidase of cunner (*Tautogolabrus adsperus* Walbaum 1792): Some characteristics relevant to hydrocarbon monitoring. *J. Fish. Res. Board Can.* **35**, 1547–1552.

Wedemeyer, G. A. and Yasutake, W. T. (1977). Clinical methods for the assessment of the effects of environmental stress in fish health. U.S. Fish and Wildl. Serv. tech. pap. 89, 18 pp.

Willford, W. A., Bergstedt, R. A., Berlin, W. H., Foster, N. R., Hesselberg, R. J., Mac, M. J., Passino, D. R. M., Reinert, R. E., and Rottiers, D. V. (1981). Introduction and summary. In: Chlorinated hydrocarbons as a factor in the reproduction and survival of lake trout (*Salvelinus namaycush*) in Lake Michigan. U.S. Fish and Wildl. Serv. tech. pap. 105, pp. 1–7.

5

HEPATIC MIXED FUNCTION OXIDASE (MFO) INDUCTION IN FISH AS A POSSIBLE BIOLOGICAL MONITORING SYSTEM

R. F. Addison

Marine Ecology Laboratory
Bedford Institute of Oceanography
Dartmouth, Nova Scotia

1. INTRODUCTION

Our current interest in mixed function oxidase (MFO) enzymes as an environmental monitoring tool derives from basic research into detoxification mechanisms begun in the late 1950s. These studies were usually carried out on the laboratory rat, with occasional excursions to the monkey or the Syrian hamster, and focused on understanding the mechanisms of drug disposition and metabolism. By the mid 1970s, the MFOs had been identified as one group, or element, of detoxification systems that possessed the following features:

1.   MFOs were found in mammalian tissues, usually though not exclusively the liver, and also in some invertebrates; they were components of the smooth

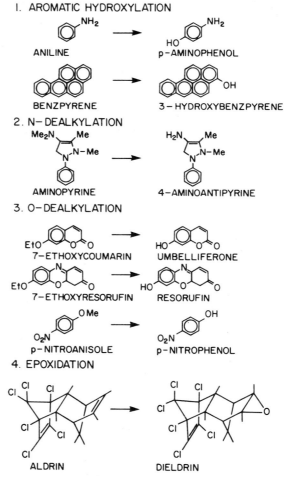

Figure 1.   Some MFO-catalyzed foreign compound degradations that have been reported in freshwater fish.

endoplasmic reticulum (SER); they catalyzed, in general, oxidation reactions involving the conversion of lipophilic substrates to more water-soluble or conjugatable products (some examples of these reactions are listed in Fig. 1).

2.   MFOs required molecular oxygen, NADPH, and a CO-sensitive cytochrome P-450 (so-called from the absorption maximum of the CO complex of its reduced form).

3.   MFOs were present with relatively low activity in normal animals, where their role seemed to be to degrade endogenous lipophilic substrates such as steroid hormones; however, if the organism were stressed by exposure to various foreign compounds, among which were environmental contaminants such as the insecticide DDT, the polychlorinated biphenyls (PCBs), or polynuclear aromatic hydrocarbons (PAHs), MFO activity would increase dramatically, apparently to enhance the degradation and clearance of the offending compounds.

This last point attracted the attention of the environmental toxicologists, for it implied that MFO activities might be used in the field as an index of the sublethal stress caused by ambient levels of environmental contaminants, that is, as a field sublethal bioassay. Since the mid-1970s various groups have explored this possibility, with varying degrees of success.

2.   MFO SYSTEMS IN AQUATIC BIOTA

A few reports have described MFO activity and induction in aquatic invertebrates, but in general activity and inducibility appear limited, and activity is reported in tissues or organs that are relatively difficult to obtain. Most studies of the potential of MFO measurements as a monitoring tool have therefore centered on fish liver, which is easier and more rewarding to investigate.

Several indices of MFO activity and induction have been studied. Many of the classic indices of MFO activity or induction that have been well-established in mammals either are not well-characterized in fish, or are not particularly responsive to inducers. For example, MFO induction in mammals is usually indicated by (1) proliferation of the SER, reflected in increases in microsomal protein concentrations; (2) increases in the amount of cytochrome(s) P-450; (3) increases in activity of microsomal enzymes that degrade various substrates. In fish, proliferation of the SER, and the accompanying increase in microsomal protein may not be clearly associated with induction (e.g., Lipsky et al., 1978); cytochrome(s) P-450 are not as well characterized as those in mammals, except in a few instances (e.g., Ahokas et al., 1976b; Chevion et al., 1977); and MFO activity is usually relatively low so that classic enzymatic reactions, such as aniline hydroxylation, proceed at rates close to their detection limit. Most studies of MFO induction in fish therefore tend to emphasize enzymatic indices, and focus on enzymatic reactions whose products can be detected at very low levels—hence the choice of fluorometric and electron capture–gas chromatographic methods, rather than colorimetric assays.

2.1.   Selectivity of Fish Hepatic MFOs to Inducing Compounds

One of the first points to emerge from studies of hepatic MFOs in fish was that these enzymes did not respond to as wide a range of inducing compounds as did those in the laboratory rat. The mammalian MFO systems were known to be induced by a wide variety of foreign compounds, including barbiturate drugs, DDT, PCBs, and PAHs (Parke, 1975). MFO systems in fish—particularly salmonids, which is the group most usually studied—were induced only by PCB mixtures or by PAHs (Buhler and Rasmusson, 1968; Addison et al., 1977, 1978; Elcombe and Lech, 1978). These observations may be interpreted as follows. MFO systems in mammals seem to fall into two major groups: (1) those that involve cytochrome(s) P-450, that catalyze reactions such as aniline hydroxylation, or aminopyrine N-demethylation, and that are induced by barbiturates or by DDT; and (2) those that involve cytochrome(s) P-448 (similar to cytochrome(s) P-450 but whose CO complex has a slightly displaced absorption maximum), that catalyze reactions such as benzpyrene hydroxylation and ethoxyresorufin O-deethylation, and that are induced by PAHs and by some PCB components, usually the planar molecules. At present, it seems that only the latter types of MFO activities are induced in fish; although MFO activity of the former type is observed, albeit at fairly low levels, this MFO activity is not induced. To further confuse the issue, there may well be species differences in fish: one histopathological study has shown some proliferation of the SER—a prelude to MFO induction—following exposure of zebrafish and guppy to DDT, although MFO induction was not measured directly (Weis, 1974).

From the evidence available now, it seems that hepatic MFO induction in fish is brought about by fewer compounds than elicit a response in the rat. Although MFO induction may therefore be less useful than was previously hoped as a general index of environmental stress, this selectivity toward a certain group of compounds may sometimes be an advantage. Table 1 summarizes induction responses that have been reported in freshwater fish exposed to various known or potential contaminants.

2.2.   Dose-Response Relationships

Dose-response relationships between MFO activity and contaminant exposure have been established only in salmonid fish.

Rainbow trout (*Salmo gairdnerii*) show generally increasing MFO activity (as illustrated by ethoxycoumarin O-deethylase, ethoxyresorufin O-deethylase, or benzpyrene hydroxylase activities) to increasing feeding levels of PCBs or poly-brominated biphenyls (PBBs) (Elcombe and Lech, 1978). The relationship between MFO activity and inducer level in the diet seemed roughly linear in dose ranges from 0–300 μg/g (approximately) (Elcombe and Lech, 1978). In contrast, ethoxycoumarin O-deethylase activity in brook trout (*Salvelinus fontinalis*) liver was related nonlinearly to the whole body residue concentration of PCBs after experimental feeding (Addison et al., 1981). (Deposition of PCBs in the body was itself related nonlinearly to the dose fed.) The most sensitive part of the dose-response curve was in the range of approximately 0 to 5 $\mu g/g$ wet weight of PCB, which is the range most significant from a human exposure viewpoint. In short, the few existing data show that a quantifiable dose-response relationship exists between MFO activities and exposure to, or accumulated burdens of, inducing compounds at environmentally realistic levels. However, the nature of the relationship seems variable and requires calibration before MFO induction could be used as a monitoring tool.

2.3.   Time Course of Induction

MFO induction in fish occurs fairly rapidly: increases in hepatic MFO activity were detected within two days of feeding a PCB mixture to brook trout (Addison et al., 1981) and other studies (e.g., Elcombe and Lech, 1978) have confirmed that induction occurs well within a week of exposure to inducing materials. MFO activity remained elevated for at least three weeks after the brook trout were fed with PCBs, and MFO levels were well-correlated with the gradually declining PCB residue concentrations in the fish over this period. Other studies on other fish have shown that MFO induction is rapid (generally occurring within a week or less of initial exposure to inducers) and may remain elevated for periods of several months (e.g., Payne and Fancey, 1982). Thus it appears that MFO activity measurements will rapidly reflect exposure to inducers, integrated over reasonable periods of time.

2.4.   Response of MFO Enzymes to Factors Other Than Foreign Compounds

2.4.1.   Reproduction

MFO activities in fish may be increased by natural factors or processes, as well as by exposure of the fish to foreign compounds. The most extensively studied of such natural factors has been sex or reproductive status. However, the effects of these conditions are not yet clear. In our experience with trout (Addison et al., 1977, 1978, 1979, 1981, 1982), sexually immature fish do not show any clear differences in MFO activity or in susceptibility to induction, and Forlin and Lidman (1979) report no sex differences in immature rainbow trout. However, sexually mature fish of various species show differences in basal MFO activity, and sometimes in cytochrome P-450 content (e.g., Forlin and Lidman, 1979; Stegeman and Chevion, 1980). Not only may there be sex differences in basal MFO activity, but the susceptibility to induction may also vary between the sexes (e.g., Forlin and Lidman, 1979; Forlin, 1980); furthermore, some MFO enzymes are more susceptible than others. As might be expected, species differences in variation of MFO activity with sex also occur (Stegeman, 1981).

The impact of sex or reproductive status on MFO activity or induction is therefore difficult to assess. However, since one function of MFOs is, apparently, to degrade certain steroid hormones, a relationship between MFO activity and sexual condition is not surprising. From the viewpoint of environmental monitoring, sexual or re-

Table 1. Responses of Freshwater Fish to Known or Potential MFO Inducers

Inducer[a]	Fish Species	Induction Observed[a]	Reference
		DDT Group	
p,p′-DDT	Salvelinus fontinalis	None	Addison et al. (1977)
p,p′-DDT	Salmo gairdneri	None	Buhler and Rasmusson (1968)
p,p′-DDE	S. fontinalis	None	Addison et al. (1977)
		PCBs	
Aroclor 1242 + 1254	S. gairdneri	BPH, ECOD, EROD	Franklin et al. (1980)
Clophen A-50	S. gairdneri	BPH, PNAD	Forlin (1980)
Aroclor 1254	S. fontinalis	ECOD, EROD	Addison et al. (1981)
Aroclor 1254	Cyprinus carpio	BPH, ECOD, EROD	Melancon et al. (1981)
Aroclor 1248	C. carpio	APND	Ito and Murata (1980)
Aroclor 1254	Ictalurus punctatus	AH, APND	Hill et al. (1976)
		Polybrominated Biphenyls	
FireMaster BP-6	S. gairdneri	BPH, ECOD, EROD	Elcombe and Lech (1978)
FireMaster BP-6	S. fontinalis	BPH	Law and Addison (1981)

Dioxins			
2,3,7,8-TCDD	*S. gairdneri*	ECOD, EROD	Vodicnik et al. 1981
Cyclodienes			
Chlordane	*S. gairdneri*	None	Buhler and Rasmusson (1968)
Chlordane	*Oncorhynchus kisutch*	None	Buhler and Rasmusson (1968)
PAHs			
3-MC	*S. fontinalis*	ECOD, AH, AE	Addison et al. (1978)
3-MC	*S. gairdneri*	BPH, PNAD	Forlin (1980)
Mixed PAHs	*Salmo trutta*	BPH	Payne and Penrose (1975)

[a] The following abbreviations, not previously identified, have been used.

AE	Aldrin epoxidase
AH	Aniline hydroxylase
APND	Aminopyrine N-demethylase
BPH	Benzpyrene hydroxylase
ECOD	Ethoxycoumarin O-deethylase
EROD	Ethoxyresorufin O-deethylase
PNAD	p-Nitroanisole O-demethylase
3-MC	3-Methylcholanthrene
2,3,7,8-TCDD	2,3,7,8-Tetrachlorodibenzodioxin.

productive status is a factor that should be eliminated as far as possible during sampling.

2.4.2. Size

A further factor that may affect uninduced MFO activity is that of size, which may be related to age or maturity. In our experience, larger fish generally have lower MFO activity than smaller ones, other factors being equal. We have tried to explain this variation in terms of a weight versus metabolic rate relationship (Addison and Willis, 1982); whether or not this is the explanation of the variation, our data indicate that weight (or age, or maturity) is another factor to be eliminated during environmental sampling.

3.   ENVIRONMENTAL MONITORING USING MFO ACTIVITY MEASUREMENTS

Most "calibrations" of the responses of fish hepatic MFOs to foreign compounds described above have been carried out on hatchery fish. These animals have the advantage of being (usually) of fairly uniform age and size, and may be from a limited genetic stock, and only one sex may be selected. Even in such homogenous samples, MFO induction by foreign compounds is a highly variable response. Fish caught in the wild are rarely as homogenous as those available from hatcheries: the natural factors and processes I have listed above as affecting MFO activity may be more pronounced, and it is more difficult to ascribe differences in MFO activity to the single cause of a difference in environmental contamination. Most studies of MFO activities in "wild" organisms have therefore compared samples from dramatically different environments, such as those contaminated by major oil spills, and those that remain relatively pristine.

  These studies have shown reasonably convincingly that fish from heavily oiled environments have appreciably higher MFO activities than those from clean environments. Burns (1976) found that *Fundulus heteroclitus* from an oil spill site had higher levels of aldrin epoxidase and benzpyrene hydroxylase activity than comparable fish from a clean site. Rather surprisingly (in view of the apparent insensitivity of fish MFOs to chlorinated hydrocarbon pesticides) *Fundulus* from a marsh sprayed with these materials also showed increased MFO activity. Payne and Penrose (1975) found similar differences in arylhydrocarbon hydroxylase activities in trout (*Salmo trutta*) from a clean lake and from one contaminated by petroleum. This was correlated with hydrocarbon content of the sediment. A more subtle effect, that of offshore hydrocarbon drilling operations, was reflected in slightly increased benzpyrene hydroxylase activities in various marine fish collected close to the drilling rig, compared to samples from a control area (Davies et al., 1981). Unfortunately, none of these studies related MFO activities to residues of petroleum in the fish, presumably because of the difficulties of the chemical analyses.

  It is worth noting that environmental stress does not always result in induction of MFO systems. Fish taken from a generally and heavily polluted environment showed some depression rather than induction in MFO activity (e.g., Ahokas et al., 1976a). Whether this reflects the effects of materials (other than MFO inducers) that may cause general liver damage is not clear.

4. CONCLUSIONS

It would be misleading to suggest that MFO induction measurements are sufficiently domesticated to be used routinely as a sensitive monitor of sublethal environmental stress. Although in the laboratory, under well-controlled conditions, MFO induction may represent a quantitative response to graded levels of contaminants, we know too little about its biochemical basis, or about its response to interfering factors to use it confidently as a field bioassay of subtle effects of specific organic contaminants. No doubt this state of affairs will improve. In the meantime, MFO induction may be useful as a laboratory sublethal bioassay of the possible effects of materials that are present or may be future environmental contaminants (e.g., Addison et al., 1978; Law and Addison, 1981; Addison et al., 1982).

REFERENCES

Addison, R. F. and Willis, D. E. (1982). Variations of hepatic ethoxycoumarin O-de-ethylase activity with body weight and other factors in brook trout (*Salvelinus fontinalis*). *Can. J. Fish. Aquat. Sci.* **29**, 924–926.

Addison, R. F., Zinck, M. E., and Willis, D. E. (1977). Mixed function oxidase enzymes in trout (*Salvelinus fontinalis*) liver: absence of induction following feeding of *p,p'*-DDT or *p,p'*-DDE. *Comp. Biochem. Physiol.* **57C**, 39–43.

Addison, R. F., Zinck, M. E., and Willis, D. E. (1978). Induction of hepatic mixed-function oxidase (MFO) enzymes in trout (*Salvelinus fontinalis*) by feeding Aroclor 1254 or 3-methylcholanthrene. *Comp.Biochem. Physiol.* **61C**, 323–325.

Addison, R. F., Zinck, M. E., Willis, D. E., and Darrow, D. C. (1979). Induction of hepatic mixed function oxidases in trout by polychlorinated biphenyls and butylated monochlorodiphenyl ethers. *Toxicol. Appl. Pharmacol.* **49**, 245–248.

Addison, R. F., Zinck, M. E., and Willis, D. E. (1981). Time- and dose-dependence of hepatic mixed function oxidase activity in brook trout *Salvelinus fontinalis* on polychlorinated biphenyl residues: implications for "biological effects monitoring." *Environ. Poll. (Ser. A)* **25**, 211–218.

Addison, R. F., Zinck, M. E., Willis, D. E., and Wrench, J. J. (1982). Induction of hepatic mixed function oxidase activity in trout (*Salvelinus fontinalis*) by Aroclor 1254 and some aromatic hydrocarbon PCB replacements. *Toxicol. Appl. Pharmacol.* **63**, 166–172.

Ahokas, J. T., Karbe, N. T., Oikari, A., and Soivio, A. (1976a). Mixed function monooxygenase of fish as an indicator of pollution of aquatic environment by industrial effluent. *Bull. Environ. Contam. Toxicol.* **16**, 270–274.

Ahokas, J. T., Pelkonen, D., and Karki, N. T. (1976b). Cytochrome P-450 and drug induced spectral interactions in the hepatic microsomes of trout, *Salmo trutta lacustris. Acta. Pharm. Toxicol.* **38**, 440–449.

Buhler, D. R. and Rasmusson, M. (1968). The oxidation of drugs by fishes. *Comp. Biochem. Physiol.* **25**, 223–239.

Burns, K. A. (1976). Microsomal mixed function oxidases in an estuarine fish, *Fundulus heteroclitus* and their induction as a result of environmental contamination. *Comp. Biochem. Physiol.* **53B**, 443–446.

Chevion, M., Stegeman, J. J., Peisach, J., and Blumberg, W. E. (1977). Electron paramagnetic resonance studies on hepatic microsomal cytochrome P-450 from a marine teleost fish. *Life Sci.* **20**, 895–900.

Davies, J. M., Hardy, R., and McIntyre, A. D. (1981). Environmental effects of North Sea oil operations. *Mar. Poll. Bull.* **12**, 412–416.

Elcombe, C. R. and Lech, J. J. (1978). Induction of monooxygenation in rainbow trout by polybrominated biphenyls: a comparative study. *Environ. Health Persp.* **23**, 309–314.

Forlin, L. (1980). Effects of Clophen A50, 3-methylcholanthrene, pregnenolone-16α-carbonitrile and phenobarbital on the hepatic microsomal cytochrome P-450-dependent monooxygenase system in rainbow trout, *Salmo gairdneri*, of different age and sex. *Toxicol. Appl. Pharmacol.* **54**, 420–430.

Forlin, L. and Lidman, N. (1979). O-Demethylation of *p*-nitroanisole by mixed function oxidase system of the rainbow trout (*Salmo gairdneri*) liver. *Gen. Pharmacol.* **10**, 411–415.

Franklin, R. B., Elcombe, C. R., Vodicnik, M. J., and Lech, J. J. (1980). Comparative aspects of the disposition and metabolism of xenobiotics in fish and mammals. *Fed. Proc.* **39**, 3144–3149.

Hill, D. W., Hejtmancik, E., and Camp, B. J. (1976). Induction of hepatic microsomal enzymes by Aroclor® 1254 in *Ictalurus punctatus* (channel catfish). *Bull. Environ. Contam. Toxicol.* **16**, 495–502.

Ito, Y., and Murata, T. (1980). Studies on the influence of polychlorinated biphenyls (PCB) on aquatic organisms. III. Influence of PCB on drug-metabolising enzyme system in carp. *Eisei Kagaku* **26**, 21–27.

Law, F. C. P. and Addison, R. F. (1981). Response of trout hepatic mixed function oxidases to experimental feeding of ten known or possible chlorinated environmental contaminants. *Bull. Environ. Contam. Toxicol.* **27**, 605–609.

Lipsky, M. J., Klaunig, J. E., and Hinton, D. E. (1978). Comparison of acute response to polychlorinated biphenyl in liver of rat and channel catfish: a biochemical and morphological study. *J. Toxicol. Environ. Health* **4**, 107–121.

Melancon, M. J., Elcombe, C. R., Vodicnik, M. J., and Lech, J. J. (1981). Induction of cytochromes P-450 and mixed function oxidase activity by polychlorinated biphenyls and β-naphthoflavone in carp (*Cyprinus carpio*). *Comp. Biochem. Physiol.* **69C**, 219–226.

Parke, D. V. (1975). Induction of the drug-metabolizing enzymes. In: D. V. Parke (Ed.), *Enzyme induction.* Plenum Press, London, pp. 207–271.

Payne, J. F. and Fancey, L. L. (1982). Effect of long term exposure to petroleum on mixed function oxygenases in fish: further support for use of the enzyme system in biological monitoring. *Chemosphere* **11**, 207–213.

Payne, J. F. and Penrose, W. R. (1975). Induction of aryl hydrocarbon (benzo[a]pyrene) hydroxylase in fish by petroleum. *Bull. Environ. Contam. Toxicol.* **14**, 112–116.

Stegeman, J. J. (1981). Polynuclear aromatic hydrocarbons and their metabolism in the marine environment. In: H. V. Gelboin and P. O. P. Ts'o (Eds.), *Polycyclic hydrocarbons and cancer, Vol 3.* Academic Press, New York, pp. 1–60.

Stegeman, J. J. and Chevion, M. (1980). Sex differences in cytochrome P-450 and mixed function oxygenase activity in gonadally mature trout. *Biochem. Pharmacol.* **28**, 553–558.

Vodicnik, M. J., Elcombe, C. R., and Lech, J. J. (1981). The effect of various types of inducing agents on hepatic microsomal monooxygenase activity in rainbow trout. *Toxicol. Appl. Pharmacol.* **59**, 364–374.

Weis, P. (1974). Ultrastructural changes induced by low concentrations of DDT in the livers of the zebrafish and the guppy. *Chem.-Biol. Interact.* **8**, 25–30.

6

PHYSIOLOGICAL RESPONSES OF FISH: PROBLEMS AND PROGRESS TOWARD USE IN ENVIRONMENTAL MONITORING

Gerald R. Bouck

Division of Fish and Wildlife
Bonneville Power Administration
U.S. Department of Energy
Portland, Oregon 97232

1. INTRODUCTION

The measurement of sublethal stress in wild fish is a problem that remains essentially unsolved today. As a result, aquatic environmental quality continues to be measured mainly with administratively established chemical parameters. In some cases environmental quality can be and is estimated by responses of ecosystems (e.g., species diversity), by populations (e.g., recruitment), or by individuals (e.g., avoidance). Except for growth, physiological parameters of fish are conspicuously absent from field studies, a fact that is in contrast to their common use in laboratory studies. This paper explores this apparent paradox, lists some problems of physiological testing in the field, and describes investigations of plasma enzymes and myogens as possible alternative measures of fish health.

The physiological approach to measuring fish health portends several advantages in environmental studies. It is widely accepted that excessive environmental stress causes a variety of detectable, recognizable changes in the blood and tissues of fish. This is because an environmental stress response begins at the molecular/cellular level and then extends to tissues and organs, whose responses occur before changes occur in populations and ecosystems. Therefore, measurement of fish health could provide early warning of excessive environmental stress, and would do so comparatively faster and cheaper than monitoring of population responses. Blood analyses also could allow health assessment without destroying the donor fish, an important consideration for endangered species or those highly valued by the public. Furthermore, domestic fish can be caged as test organisms when wild fish are impractical to collect. Finally there are some cases when it is imperative to know the current health status of fish prior to selecting alternatives in hydromanagement.

2. FACTORS THAT AFFECT THE DIAGNOSTIC USE OF PHYSIOLOGICAL TESTS

2.1. Field Application of Tests

Activity in the field is controlled mainly by the weather and other physical factors. Collection of fish can be dangerous and strenuous both for investigator and for fish. Further limitations are imposed by unstable surfaces, cold temperatures, and protective clothing that can destroy manual dexterity and render testing impossible. For example, inclement weather can transform otherwise simple tasks such as weighing fish into extremely challenging undertakings.

Unfortunately there are several other serious problems limiting the potential usefulness of physiological tests in field work (Wydoski and Wedemeyer, 1976), and most of these are still unresolved. Sensitive indices of stress can be used on fish in laboratory tanks, but wild fish must be collected before they can be tested. This creates two problems: (1) how to collect and test statistically valid numbers of fish in the field; and (2) recognizing that capture may stress the fish, which tests should be used on wild fish? (As a corollary to the latter, is it possible to develop new tests that are insensitive to capture stress, but responsive to chronic environmental stress?)

Both samples and reagents require greater preparation for the field and must be protected from adverse temperatures, contamination, deterioration, and breakage. For example, assume one wants to test the level of leucine aminonaphthylamidase (LAN) in the blood of wild fish; this requires either serum or plasma. Whole fish blood from stressed fish clots almost immediately (Bouck and Ball, 1965), which delays further action (overnight) until clot retraction expresses the serum. Additionally, sunshine and warm temperatures cause hemolysis, which alone would destroy the sample for LAN analysis. A preferable course of action is to use an anticoagulant, and harvest and freeze the plasma immediately after a short centrifugation. The choice of anticoagulants is again influenced by field conditions: heparin is too unreliable to risk its usage, and oxalate tends to cause hemolysis; sodium citrate is highly effective and reliable but precludes the determination of plasma sodium; citrate also chelates divalent cations that may influence plasma enzyme activities. At room temperatures, about 80% of the LAN will be lost in 24 hr (Bouck et al., 1975). Obviously the effects of sample handling can be influential and must be understood prior to working in the field.

A related problem concerns the development and application of new tests for stress in fish, a process that requires long and difficult investigations before routine clinical use is warranted. Therefore, it is important to distinguish between the *investigative use* and the *clinical use* of tests on fish. Investigative use is the development and testing of potentially useful analyses. Clinical use implies that the method has been validated by extensive testing—both to identify the biological significance of the parameter and the limitations in applying the test to the species in question. Given those definitions, the following must be recognized: (1) relatively few tests for stress in fish have progressed beyond investigative stages; and (2) relatively few tests currently lend themselves to field application. The result of these two factors often produces data whose meaning is not currently apparent or relevant to environmental quality.

2.2. Fish Capture Methods

Bouck and Ball (1966) tested and rejected the null hypothesis that blood characteristics would not be affected by three capture methods: hooking, seining, or electroshocking. The fish were all collected at the same time, confined in flowing water, anesthetized with buffered tricaine methane sulfonate (MS-222), and sampled at 15-minute intervals. Capture methods affected the concentrations of hemoglobin, plasma protein, erythrocytes, and five electrophoretic fractions of plasma protein, consistent with subsequent reports (Wedemeyer, 1972; Schreck et al., 1976; Wydoski et al., 1976).

The above problems might be solved by collecting blood more quickly after capture (or by measuring slower responses to stress). This hypothesis was tested by comparing fish that were hand-netted individually from aquaria to fish that were sampled immediately after capture by seining, electroshocking, or angling (Bouck et al., 1978). As before (Bouck and Ball, 1966), hematocrits and hemoglobin were disturbed by capture (Table 1). Plasma acid phosphatase (AP) activity was too low to measure accurately. Creatine phosphokinase (CPK) responded significantly to handling stress. The mean activity of the other two enzymes, lactate dehydrogenase (LDH) and LAN, did not differ significantly between capture methods. This work demonstrated that some blood parameters respond rapidly to stress (see also Chavin, 1973) and are unacceptable for detecting low levels of environmental stress in wild fish. For others, the time interval between capture and blood collection must be very short, probably no more than 1 minute. Therefore, fish must not be collected faster than they can be sampled.

The required number of fish to be collected at each station depends on the sample variance of each parameter and was estimated for LDH (Bouck et al., 1978). Twenty fish would produce a mean LDH level with a 95% confidence interval of 10 international units (about 3% of control activity, Table 1). However, the statistical comparison of enzyme levels may be less necessary than with other characteristics. Tissue damage generally causes such a great increase in some plasma enzymes (Hess, 1963) that "stressed" individuals might be identified on an all-or-none (quantal) basis.

2.3. Repeated Bleeding of Fish

An alternative approach to testing environmental quality is by repeated sampling of blood from caged fish. Wild fish might also be resampled, either by intention or accident. Cairns and Christian (1978) reported that repeated bleeding seriously disturbed the blood of rainbow trout (*Salmo gairdneri*). However, their sampling method was nonsterile and included stresses such as electroshocking and starvation, which probably exceeded those of a realistic blood sampling program. Therefore, we studied the effect of repeated bleeding on several blood parameters of rainbow trout as described below.

Each fish was in its own aquarium, supplied with flowing water, and fed Oregon Moist Diet at 7% of its body weight per week. After anesthesia with MS-222, blood was drawn aseptically from the dorsal aorta once a week for 6 weeks. There levels of bleeding were used: 0.3, 0.6, and 0.9% of the body weight. These amounts equalled 50, 100, and 150% of their estimated safe bleeding level (Klontz and Smith, 1968), assuming that the total blood volume was about 6% of body weight (Smith, 1966). Controls included undisturbed fish, which were sampled only once, and sham lots, which were punctured frequently with the hypodermic needle but were bled only once. Blood analyses measured hematocrits, the levels of hemoglobin, plasma protein, plasma glucose, and plasma chloride, and the activities of LDH and LAN. The analyses were done as before (Bouck et al., 1978) and methods were comparable to those of Cairns and Christian (1978).

Weekly bleeding at 0.3, 0.6, and 0.9% of the body weight each produced significant changes in the blood of rainbow trout. The most profound changes were

Table 1. Comparison of Hematocrit, Plasma protein, and Some Plasma Enzyme Levels of Rainbow Trout (*Salmo gairdneri*) upon Capture by Four Methods. Mean Values + SE with Numbers of Fish in Parentheses[a]

Fork Length (cm)	Total Weight (g)	Hematocrit (%)	Plasma Protein Concentration (g/100 ml)	Enzyme Activity Units Per Gram of Protein			
				Plasma Acid Phosphatase	Plasma Lactate Dehydrogenase	Plasma Creatine Phosphokinase	Plasma Leucine Aminonaphthyl-amidase
Hand-netted individually from aquaria							
25.5 (9)	180 (9)	28.6 ± 0.8 (10)	3.52 ± 0.16 (10)	0.029 ± 0.003 (9)	339.2 ± 14.7 (9)	1.35 ± 0.25 (9)	20.20 ± 1.19 (10)
Seined from large tank							
25.5 (10)	185 (10)	34.6 ± 0.6[b] (8)	3.60 ± 0.13 (10)	0.023 ± 0.002 (9)	370.2 ± 21.6 (10)	1.05 ± 0.14[b] (10)	18.16 ± 1.22 (10)
Electroshocked in individual aquaria							
25.4 (10)	193 (10)	29.5 ± 1.9 (9)	3.34 ± 0.12 (9)	0.0023 ± 0.003 (9)	367.8 ± 24.2 (9)	2.91 ± 0.33[b] (9)	18.28 ± 1.53 (9)
Caught by hook and line from large tank							
24.7 (10)	168 (10)	33.6 ± 2.9[b] (9)	3.46 ± 0.21 (10)	0.0021 ± 0.002 (10)	369.8 ± 26.8 (10)	1.61 ± 0.50[b] (10)	19.50 ± 0.94 (10)

[a] After Bouck et al., 1978.

[b] Significantly different ($p < .05$) from control (hand-netted) by analysis of variance.

in hematocrit and hemoglobin. Plasma protein and glucose were lower in bled than in unbled control groups, while chlorides were relatively stable in all groups over the 6 week period. The activity of LDH was too variable for interpretation. Conversely, plasma LAN decreased in bled fish, both in activity per 100 ml of plasma and per gram of plasma protein. Unfortunately, the experimental fish began reproductive development during the study, which could have affected the results. However, weekly bleeding did change some blood characteristics, even at half the level recommended by Klontz and Smith (1968). The use of caged fish or repeated blood samples is not ruled out, but further study must define the proper combination of bleeding volume and frequency.

3. PLASMA ENZYMES AS INDICATORS OF STRESS

The measurement of blood characteristics that respond slowly to capture stress was recommended in Section 2.2. One obvious possibility was the assay of changes in plasma enzyme levels, which can be classified as either *blood specific* (those having a function and purpose in the blood) or *blood nonspecific* (those derived from moribund cells). Blood nonspecific enzymes are preferred for this purpose because their presence may indicate tissue damage possibly within a single cell type of one organ or tissue.

3.1. Leucine Amino Naphthylamidase (LAN)

This proteolytic enzyme is located mainly in lysosomes (Kaulen et al., 1970). On cell death, LAN is released from the lysosomes and assists in autolysis. Once enzymes cross cell membranes into the interstitial space, they are carried passively by the lymphatic system to the blood and circulate with the other constituents until eliminated. Consequently, LAN may be a measure of cellular death: its activity rises in human plasma during periods of stress, particularly during wasting diseases (Hess, 1963).

3.1.1.   LAN Assays in Fish Blood

The assay of LAN activity in rainbow trout was reported by Bouck et al. (1975), along with several of its biochemical characteristics. Maximum activity in vitro occurred at 55°C, diminishing above or below this temperature; 95% of the enzyme activity was lost at 70°C. Therefore, sample storage is a problem, since 80% of the activity could be lost at room temperature in 1 day.

Maximum LAN activity in vitro was at pH 8.0, well above proposed physiological limits. Although citrate at anticoagulant levels did not inhibit LAN, both cyanide and ethylene diamine tetraacetic acid (EDTA) did. Centrifugation for 10 minutes at $1200 \times g$ sedimented about 50% of the LAN activity in disrupted liver; this fraction contained cellular debris and lysosomes. Considerable LAN activity occurred in cellular sap in vitro, but it is not clear whether it was active in vivo.

At least three isozymes of LAN were identified after electrophoresis of tissue extracts: LAN-I in liver, gill, and kidney; LAN-II in midgut, ceca, and hindgut; and LAN-III (isoelectric) in liver, kidney, midgut, ceca, and hindgut. Only LAN-I was found in the plasma of control rainbow trout. Plasma LAN activity was strongly linked to several blood or body characteristics at two acclimation temperatures.

3.1.2. Tissue Distribution of LAN

LAN has a discrete distribution within tissues of rainbow trout, a property that may have some diagnostic advantages. Bouck (1980), using a histochemical method, reported high levels of LAN-II and III in the epithelial mucosa of midgut, ceca, and hindgut. Some LAN was in hepatocytes, but the highest levels in the liver were in the Kupffer cells. Kidney LAN (LAN-I and LAN-II) was located in the epithelium of parts of the renal tubule. A small amount of LAN was present in the gills. In summary, LAN occurred mainly in those tissues most frequently affected by toxicants: gill, liver, kidney, and gut.

The distribution of LAN in rainbow trout was also measured biochemically by methods more sensitive and quantitative than histochemistry (Bouck, 1980). LAN was detected in all tissues except brain. The lowest activity in healthy fish was in blood plasma, where it may represent the residue from moribund and dead cells. All other tissues had considerably higher activities, ranging from 2.2 to 36 times the level in blood plasma. Given this distribution and concentration gradient, it would be difficult to damage any tissue except brain without increasing the LAN level in blood. Some cellular LAN might be removed by the lymphatic system or excreted via urine and/or feces, but such removal has not been proved. A large but unknown proportion of the LAN from dead cells passes from the interstitial space into the blood. However, damage to an equal weight of several tissues would not necessarily increase plasma LAN activity equally, since LAN activities are not evenly distributed.

3.1.3. Effect of Different Stressors on Plasma LAN Activity

The responses of LAN to various stresses have been measured over a decade. The tests included fish from three different gene pools and (Table 2) modifying factors such as age, acclimation temperature, water supply, and diet. Despite the above factors, control fish from a given gene pool had remarkably similar levels of LAN activity. Levels of plasma LAN were higher in older than in younger fish, as is true for plasma protein concentrations.

Nutrition had a very important effect on plasma LAN activity over a 6-week period (Table 2). Compared with fish fed Oregon Moist Diet (59.6 units/100 ml), fasting fish had an activity of 103.4 units/100 ml and fish fed sweet corn had 73.9. Increased LAN may indicate cell destruction due to protein deficiency, similar to wasting disorders of nutritionally deficient mammals (Hess, 1963) and fish (Sauer and Haider, 1979). Malnutrition may account for the elevated LAN levels reported by Cairns and Christian (1978), whose fish were both starved and bled. In my study, the plasma LAN level dropped among well fed rainbow trout that were bled repeatedly (Table 2). Plasma activity fell about 36% over a 6 week period, indicating

Table 2. Summary of Leucine Aminonaphthylamidase (LAN) Levels in Plasma of Rainbow Trout (*Salmo gairdneri*) Under Various Conditions

	Mean LAN Activities	
Condition	Units/100 ml Plasma	Units/g Plasma Protein
Capture Stress (Bouck et al., 1978)[a]		
Handnetted (control)	60.4	20.2
Electroshocked	70.6	18.3
Hooking	67.0	19.5
Seining	65.6	18.2
Physiological Stress		
Repeated bleeding (Cairns and Christian, 1978)[a]		
Initial	68.0	—
Ninth bleeding	125.0	—
Repeated bleeding (Bouck, unpublished data)		
Control	84.7	35.2
0.3% of body wt per wk for 6 wks	80.0	22.4
0.6% of body wt per wk for 6 wks	81.5	21.1
0.9% of body wt per wk for 6 wks	74.7	21.8
Nutritional Stress (Bouck, unpublished data)[b]		
Control (OMP Diet)	59.6	21.2
Fasting	103.4	27.8
Sweet corn diet	73.9	34.4
Environmental Stress		
Temperature (Bouck et al., 1975)[a] 7°C =	53.4	24.3
17°C =	39.1	17.5
Salinity (Bouck, unpublished data)[c] control =	80.0	22.1
216 hr in seawater =	92.3	24.1
Toxic Stress (Bouck, unpublished data)[b]		
OMP	59.6	21.2
OMP + 10% borate	88.8	24.0

[a] Wisconsin rainbow trout studied in Michigan.

[b] Roaring rainbow trout studied in Oregon.

[c] Puyallup rainbow trout studied in Washington.

a loss due to bleeding that exceeded the input from all other sources. Therefore, bleeding alone may not increase LAN levels in blood.

I used salinity as a model environmental stress for rainbow trout (Table 2). Freshwater trout can adjust to seawater in the spring, but they generally do not thrive in high salinity seawater. Changes in their blood were followed over a 10-day acclimation period in 29‰ seawater. Plasma chlorides rose rapidly in the first

2 days, but returned to normal by day 10. Some dehydration and hemoconcentration occurred at first, as judged by hematrocrits and hemoglogin. Plasma protein and plasma glucose rose slowly, peaked on day 3, and returned to normal soon thereafter. Plasma LAN and LDH activities also showed an initial transitory rise as a result of hemoconcentration. However, LAN activity continued to rise and appeared to be still rising on day 10 when the study ended. Sauer and Haider (1979) reported that a related enzyme, leucine amino peptidase, in rainbow trout was elevated at 10‰ salinity, but depressed at 20‰.

Changes in plasma LAN activity were also investigated in rainbow trout fed 6 weeks on Oregon Moist Diet containing 10% tetrahydroborate. LAN activity in plasma rose in fish fed borate, but it is not clear whether the increase represented intoxication or diminished feed consumption. However, these results were consistent with those of Racicot et al. (1975). They reported that both carbon tetrachloride toxicity and infection increased the activities of several plasma-nonspecific enzymes of rainbow trout.

4. PROTEIN COMPOSITION AS A POTENTIAL INDICATOR OF STRESS

Although blood composition may reflect environmental stress, it is clear that additional methods must be developed and tested. An alternative to blood assays is electrophoresis of soluble proteins in skeletal muscle (myogens), which are easier to collect and assay than plasma/serum protein.

Rock bass (*Ambloplites rupestris*) were sampled by Bouck (1972) after diurnal exposure to low oxygen for 9 days (3 ppm oxygen for 8 hr each day). Muscle was excised and frozen under simulated field conditions, but capture stress was excluded. Separation of myogens, hemogens, and plasma proteins by electrophoresis showed that the amount of proteins in each fraction changed significantly in response to sublethal hypoxia; generally the amount of low mobility proteins increased. Judging from other associated changes (hematocrits, hemoglobin, plasma protein concentration, body weight gain, ventilation activity, and feeding behavior), the changes in protein were responses to adverse environmental conditions. The psysiological significance of these protein changes is uncertain; however, since they involve about 80% of the fish's body weight and composition, they should not be dismissed lightly. Equally important, muscle samples can be collected from organisms too small for blood sampling.

5. SUMMARY

The use of physiological responses of wild fish to assess environmental quality is seldom successful, in part because wild fish are usually stressed during collection. Capture stress precludes the use of sensitive, early indicators of environmental stress. An alternative strategy diminishes the effects of capture by the use of more slowly responding indicators, such as leakage of enzymes from cells into the blood. Studies were conducted on plasma and tissue leucine amino naphthylamidase activity (LAN) in rainbow trout. Biochemical studies validated the methodology and demonstrated the presence of LAN in all tissues except brain. LAN is apparently liberated from lysosomes of dead cells and passes indirectly into the blood as debris. Plasma LAN can be changed by environmental stress (salinity), nutritional stress (fasting), physiological stress (reproduction, fasting, and bleeding), and toxicant stress (borate). Therefore, LAN may be useful in assessing wild fish, provided they are sampled immediately upon capture.

REFERENCES

Bouck, G. R. (1972). Effects of diurnal hypoxia on electrophoretic protein fractions and other health parameters of rock bass (*Ambloplites rupestris*). *Trans. Amer. Fish. Soc.* **101**, 488–493.

Bouck, G. R. (1980). Concentrations of leucine aminonaphthylamidase (LAN) and soluble proteins in tissues of rainbow trout (*Salmo gairdneri*). *Can. J. Fish. Aquat. Sci.* **37**, 116–120.

Bouck, G. R. and Ball, R. C. (1965). Influence of a diurnal oxygen pulse on fish serum proteins. *Trans. Amer. Fish. Soc.* **94**, 363–370.

Bouck, G. R. and Ball, R. C. (1966). Influence of capture methods and blood characteristics and mortality in the rainbow trout (*Salmo gairdneri*). *Trans. Amer. Fish. Soc.* **95**, 170–176.

Bouck, G. R., Schneider, P. W., Jr., Jacobson, J., and Ball, R. C. (1975). Characterization and subcellular localization of leucine aminonaphthylamidase (LAN) in rainbow trout (*Salmo gairdneri*). *J. Fish. Res. Board Can.* **32**, 1289–1295.

Bouck, G. R., Cairns, M. A., and Christian, A. R. (1978). Effect of capture stress on plasma enzyme activities in rainbow trout (*Salmo gairdneri*). *J. Fish. Res. Board Can.* **35**, 1485–1488.

Cairns, M. A. and Christian, A. R. (1978). Effects of hemorrhagic stress on several blood parameters in adult rainbow trout (*Salmo gairdneri*). *Trans. Amer. Fish. Soc.*, **107**, 334–340.

Chavin, W. (1973). Teleostean endocrine and para-endocrine alterations of utility in environmental studies. In: W. Chavin (Ed.), *Responses of fish to environmental changes*. Thomas Publishing Co. Springfield, IL, pp. 199–239.

Hess, B. (1963). *Enzymes in blood plasma*. Academic Press, New York, 167 pp.

Kaulen, H. D., Henning, R., and Stoffel, W. (1970). Biochemical analyses of the pinocytotic process: II. Comparison of some enzymes of the lysosomal and the plasma membrane of the rat lower cell. *Hoppe-Seyler's Physiolog. Chem.* **351**, 1551–1563.

Klontz, G. W. and Smith, L. S. (1968). Methods of using fish as biological research subjects. In: W. I. Gay (Ed.), *Methods of animal experimentation*, Vol. 3. Academic Press, New York, pp. 323–385.

Racicot, J. G., Gaudet, M., and Leray, C. (1975). Blood and liver enzymes in rainbow trout (*Salmo gairdneri*) emphasis on their diagnostic use: Study of CCl_4 toxicity and a case of Aeromonas infection. *J. Fish Biol.* **7**, 1–12.

Sauer, D. M. and Haider, G. (1979). Enzyme activities in the plasma of rainbow trout, *Salmo gairdneri* Richardson, the effects of nutritional status and salinity. *J. Fish Biol.* **14**, 407–412.

Schreck, C. B., Whaley, R. A., Bass, M. L., Maughan, O. E., and Solazzi, M. (1976). Physiological responses of rainbow trout (*Salmo gairdneri*) to electroshock. *J. Fish. Res. Board Can.* **33**, 76–84.

Smith, L. S. (1966). Blood volumes of three salmonids. *J. Fish. Res. Board Can.* **23**(9), 1439–1446.

Wedemeyer, G. (1972). Some physiological consequences of handling stress in the juvenile coho salmon (*Oncorhynchus kisutch*) and steelhead trout (*Salmo gairdneri*). *J. Fish. Res. Board Can.* **29**, 1730–1738.

Wydoski, R. S. and Wedemeyer, G. A. (1976). Problems in the physiological monitoring of wild fish populations. In: *Proc. of the annu. conf. of the Western Assoc. of Game and Fish Commissioners*, vol. 56, pp. 200–214.

Wydoski, R. S., Wedemeyer, G. A., and Nelson, N. C. (1976). Physiological response to hooking stress in hatchery and wild rainbow trout (*Salmo gairdneri*). *Trans. Amer. Fish. Soc.* **105**, 601–606.

7

FISH SERUM CHEMISTRY AS A PATHOLOGY TOOL

W. L. Lockhart and D. A. Metner

Department of Fisheries and Oceans
Freshwater Institute
Winnipeg, Manitoba

1.   INTRODUCTION

Several authors have reported chemical analyses of blood taken from fish under a variety of circumstances (Hunn, 1967; Love, 1970). One approach has been similar to that of hospital laboratories in which a series of measurements are made in an attempt to characterize an individual. The individual can then be compared with reference data from other experience and any differences from "normal" assay values can be recognized. While this approach has some statistical limitations in its application to fish, it is often successful in discriminating groups of fish exposed to some experimental stress from similar groups not exposed to the stress. In this paper we relate some of our experience in applying biochemical analyses to fish from laboratory experiments and from field collections.

There is no attempt to provide a detailed literature review of fish blood biochemistry; this has been provided recently for rainbow trout by Hille (1982). Similarly, there is no attempt to review effects of environmental stressors on blood biochemistry; Wedemeyer and McLeay (1981) have provided a timely review of this subject.

2. MATERIALS AND METHODS

2.1. Fish Species

Species included in this paper are rainbow trout (*Salmo gairdneri*), white sucker (*Catostomus commersoni*), and brook stickleback (*Culaea inconstans*).

2.2. Fish Sampling

Our intent with experimental groups of fish in the laboratory has been to maintain treated and untreated groups of fish under similar circumstances so that any sampling or experimental artifacts (Bouck and Ball, 1966) should be common to both groups and should therefore not obscure legitimate treatment effects. Live capture of wild fish from field sites was accomplished with trap nets, hoop nets, seine nets, or minnow traps set in shallow (< 3 m) water.

Blood was obtained from large rainbow trout by restraining them immediately upon removal from the water and generally by inserting a needle to puncture caudal vessels (Steuke and Schoetteger, 1967). With white suckers, however, we have been more successful with heart puncture than with caudal puncture. Small fish (about 10 g or smaller) have been sampled by severing the caudal peduncle and collecting blood in a microhematocrit tube.

Once obtained, blood was kept chilled on ice and allowed to clot for at least 1 hour; it was then centrifuged so that serum could be conveniently withdrawn. When sampling at field locations sera were frozen on dry ice for transport to the laboratory; when sampling at the laboratory sera could be analyzed fresh starting on the day of collection.

2.3. Biochemical Analyses

Methods employed were generally adaptations of procedures described for microliter quantities of human serum (Mattenheimer, 1970) as listed below:

Analysis	Method
Hemoglobin (whole blood)	Cyanmethemoglobin
Hematocrit (whole blood)	Centrifugation
Sodium, potassium	Flame emission
Calcium, magnesium	Atomic absorption
Chloride	Microtitration
Protein	Biuret
Creatinine	Alkaline picrate
Urea nitrogen	Urease, then ammonia assay
Glucose	Glucose oxidase
Total lipid	Sulfophosphovanillin
Triglyceride	Glycerol kinase
Alkaline phosphatase (AP)	p-Nitrophenyl phosphate
Lactate dehydrogenase (LDH)	Oxidation of NADH
Glutamate oxalacetic transaminase (GOT)	MDH coupled oxidation of NADH
Cholinesterase	Acetylthiocholine
Uric acid	Uricase

As much as possible, reagents were obtained as prepackaged kits designed for use in clinical chemistry laboratories and were scaled down to suit amounts of fish blood available.

2.4. Sources of Fish

Fish were obtained from a variety of laboratory experiments ranging from exposures to chemicals in glass aquaria maintained at constant temperature and photoperiod to outdoor polyethylene pools with no physical control other than the supply of water. Fish were also captured from several lakes, rivers, and streams undergoing different pollutant loadings.

For laboratory tests a control group was always included in the experimental design, but with field samples this was not always possible. In the case of samples from flowing waters following single-event pollutant inputs, the best control samples were considered to be from the stream before the input event, and from upstream of the input during and after the event. With standing water where the pollutant concentration changed rapidly, as in forest spraying, pretreatment samples were considered to be the best controls.

3. RESULTS AND DISCUSSION

3.1. Fish Exposed in Outdoor Polyethylene Pools

Rainbow trout were maintained in 12 outdoor polyethylene pools as described by Lockhart et al. (1975) and fish in 2 of these pools were exposed to a synthetic triaryl phosphate oil, IMOL-S-140. For a 4 month period from July 9 to November 13 one of these groups of fish was monitored for biochemical responses to this treatment, and results were compared with the same fish from before treatment and with other groups of fish not receiving the treatment. During this period the blood biochemistry of fish exposed to this toxicant became highly abnormal with respect to the enzymes LDH and GOT. Table 1 shows some of the values obtained for the exposed group. Apparently, fish treated with IMOL-S-140 experienced a degree of stress since some began to refuse floating pellets of food (Purina Trout Chow) 8 days after the start of treatment, and all completely rejected food after 14 days. It was also noted that treated fish remained in a covered portion of their pool during daylight hours, while untreated fish did not. Early in November some fish in the treated pool died, and internal fatty tissues were found to be discolored. The measurements of LDH and GOT had shown a response to treatment within 16 days, and this was even more striking with the final samples in November. The mean LDH and GOT values for this group of fish exceeded by severalfold any we observed with the other groups of fish held in these outdoor pools.

In this instance when the fish were clearly experiencing a stress condition, the blood biochemistry did in fact distinguish them from other fish not experiencing the stress.

3.2. Fish Exposed in Laboratory Vessels

We have also exposed rainbow trout to chemicals under laboratory conditions with better control over variables such as temperature and photoperiod than in the case of outdoor pools. For example, fingerlings were exposed to the insect growth regulator, methoprene, at 10 mg/l, in a continuous flow diluter at 15° C for 96 hr (Madder and Lockhart, 1978). Blood analysis indicated a significant fall in serum glucose, with no measurable effect on hematocrit, lipid, GOT, or sodium. In view of this effect on serum glucose, the exposure was repeated using methoprene concentrations from 0.625 mg/l to 10 mg/l. A dose-dependent fall in serum glucose values was noted for concentrations below the solubility of methoprene (1.39 mg/l) with no further drop in glucose at concentrations over saturation (Fig. 1).

This observation is difficult to explain in terms of a fish "equivalent" of a general adaptation syndrome, since the response to a stressor would typically be an increase rather than a drop in serum glucose values. It seems possible that methoprene acted specifically on sugar regulation rather than eliciting a generalized response to stress; the fish showed no obvious indication of irritation. However, in this case also, the

Table 1. **Mean Blood Chemistry Data for Rainbow Trout Exposed to a Mixed Triaryl Phosphate Oil, IMOL-S-140, from July 9 to November 6**[a]

		May 22	July 25	November 13	Control (July 25)
Number Sampled		7	8	11	8
Hematocrit (%)	Mean	37	[29]	[22]	41
	S.D.	4.2	5.6	7.6	4.6
Hemoglobin (g/100 ml)	Mean	8.1	[6.8]	[6.7]	9.4
	S.D.	0.96	1.13	2.43	1.24
Glucose (mg/100 ml)	Mean	100	73	140	114
	S.D.	12.6	36.1	59.9	48.6
Calcium (mg/100 ml)	Mean	10.8	15.2	13.1	14.0
	S.D.	1.02	4.63	1.51	2.36
Magnesium (mg/100 ml)	Mean	2.1	3.1	2.4	3.4
	S.D.	0.18	0.68	0.29	0.38
Chloride (mg/100 ml)	Mean	493	547	491	588
	S.D.	15.3	21.0	18.5	42.3
LDH (mU/ml)	Mean	721	[4730]	[34470]	1550
	S.D.	397	2205	59970	993
GOT (mU/ml)	Mean	175	[410]	[5200]	145
	S.D.	51.5	82.4	5402	57.0
AP (mU/ml)	Mean	37	12	20	19
	S.D.	12.1	4.14	12.2	7.26
Protein (g/100 ml)	Mean	4.30	4.21	4.07	4.12
	S.D.	0.14	0.47	0.89	0.78
Creatinine (mg/100 ml)	Mean	0.43	0.42	0.21	0.34
	S.D.	0.07	0.15	0.03	0.07
Urea N (mg N/100 ml)	Mean	0.86	0.49	0.91	0.63
	S.D.	0.47	0.23	1.03	0.17
Sodium (mg/100 ml)	Mean	391	406	394	426
	S.D.	28.8	14.7	24.6	25.2
Potassium (mg/100 ml)	Mean	11.9	15.6	9.92	9.89
	S.D.	4.05	4.61	2.65	1.81
Cholinesterase (mU/ml)	Mean	—	52.9	35.3	60
	S.D.		10.2	7.92	20.3

[a] Values considered abnormal are shown in brackets. Comparisons were by t test using values transformed to natural logarithms, and the criterion for significance was $p < .05$.

blood biochemistry discriminated among control and exposure groups, although not in the pattern that might have been expected.

3.3. Fish with Diseases

Rainbow trout under our hatchery conditions have sometimes developed a kidney condition described as nephrocalcinosis (Gillespie and Evans, 1979). Biochemical

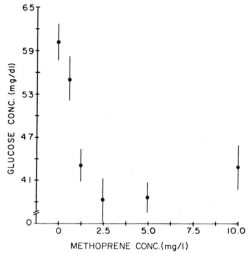

Figure 1.   Serum glucose concentrations in rainbow trout fingerlings exposed to several concentrations of methoprene (Altosid, Zoecon Corp.). The solubility of methoprene is 1.39 mg/l and concentrations nominally above that were presumably at saturation of 1.39 mg/l.

analysis of the blood of these fish (Yurkowski et al., 1983) has revealed surprisingly few major effects on serum chemistry in view of the gross changes physically apparent in kidney morphology. The clearest indication of any failure in kidney function may have been the creatinine values, which increased from about 0.2 mg/100 ml in fish with the least severe symptoms to about 0.8 mg/100 ml in fish with the most severe symptoms. A more complete description of the blood chemistry of these fish can be found in Yurkowski et al. (1983).

Another group of rainbow trout maintained in outdoor pools developed a severe case of columnaris disease (Lockhart et al., 1974) with about a 50% mortality. The causative agent was identified as a strain of *Chrondrococcus* of intermediate virulence since individuals with severe gross lesions sometimes recovered completely. Fish showing clinical signs were characterized by striking elevations in blood glucose, and by a doubling of urea nitrogen with no change in uric acid and creatinine, the other nitrogenous compounds measured. Urea is excreted mainly by the gills, while creatinine and uric acid are excreted by the kidneys (Smith, 1929). Columnaris disease is thought to damage the respiratory functional capacity of gills (Pacha and Ordal, 1970) and so it would not be surprising if the anatomical damage to gills also diminished the clearance of urea and resulted in some uremia. The elevated serum glucose may be interpreted either as a response to respiratory insufficiency (Menton, 1927), or perhaps more likely, as a secondary aspect of a generalized response due to the disease.

In comparing the two diseases, nephrocalcinosis with its gross effects on kidney morphology had its major effect on a metabolic product (creatinine) excreted by the kidney (Table 2) with little or no effect on one excreted by the gills (urea). Columnaris disease, with an obvious effect on the gills, was characterized by an elevation in a product excreted by the gills (urea), with little or no change in creatinine or uric acid, which are excreted by the kidneys. Based on these observations it seems reasonable to suggest that measures of these nitrogenous compounds, and perhaps ratios of them, may offer promise as a diagnostic test of the functional integrity of kidneys and gills.

3.4.   Fish Sampled from Field Populations

Field samples of fish often show a greater statistical variation than laboratory samples subjected to the same analyses. For example, cholinesterase enzymes are often used as tools to monitor for sublethal poisoning after applications of organophosphorus and carbamate pesticides. In the laboratory it is straightforward to show the enzyme inhibiting effect of exposure to these materials by comparing a group of exposed fish with a control group. For example, a group of rainbow trout killed by exposure to fenitrothion had a mean brain cholinesterase value of 124 ($\pm$15) mU/mg brain protein as compared to control values of 181 ($\pm$13). Standard deviations were of the order of 10% of the mean values, enabling relatively small differences to be detected with simple statistical tests. Fig. 2 illustrates the problem of applying the cholinesterase technique to a field sample of brook sticklebacks captured at intervals before and after spraying over a stream in a forest area near Spruce Woods Park, Manitoba (Lockhart et al., 1977). From these data it was concluded that little or no sublethal poisoning resulted from the spray operation. It is evident from the spread of values (standard deviations about one third of means) that larger sample numbers would be needed to detect the same degree of effect demonstrated with laboratory experiments with uniform groups of fish. As an example of this problem, a laboratory group with a mean of 100 and a standard deviation of 10 might be compared with a field group having a mean of 100 and a standard deviation of 30.

Table 2.   Blood Serum Creatinine Values in Hatchery Rainbow Trout with Different Severities of Nephrocalcinosis

| | | Creatinine (mg/100 ml) | |
		Few or No Symptoms of Disease	Severe Symptoms of Disease
Post spawning males	Mean	0.20	0.75
	S.D.	0.115	0.238
	n	7	4
Post spawning females	Mean	0.20	0.78
	S.D.	0.074	0.148
	n	12	5
Spawning females	Mean	0.23	0.90
	S.D.	0.096	0.219
	n	4	6

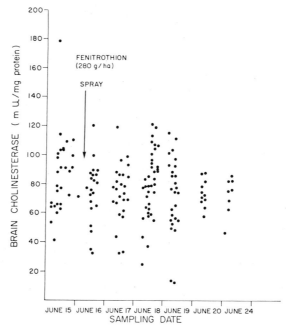

Figure 2.   Brain cholinesterase values (mU/mg protein) for brook sticklebacks captured before and after spraying a forested area in southern Manitoba with fenitrothion at 280 g/ha in June, 1975.

In order to detect the same reduction in mean values due to poisoning, the field sample must be larger by a factor equal to the ratio of the two variances; hence the field sample would have to be about nine times larger. Practically this means that the amount of work required to show even a well-established response with field populations is formidable. In order for biochemical or other methods to be practical tools in fisheries biology, however, it is essential that they be applicable to field populations.

Field samples of several species of fish were captured from the Athabasca River, Alberta, following applications of methoxychlor for control of blackfly larvae in 1974 and 1975. Examination of serum from these fish revealed an unusual group of white suckers captured about 75 km downstream from the treatment site on July 24, several weeks after treatment. For example, this group, although only four fish, had serum protein values consistently under 2 mg/100 ml (Fig. 3). Other white suckers captured during the sampling program did not develop such extremely low protein values. For comparison, several groups of white suckers from a number of locations have been analyzed for serum protein, with results shown in Table 3. Generally suckers have protein values well over 3 g/100 ml of serum, but those from the 75 km site were consistently low, between 2 and 4 g/100ml, with the most extreme values under 2 g/100 ml on July 24. This Athabasca River sample from late July was outside the range of all our previous experience with field or laboratory samples. We cannot be certain what caused an unusual set of values like

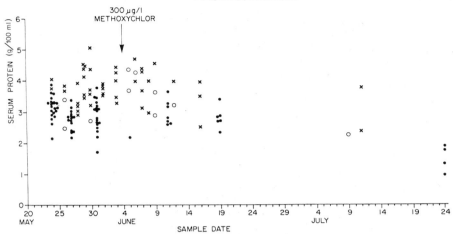

Figure 3. Blood serum protein from white suckers captured in the Athabasca River, Alberta, in 1974 at several sites upstream and downstream from an injection of methoxychlor on June 4.

these, and the possibility of some artifact of sampling or analysis always arises, although we believe that to be unlikely. Experiments with methoxychlor poisoning have led us to rule out the possibility that methoxychlor exposure was responsible by any direct pharmacological action on the fish. Examination of fish from the same site in July of the following year revealed no unusually low values, and so the phenomenon was probably not just an extreme in seasonal variation.

During our field sample collections from the Athabasca River similar efforts were being made by entomologists to understand the impact of methoxychlor treatment

Table 3. Blood Serum Protein Values for White Suckers Captured at Various Locations

Source of Fish	Number of Fish	Mean Serum Protein (g/100 ml ($\pm$ S.D.))	Range
George Lake, Ont.	22	4.5 (0.69)	2.9–6.0
Georgian Bay, Ont.	11	5.0 (0.85)	3.8–6.9
Lake Manitoba	28	4.4 (0.73)	3.2–5.7
Prince Albert, Sask. (1973)	6	3.3 (0.41)	2.6–3.6
Prince Albert, Sask. (1974)	12	3.4 (0.48)	2.7–4.2
Lake 224, Kenora, Ont.	12	5.4 (0.49)	4.9–6.4
Lake 240, Kenora, Ont.	12	4.6 (0.43)	4.0–5.4
Roddy Lake, Kenora, Ont.	12	3.9 (0.49)	3.1–4.7
Roddy Lake, Kenora, Ont.	14	3.9 (0.51)	3.0–4.6
Lake 223, Kenora, Ont.	10	4.5 (0.37)	3.9–5.2

on populations of benthic invertebrate animals. Flannagan et al. (1979) observed that invertebrate populations were decreased substantially for a long period after treatment at the site where fish with low protein values were taken, giving rise to the hypothesis that fish may have undergone a period with inadequate food available. Invertebrate populations nearer the treatment site appeared to recover more readily, presumably as a result of recolonization from untreated reaches upstream.

The entomological observations gave rise to speculation that white sucker serum protein values might indicate chronic food shortage, and this hypothesis was tested on a group of white suckers captured in the Red River, Manitoba, and maintained in the laboratory for several months. These fish all fed readily on Silver Cup trout food until they were divided into two groups; one group continued to receive a daily ration of food and the other received none. At intervals over several weeks five fish were sampled from each group, and serums were analyzed for protein, total lipid, and triglyceride (Table 4). The results indicated (analysis of variance) that all three serum measurements were sensitive to the starvation, but it is clear that a period somewhat longer than 8 weeks was required to produce the responses in serum values.

These results are not presented as evidence that methoxychlor treatment destroyed invertebrate animals, and that fish populations feeding on those invertebrates were short of food, although the results may be consistent with such an interpretation. Rather they are presented to show that a relatively simple serum measure like serum protein may have responded in some unsuspected way to the history of fish. It may be possible, in some cases, to use the biochemistry to extrapolate back from the fish to some feature of habitat quality. In the case cited here we suspect that habitat carrying capacity was reduced due to effects of methoxychlor on fish food organisms, and that the fish population had not yet adjusted to the new circumstances.

If indeed the serum protein measurement reflected the history of the fish with respect to some critical resource like food, and further if that same resource were also limiting to growth, then we might expect some relationship between body size measurements and serum protein measurements. Fig. 4 shows a preliminary plot of the relationship between the length-weight measurements and serum protein

Table 4.   Mean Values for White Sucker Blood Sera Analyzed for Protein, Total Lipid, and Triglyceride, After Periods of Feeding or Starvation

Time After Start (weeks)	Protein (g/100 ml)	Lipid (mg/100 ml)	Triglyceride (mg/100 ml)
0 Feeding	4.20 (0.89)	1650 (572)	331 (110)
0 Starving	3.90 (0.74)	1220 (418)	214 (93.6)
4 Feeding	3.95 (0.70)	1420 (288)	344 (89.7)
4 Starving	3.82 (0.45)	1110 (340)	254 (98.8)
8 Feeding	4.10 (0.91)	1390 (571)	353 (230)
8 Starving	4.17 (0.54)	1260 (193)	259 (66.3)
16 Feeding	4.81 (0.48)	1550 (294)	416 (72.1)
16 Starving	3.06 (0.57)	662 (362)	118 (121)

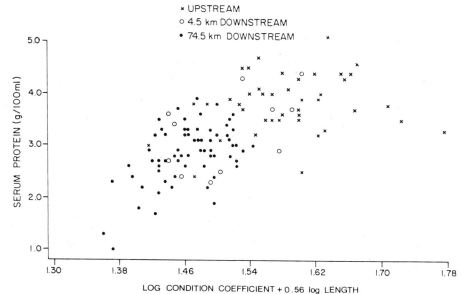

Figure 4. Blood serum protein of Athabasca River white suckers plotted as a function of body size measurements. The abscissa is the log "condition factor" plus 0.56 log length. The condition factor was calculated as the (weight × 100000)/length^3.

measurements of white suckers captured from the Athabasca River in 1974. We used a one-way analysis of variance (Statistical Analysis Services' general linear models procedure, SAS Institute, 1979) to compare serum protein between sites with log length and log weight as covariates. The site means were significantly different, and contributions of the covariates were also significant ($p < .02$). The solution was

serum protein (g/100 ml) = 3.704 log weight (g) -9.023 log length (mm) $+K$

($K = 16.4096$ for site 1, upstream; $K = 16.0666$ for site 2, 4.5 km downstream; $K = 15.8376$ for site 3, 74.5 km downstream.)

$$R^2 = 0.470$$

Since log condition factor (CF) = (5 + log weight) $-$ (3 log length), the solution equation can be rewritten as

serum protein (g/100 ml) = 3.704 (log CF + 0.564 log length) $-18.5185 + K$

(K values are as shown above). In Fig. 4 the relationship between serum protein and the expression (log CF + 0.564 log length) is shown.

It would be simplistic to argue that the statistical relationship among these measures resulted from the brief interval between methoxychlor treatment on June 4 and sample collection on July 24. Even if serum protein values responded in that time interval, it is hard to imagine fish length making such a rapid corresponding adjustment. Inspection of Fig. 3 suggests that the protein values differed among sites even before treatment. At most then, the treatment may have augmented differences that already existed among groups of fish taken at the different points in the river. It is tempting to speculate that the site 74.5 km downstream showing the low protein values in July was naturally limiting to fish in terms of food and that this situation was exacerbated by methoxychlor treatment.

Clearly a single set of data such as that in Fig. 4 could not establish a general relationship; however, the high level of statistical significance suggests that serum analysis might have some unexploited potential to reveal aspects of the history of fish in relation to habitat quality. As a general hypothesis this seems amenable to field and laboratory testing.

4. SUMMARY

Fish from a variety of sources ranging from laboratory experiments to field capture have been examined for serum biochemical components. In cases where some type of toxic chemical or disease stressor was known to be acting on the fish, the biochemical measurements consistently separated exposed fish from others not experiencing the known stressor. It has been assumed that the many unknown stressors acting on experimental groups (crowding, handling, parasites, undiagnosed diseases, etc.) have been acting on both experimental and control groups, and so should not obscure treatment effects. If the nature of the applied stressor is known, then biochemical responses can suggest possible modes of toxic action; however, if the applied stressor is unknown, then the biochemical analyses are unlikely to allow its unequivocal identification.

In conclusion, the use of biochemical methods offers promise in three areas, namely detection of states of stress, suggestion of modes of action, and tentatively as tools to explain the metabolic basis for conventional fishery statistics like growth. A great deal of further work will be needed to derive the statistical and mechanistic relationships linking biochemical characteristics with exposures to specific groups of chemicals or to population characteristics, but the potential seems clearly established. These procedures could be applied as readily to Great Lakes populations as to those reported here, and could provide a basis for agreement on the presence or absence of states of stress, as well as providing clues to explain more conventional fishery production statistics.

REFERENCES

Bouck, G. R. and Ball, R. C. (1966). Influence of capture methods on blood characteristics and mortality in the rainbow trout (*Salmo gairdneri*). *Trans. Amer. Fish. Soc.* **95**, 170–176.

Flannagan, J. F., Townsend, B. E., DeMarch, B. G. E., Friesen, M. K. and Leonhard, S. L. (1979). The effects of an experimental injection of methoxychlor on aquatic invertebrates: Accumulation, standing crop, and drift. *Can. Entomol.* **111**, 73–89.

Gillespie, D. C. and Evans, R. E. (1979). Composition of granules from kidneys of rainbow trout (*Salmo gairdneri*) with nephrocalcinosis. *J. Fish. Res. Board Can.* **36**, 683–685.

Hille, S. (1982). A literature review of blood chemistry of rainbow trout, *Salmo gairdneri* Rich. *J. Fish. Biol.* **20**, 535–569.

Hunn, J. B. (1967). Bibliography on the blood chemistry of fishes. U.S. Dept. of the Interior, Fish and Wildl. Serv. Bureau of Sport Fish. and Wildl., Res. rep. 72, 32 pp.

Lockhart, W. L., Gillespie, D. C., and Metner, D. A. (1974). Changes in blood chemistry associated with Columnaris disease in rainbow trout. Unpublished manuscript, 12 pp.

Lockhart, W. L., Wagemann, R., Clayton, J. W., Graham, B., and Murray, D. (1975). Chronic toxicity of a synthetic tri-aryl phosphate oil to fish. *Environ. Physiol. Biochem.* **5**, 361–369.

Lockhart, W. L., Flannagan, J. F., Moody, R. P., Weinberger, P., and Greenhalgh, R. (1977). Fenitrothion monitoring in southern Manitoba. In: *Proc. Symp. Fenitrothion*, Nat. Res. Council Canada, NRCC no. 16073, pp. 233–252.

Love, R. M. (1970). *The chemical biology of fishes.* Academic Press, London.

Madder, D. J. and Lockhart, W. L. (1978). A preliminary study of the effects of diflubenzuron and methoprene on rainbow trout (*Salmo gairdneri* Richardson). *Bull. Environ. Contam. Toxicol.* **20**, 66–70.

Mattenheimer, H. (1970). *Micromethods for the clinical and biochemical laboratory.* Ann Arbor Science Publishers, Ann Arbor.

Menton, M. L. (1927). Changes in the blood sugar of the cod, sculpin, and pollock during asphyxia. *J. Biol. Chem.* **72**, 249–253.

Pacha, R. E. and Ordal, E. J. (1970). Myxobacterial diseases of salmonids. In: *Symp. on diseases of fishes and shellfishes.* Amer. Fish. Soc. spec. pub. 5, pp. 243–257.

SAS Institute (1979). *SAS Users Guide*, 1979 ed. SAS Inst., Raleigh, NC.

Smith, H. W. (1929). The excretion of ammonia and urea by the gills of fish. *J. Biol. Chem.* **81**, 727–742.

Steuke, E. W. Jr. and Schoetteger, R. A. (1967). Comparison of three methods of sampling trout blood for measurement of hematocrit. *Prog. Fish Cult.* **29**, 98–101.

Wedemeyer, G. A. and D. J. McLeay (1981). Methods for determining the tolerance of fishes to environmental stressors. In: A. D. Pickering (Ed.), *Stress in fish.* Academic Press, New York, pp. 247–275.

Yurkowski, M., Gillespie, D. C., Metner, D. A., and Lockhart, W. L. (1983). Description of nephrocalcinosis stages and their influence on the biochemical parameters of blood and blood serum of mature rainbow trout. Manuscript in preparation.

8

MONITORING LEAD
EXPOSURE OF FISH

Peter V. Hodson, Beverley R. Blunt, and D. Michael Whittle

Great Lakes Fisheries Research Branch
Department of Fisheries and Oceans
Canada Centre for Inland Waters
Burlington, Ontario L7R 4A6

1. INTRODUCTION

Acute and chronic toxicity of lead to fish includes histopathologies, deformities, neurotoxicity, hemolytic anemia, and inhibition of hemoglobin synthesis (Hodson et al., 1984). These effects occur at waterborne lead concentrations as low as 8 μg/ l (Davies et al., 1976); water quality objectives for Lakes Superior, Huron, Michigan, Erie, and Ontario are 2, 3, 4, 4, and 5 μg Pb/l, respectively (AEOC, 1980). However, the measurement of lead levels in water to determine if objectives are met does not mean that lead effects on fish can be predicted. Current methods for water analysis cannot measure biologically available lead. Also, the frequency of sampling is usually inadequate to estimate the dose accumulated by fish over space and time. Since it is impossible to predict the effects on fish of an unknown exposure, biological surveys are required to detect lead toxicity.

This paper reviews methods for measuring the exposure of Great Lakes fish to lead so that effects may be predicted more accurately.

1.1. Techniques for Assessing Lead Exposure

Several methods are used in human medicine to diagnose lead exposure. The most common are measurements of the levels of blood lead, blood porphyrins, and urine amino levulinic acid (ALA), and of the inhibition of erythrocyte δ-amino levulinic acid dehydratase (ALA-D, EC 4.2.1.24) (Granick et al., 1972; Posner, 1977).

1.1.1. Blood Lead

Blood lead levels in fish directly reflect lead exposure and are related to lead neurotoxicity. They increase proportionally with waterborne lead levels in both rainbow trout (*Salmo gairdneri*) and brook trout (*Salvelinus fontinalis*) and the slope of this relationship is similar to that of most other tissues (Hodson et al., 1978b; Holcombe et al., 1976). The prevalence of black tails, a sign of lead neurotoxicity in trout, increases when blood lead levels exceed 300 μg/l (Holcombe et al., 1976; Hodson et al., 1978b, 1979, 1980b, 1982).

Blood lead levels of fish are more sensitive indicators of lead exposure than whole-body lead analyses. The detection limit for blood lead assays by furnace atomic absorption spectrophotometry (AAS) (method of Hodson et al., 1977) is 10 μg/liter, while that for whole-body lead by flame AAS after acid digestion is 30 μg/kg (Hodson et al., 1982).

Fish exposed to lead experimentally may have up to 10,000 μg/l of blood lead (Hodson et al., 1978b, 1983b; Krajnović-Ozretić and Ozretić, 1980). The "normal" values for unexposed control fish are ≤ 100 μg/l. However, many fish from clean offshore waters have levels < 10 μg/l (Hodson et al., 1983). Therefore the laboratory "normal" may represent some contamination, either of the fish or of the sample during handling.

The variability of blood lead tests is due to analytical error and variation between fish. We sampled a large volume of blood from one rainbow trout exposed to lead in the laboratory. Subsamples were freeze-dried and analyzed for lead on several occasions with samples taken in the field; the arithmetic mean and standard deviation were 1460 ± 162 μg Pb/l ($N = 10$; relative standard deviation = 11.1%) (Hodson and Blunt, unpublished data). Blood lead levels are usually reported as "μg of lead per volume of whole blood," and we have adopted this convention. However, more than 90% of blood lead is bound to red blood cells (RBCs) (Posner, 1977). Therefore, the variability of results between fish may be reduced by dividing whole blood concentrations by the hematocrit to give "μg of lead per volume of RBCs" (Hodson. unpublished data).

In our experience, the major problems of this technique are sample contamination and the expense and complexity of furnace AAS.

1.1.2. ALA-D Activity

ALA-D catalyzes the synthesis of one molecule of porphobilinogen (PBG) from two molecules of ALA. PBG is a precursor of hemoglobin, catalase, various cytochromes, and chlorophyll a. Therefore ALA-D occurs in most aerobic bacteria, plants, and animals (Hernberg and Nikkanen, 1972). In fish, ALA-D is a very good indicator of hemopoiesis in the spleen and kidney (Johansson-Sjöbeck, 1979; Johansson-Sjöbeck and Larsson, 1978), while in dogfish (*Scyliorhinus canicula*), high activity is found in the kidney, epigonal organ, and Leydig's organ (Fänge and Johansson-Sjöbeck, 1975).

ALA-D activity is very sensitive in vitro to a variety of metals due to its high dependence on sulfhydryl groups. However, only lead is effective in vivo. Activity is inhibited in mammals and birds at blood lead levels far below those causing hemoglobinemia (Hernberg and Nikkanen, 1972; Dieter and Finley, 1979). Lead-induced inhibition of ALA-D occurs in liver, kidney, spleen, and blood of fish (Jackim, 1973; Johansson-Sjöbeck and Larsson, 1979); the most convenient rich source appears to be the blood. ALA-D may be a "relict" enzyme in the blood of mammals because hemoglobin formation is usually complete before RBCs circulate. As a result erythrocyte ALA-D can be greatly inhibited without hemoglobinemia (Posner, 1977).

The assay of erythrocyte ALA-D activity is quite simple. Erythrocytes are hemolyzed in a soapy, buffered medium with (test) or without (blank) substrate (ALA) and incubated for a fixed time at a fixed temperature. The reaction is stopped by deproteination. Following centrifugation, the amount of PBG in the supernatant is measured by adding dimethylaminobenzaldehyde. The reaction product is the "Ehrlich salt," and its absorption is measured at 620 nm. Activity is calculated from the nanomolar extinction coefficient for the Ehrlich salt, corrected for incubation time and hematocrit (the enzyme is intracellular). Activity is expressed as "nanomoles

PBG produced per ml of red blood cells per hour" (nm PBG/ml RBCs·hr) (Hodson et al., 1977).

1.1.3. Urinary ALA

When ALA-D is inhibited in vivo by lead, excess substrate (ALA) is excreted in the urine (Posner, 1977). Assays of ALA in fish urine have not been tried due to the problems of reliably collecting samples of urine.

1.1.4. Blood Porphyrins

Levels of protoporphyrin, zinc protoporphyrin, and coproporphyrin increase in the blood of lead-exposed mammals due to the effects of lead on the break down of hemoglobin and the inhibition of porphyrin synthetase (Granick et al., 1972). Porphyrins fluoresce when exposed to light, and the wavelength at which each fluoresces is distinctive (Granick et al., 1972). We used scanning spectrofluorometry on acid extracts of fish blood to look for porphyrins in the blood of chronically lead-exposed and control rainbow trout. Although the lead exposure was sufficient to cause neurotoxicity and to inhibit ALA by more than 50%, we found only very low levels of porphyrins (Hodson, Whittle, and Blunt, unpublished data). Therefore, either the method described by Granick et al. (1972) was inadequate, or the interaction between lead and hemoglobin metabolism in fish differs from that in humans.

2. LABORATORY STUDIES OF BLOOD LEAD AND ALA-D ACTIVITY OF FISH

2.1. Species

ALA-D inhibition in vivo was first seen in the livers and kidneys of lead-exposed killifish (*Fundulus heteroclitus*) and flounder (*Pleuronectes platessa*) (Jackim, 1973). Later in vivo studies have shown lead inhibition of erythrocyte ALA-D in goldfish (*Carassius auratus*), rainbow trout, pumpkinseed (*Lepomis gibbosus*), brook trout, carp (*Cyprinus carpio*), dogfish (*Scyliorhinus canicula*), and mullet (*Mugil auratus; M. chelo*) (D'Amelio et al., 1974; Hodson, 1976; Hodson et al., 1977; Johansson-Sjöbeck and Larsson, 1979; Beritić et al., 1980; Krajnović-Ozretić and Ozretić, 1980).

2.2. Relationship to Lead Exposure and Lead Effects

ALA-D activity is negatively correlated to total waterborne lead levels (Hodson et al., 1977; Johansson-Sjöbeck and Larsson, 1978; Haux et al., 1980). However, the link is not very consistent due to many biological and environmental factors that affect the amount and uptake of available lead (Hodson et al., 1983). In contrast, there is a strong negative correlation between the ALA-D activity of rainbow trout and their blood lead levels—the slopes of regression lines from a variety of experiments are parallel and repeatable (Hodson et al., 1979). Therefore, blood lead levels and ALA-D activities are better measures of biologically available lead than analyses of waterborne lead.

A relationship also exists in lead-exposed fish between ALA-D activity and neurotoxicity. Black tails are more prevalent in lead-exposed rainbow trout when ALA-D is inhibited by 50% or more (Hodson et al., 1978b; 1979; 1980a; 1982).

2.3. Sensitivity

ALA-D activity of rainbow trout is definitely inhibited at low levels of waterborne (5 μg/l) and blood lead (300 μg/l) (Haux et al., 1980; Hodson et al., 1982). Inhibition occurs within 1–2 days of the onset of exposure and clearly follows the increase in blood lead during exposure (Hodson et al., 1982).

Blood lead levels and ALA-D activity are also good early-warning indicators of lead toxicity. They respond quickly ($\leq$ 1 week), before neurotoxicity is evident ($\geq$ 8 weeks) (Hodson et al., 1978b, 1979).

Blood lead is lost and ALA-D activity regained if fish are placed in clean water, and these processes speed up as temperatures increase (Hodson, unpublished data). The depuration rate also varies with exposure time. After 1 week of lead exposure, the ALA-D activity of rainbow trout returned to normal in 8 weeks (Hodson, et al., 1977). After 30 days exposure, there was little recovery after 7 weeks in clean water (Johansson-Sjöbeck and Larsson, 1979). Long-term exposures increase the levels of lead in tissues other than blood. Hence, in clean water, blood lead levels may be maintained as lead is transferred from other tissues (Varanasi and Gmur, 1978). Therefore high levels of blood lead or ALA-D inhibition in feral fish indicate only that lead exposure has occurred. The timing of that exposure could be estimated by comparing the ratios of blood lead levels to levels in other soft and hard tissues. The recovery of ALA-D activity may also depend on the turnover rate of RBCs, which bind more than 90% of whole blood lead (Posner, 1977).

2.4. Specificity

ALA-D inhibition is a highly specific response to environmental lead contamination. Extended in vivo exposures of rainbow trout to near-lethal levels of copper, mercury, cadmium, zinc, and polychlorinated biphenyls (PCBs) did not inhibit activity (Westman et al., 1975; Hodson et al., 1977; Johansson-Sjöbeck and Larsson, 1978). Similar results occurred in livers of mummichog exposed in vivo to copper, cadmium, mercury, silver, and zinc (Jackim, 1973).

Zinc caused a slight (8–10%) but inconsistent increase in erythrocyte ALA-D activity (Hodson et al., 1977). This resembles the activation in mammalian or fish erythrocytes exposed to zinc in vitro (Davis and Avram, 1980; Krajnović-Ozretić and Ozretić, 1980). Therefore, estimates of lead exposure by measurement of ALA-D activity may be biased by the simultaneous exposure of fish to zinc. Similarly, chronic exposures of fish to cadmium cause anemia. A compensatory rise in renal ALA-D activity occurred after a 9 week exposure of flounder to 50 µg Cd/l. There was, however, no change in erythrocyte and splenic activity (Johansson-Sjöbeck and Larsson, 1978).

In vitro exposure to cadmium, mercury, zinc, copper, lead, and aluminum all affected mullet ALA-D (Krajnović-Ozretić and Ozretić, 1980). The difference between in vivo and in vitro exposures could mean that most metals do not normally come in contact with ALA-D in vivo.

2.5. Interferences

2.5.1. Assay Conditions

Measurements of ALA-D activity can be affected by experimental methods. Activity increases with substrate concentration to a peak at 4 mM ALA (670 µg/ml) (Krajnović-Ozretić and Ozretić, 1980). The pH of the reaction must be held at 6.2, the optimum for rainbow trout (Hodson et al., 1977), mullet (Krajnović-Ozretić and Ozretić, 1980), white sucker (Catostomus commersoni), carp, and lake trout (Hodson, unpublished data).

ALA-D activity increases with reaction temperature up to at least 47°C (Krajnović-Ozretić and Ozretić, 1980). However, this temperature effect is not constant— activity of rainbow trout varied with acclimation temperature at reaction temperatures above 15°C (Hodson, unpublished data). Therefore, a constant, common reaction temperature is required to prevent bias when comparing activities between fish of different thermal histories. The temperature at which fish are exposed to lead also dramatically changes their lead accumulation (Hodson, unpublished data), changes that this enzymatic activity can easily detect.

Sample storage is also important. Fresh blood must be heparinized to avoid clotting; blood can then be held on ice for up to 24 hr without loss of ALA-D activity. Deep freezing whole blood in liquid nitrogen maintains activity indefinitely. Similarly, the supernatant produced after deproteination and centrifugation of the reaction mixture can be deep-frozen without losing PBG or color development with the Ehrlich salt. Whenever samples are stored however, hematocrit must be measured within 24 hr of sampling. Hematocrits increase in stored blood because RBCs swell as oxygen is consumed (Soivio and Nyholm, 1973). Therefore the accuracy of ALA-D assays should increase if the blood is aerated before subsampling for hematocrits.

Blood lead levels should be unaffected by storage if the sample container is tightly capped to limit evaporation and contamination. Blood stored for even short intervals must be thoroughly mixed before subsampling to avoid bias caused by RBC settling. We normally freeze-dry 50 µl aliquots of whole blood in small plastic vials for storage before lead analysis (Hodson et al., 1977).

2.5.2. Biological Factors

Biological factors affecting ALA-D activity must be recognized in monitoring; for example "normal" activity (activity at blood lead levels ≤ 100 µg/l) varies between species (Hodson et al., 1983). Normal ALA-D activities (nm PBG/ml RBCs·hr, ± one standard deviation) were measured in field studies for the following species: lake trout (272 ± 42), carp (175 ± 56), white sucker (182 ± 63), and white perch (*Roccus americanus*) (131 ± 54) (Hodson et al., 1983). In laboratory studies, the mean ALA-D activities of "control" rainbow trout with normal blood lead levels range from 240 to 410 (overall mean of 305) nm PBG/ml RBCs·hr (Hodson et al., 1977; 1978b; 1982).

In other studies where blood lead was not measured, the activities of "control" rainbow trout, flounder, and dogfish were 3000, 1080, and 483 nm PBG/ml RBCs·hr, respectively (Fänge and Johansson-Sjöbeck, 1975; Johansson-Sjöbeck and Larsson, 1978; 1979).These latter values were calculated by converting the author's results (as μg PBG/ml whole blood·hr) to nanomoles PBG, and dividing by hematocrit to express activity as nm/ml RBCs. The large differences in activity between studies are due to temperature; the assays by Hodson et al. were at 15°C while those by Johansson-Sjöbeck et al. were at 37°C; human ALA-D activity, measured at 37°C, is about 980–1200 nm PBG/ml RBCs·hr (Hernberg and Nikkanen, 1972).

Blood lead levels also vary between fish species. For example, during identical exposures to lead, rainbow trout accumulated more than did brook trout, which may explain their greater sensitivity to lead toxicity (Hodson et al., 1977). The species effect on lead accumulation will affect the link between ALA-D inhibition and waterborne lead levels—species that accumulate lead readily will be better indicators of environmental contamination than will those that accumulate little.

The size of fish also affects lead accumulation, but the effect is not consistent. ALA-D activity increased with weight in carp and white suckers, which suggests that blood lead levels decreased with weight (Hodson et al., 1983). Similarly, whole-body lead levels of feral smelt (*Osmerus mordax*) decreased with increasing weight (Hodson et al., 1982). However, size did not affect lead accumulation by rainbow trout during experimental exposures, and whole-body lead levels of feral lake trout did not vary with size (Hodson et al., 1982). Therefore, size may not directly affect lead metabolism—instead, size may control habitat selection, and hence lead exposure.

Growth rate (or ration level) has no affect on either lead accumulation or ALA-D activity of rainbow trout, although high growth rates may accentuate lead toxicity (Hodson et al., 1982). In contrast, diet composition may have important effects. Small amounts of dietary calcium reduced the uptake of waterborne lead by fish, presumably by affecting gill permeability (Varanasi and Gmur, 1978). Dietary ascorbic acid chelates lead and increases its depuration rate in humans. However, high levels in rainbow trout diets did not affect their accumulation of waterborne lead (Hodson et al., 1980b).

3. FIELD STUDIES

3.1. Sweden

Haux et al. (1981) found a great reduction (87–88%) in ALA-D activities of white-fish (*Coregonus peled*) from two lead-contaminated lakes in Sweden relative to the activity of whitefish from a clean lake. The source of lead contamination was a mine, and waterborne levels during sampling were 0.5–4.5 μg/l. While these waterborne levels appear rather low, it is possible that inhibition was caused by earlier and higher levels of exposure (see Section 2.3).

3.2. Lake Ontario

The lead content of lake trout in Lake Ontario was measured in 1979. Blood lead levels and ALA-D inhibition were least in fish from the eastern end and greatest in the west (Hodson et al., 1980a). Both indicators gave the same results, except at Point Traverse, in the eastern basin. Blood lead levels were higher than expected, considering the high ALA-D activities of the same fish and the low blood lead levels of fish from nearby Main Duck Island. The cause of these high values may have been sample contamination. Most measured values were quite low ($<<$ 100 μg/l so that the slightest amount of added lead would affect the results (Hodson et al., 1980a).

In 1980 the surveys of Lake Ontario were repeated and included lake trout, carp, white suckers, and white perch (Hodson et al., 1983). The results were as follows:

1. The east-west trend in lead contamination of lake trout was no longer evident. Blood lead levels were uniformly low (geometric means $\leq$ 30 μg/liter) and ALA-D activities uniformly high ($\geq$ 220 nm PBG/ml RBCs·hr).

2. Inshore species (carp, white sucker, white perch) had higher blood lead levels than did lake trout (range of geometric means = 51–456 μg/l).

3. Of fish caught in the same nets in Hamilton Harbour, benthic feeders (carp, white suckers) had higher blood lead levels than did pelagic feeders (white perch) (geometric means of 127 and 79 compared to 59 μg/l).

4. The relationship between mean blood lead levels and mean ALA-D activities of lake trout, carp, and white perch was weak. The most likely reason was the very narrow range of mean blood lead concentrations between each site. However, when ALA-D activities of individual fish within each species were compared to their individual blood lead levels, there were significant ($P < .05$) negative regressions. Nevertheless, the ALA-D activity of white perch was not a good indicator of their lead contamination because no pH optimum was found for the enzyme assay.

5. For white suckers the range of mean blood lead levels between sites was broader than that for the other species. The mean ALA-D activities at each site tended to decrease as mean levels increased. On a fish-by-fish basis, this regression was again significant ($P < .05$).

3.3. St. Lawrence River

In 1981, sampling was restricted to carp and white sucker from the St. Lawrence River near an alkyllead industry (Hodson et al., 1983). Samples taken beside the industry, 11–12 km downstream and 16–18 km upstream were expected to show high, medium, and low levels of lead contamination, respectively. This situation should have given a good test of the utility of the ALA-D assay. The results of this survey are as follows:

1. The variability of blood lead levels and ALA-D activities within a site was very high. There were both contaminated and uncontaminated fish at each site; however, the proportion that were contaminated were least upstream, greatest beside the industry, and intermediate downstream.
2. Mean blood lead levels of both species followed the expected pattern.
3. Carp had much higher blood lead levels than did white suckers.
4. The mean blood lead levels of white suckers from the site beside the industry were fairly consistent between years: 456 μg/l in 1980 and 694 μg/l in 1981.
5. The geometric mean levels of blood lead covered a wide range: 175–4864 μg/l for carp and 35–694 μg/l for suckers.
6. ALA-D activities were highest at the upstream site. However, the inhibition downstream (28–37%) seemed quite low relative to the degree and range of lead contamination.

This survey suggests that only blood lead measurements may be useful in assessing the lead exposure of feral fish. The degree of inhibition of ALA-D was generally inadequate to separate contaminated from uncontaminated fish. However, subsequent whole-body analyses of these same fish showed that most tissue lead was alkylated. Alkyllead compounds inhibit ALA-D activity only to a small extent, even at extremely high blood lead levels (Grandjean and Nielson, 1979). Therefore, these fish did not represent a reasonable test of the utility of the ALA-D assay. Confirmation of its utility will require sites highly contaminated with inorganic lead.

These results were not entirely negative. If the ALA-D assay can be successfully applied, it might be used to diagnose both lead exposure and lead form. Blood lead levels and enzyme inhibition could be interpreted as follows:

1. Low blood lead levels, no ALA-D inhibition = unexposed fish.
2. High blood lead levels, slight ALA-D inhibition = fish exposed to alkyllead compounds.
3. High blood lead levels, high ALA-D inhibition = fish exposed to inorganic lead.

The field studies in Lake Ontario and the St. Lawrence River gave some useful statistics about ALA-D activity (Hodson et al., 1983). The relative standard deviations of activity within any location averaged 12.9% (lake trout), 28.5% (carp), 31.8% (white sucker), and 24.8% (white perch). The higher relative standard deviations for inshore species reflects both higher and more variable blood lead levels. The variation arose from catching both contaminated and uncontaminated fish in the same net, probably as a result of fish migration. Sample contamination during analysis and simultaneous exposure of fish to lead and zinc may also increase ALA-D variability.

4. SUMMARY

Blood lead levels provide a direct measure of the dose accumulated by fish, and hence a reliable measure of their lead exposure. They are also strongly linked to the severity of lead neurotoxicity and hence indicate lead effects. The problems with blood lead monitoring are those of expense, complexity, and sensitivity to sample contamination.

Measurements of the inhibition of erythrocyte ALA-D activity of fish are inexpensive, technically simple, rapid, and specific for lead. The extent of inhibition is strongly linked to inorganic lead exposure and to the neurotoxic effects of lead. There are three problems with the measurement of ALA-D activity. First, the enzyme is inhibited by blood lead, which in turn varies with lead exposure; hence ALA-D activity is an indirect, or secondary, indicator of lead exposure. Second, ALA-D activity is inhibited only slightly by alkyllead compounds; therefore activity will not reflect exposure to these compounds. Third, monitoring of ALA-D activity of feral fish has proven only partially successful because: (1) it is necessary to describe the optimum conditions for assaying activity in each species examined; (2) migration of fish confounds clear-cut comparisons between contaminated and uncontaminated sites; (3) contamination by inorganic lead is rarely severe enough to provide a good test of the ALA-D method. Nevertheless, field trials are promising and further research is warranted to realize the advantages of this technique.

Other possible indicators of lead exposure are assays of blood porphyrins. While these methods are suitable in human toxicology, we were unable to find high porphyrin levels in the blood of control or lead-exposed rainbow trout. Therefore the effect of lead on hemoglobin metabolism in fish may differ from that in humans.

ACKNOWLEDGMENTS

We thank Dr. G. D. Sweeney, McMaster University, Hamilton and Dr. Bruce Burnham, Porphyrin Products, Logan, Utah for their help in studies of fish porphyrins, and Caryl Fawcett, who typed the manuscript.

REFERENCES

AEOC (1980). Lead. In: *Annual report of the aquatic ecosystems objectives committee*. Int. Joint Commission, Regional Office, Windsor, Ont.

Beritić, T., Zibar-Šikić, J., Prpić-Majić, D., and Tudor, M. (1980). Some morphological, biochemical, and hematological parameters of abnormal lead absorption in fish. *Thallassia Jugosl.* **16**, 263–269.

D'Amelio, V., Russo, G., and Ferraro, D. (1974). The effect of heavy metals on protein synthesis in crustaceans and fish. *Rev. Int. Oceanogr. Med.* **33**, 111–118.

Davies, P. H., Goettl, J. P., Jr., Sinley, J. R., and Smith, N. F. (1976). Acute and chronic toxicity of lead to rainbow trout, *Salmo gairdneri*, in hard and soft water. *Water Res.* **10**, 199–206.

Davis, J. R. and Avram, M. (1980). Correlation of the physicochemical properties of metal ions with their activation and inhibition of human erythrocytic δ-aminolevulinic acid dehydratase (ALA-D) *in vitro*. *Toxicol. Appl. Pharmacol.* **55**, 281–290.

Dieter, M. P. and Finley, M. T. (1979). Delta-aminolevulinic acid dehydratase enzyme activity in blood, brain and liver of lead-dosed ducks. *Environ. Res.* **19**(1), 127–135.

Fänge, R. and Johansson-Sjöbeck, M. (1975). The effect of splenectomy on the hematology and on the activity of δ-aminolevulinic acid dehydratase (ALA-D) in hemopoietic tissues of the dogfish, *Scyliorhinus canicula*. (Elasmobranchii). *Comp. Biochem. Physiol.* **52A**, 577–580.

Grandjean, P. and Nielsen T. (1979). Organo-lead compounds: Environmental Health Aspects. *Residue Rev.* **72**, 97–148.

Granick, S., Sassa, S., Granick, J. L., Levere, R. D., and Kappas, A. (1972). Assays for porphyrins, δ-amino levulinic-acid dehydratase, and porphyrinogen synthetase in microliter samples of whole blood: Applications to metabolic defects involving the heme pathway. *Proc. Nat. Acad. Sci. USA* **69** (9), 2381–2385.

Haux, C., Sjöbeck, M. L., and Larsson, A. (1980). Some toxic effects of lead on fish. Poster abstract for "Stress in fish." FSBI int. symp., Univ. of East Anglia, Norwich, England, September 9–12, 1980.

Haux, C., Larsson, A., Lithner, G., and Sjöbeck, M.-L (1981). Physiological studies on fish in lead-contaminated lakes. Abstract of presentation at the 3rd symp. on fish physiol. Bangor, Wales, U.K., September 9–11, 1981.

Hernberg, S. and Nikkanen, J. (1972). Effect of lead on δ-aminolevulinic acid dehydratase. A selective review. *Pacov. Lek.* **24**, 77–83.

Hodson, P. V. (1976). δ-Aminolevulinic acid dehydratase activity of fish blood as an indicator of a harmful exposure to lead. *J. Fish. Res. Board Can.* **33**, 268–271.

Hodson, P. V., Blunt, B. R., Spry, D. J., and Austen, K. (1977). Evaluation of erythrocyte δ-aminolevulinic acid dehydratase activity as a short-term indicator in fish of a harmful exposure to lead. *J. Fish. Res. Board Can.* **34**, 501–508.

Hodson, P. V., Blunt, B. R., and Spry, D. J. (1978a). pH-Induced changes in blood lead of lead-exposed rainbow trout (*Salmo gairdneri*). *J. Fish. Res. Board Can.* **35**, 437–445.

Hodson, P. V., Blunt, B. R., and Spry, D. J. (1978b) Chronic toxicity of water-borne and dietary lead to rainbow trout (*Salmo gairdneri*) in Lake Ontario water. *Water Res.* **12**, 869–878.

Hodson, P. V., Blunt, B. R., Jensen, D., and Morgan, S. (1979). Effect of fish age on predicted and observed chronic toxicity of lead to rainbow trout in Lake Ontario water. *J. Great Lakes Res.* **5**, 84–89.

Hodson, P. V., Blunt, B. R., and Whittle, D. M. (1980a). Biochemical monitoring of fish blood as an indicator of biologically available lead. *Thalassia Jugosl.* **16**, 389–396.

Hodson, P. V., Hilton, J. W., Blunt, B. R., and Slinger, S. J. (1980b). Effects of dietary ascorbic acid on chronic lead toxicity to young rainbow trout (*Salmo gairdneri*). *Can. J. Fish. Aquat. Sci.* **34**, 170–176.

Hodson, P. V., Dixon, D. G., Spry, D. J., Whittle, D. M., and Sprague, J. B. (1982). Effect of growth rate and size of fish on rate of intoxication by waterborne lead. *Can. J. Fish. Aquat. Sci.* **39** (9), 1243–1251.

Hodson, P. V., Blunt, B. R., and Whittle, D. M. (1983). Suitability of a biochemical method for assessing the exposure of feral fish to lead. In: W. E. Bishop, R. D. Cardwell, and B. B. Heidolph (Eds.), *Aquatic toxicology and hazard assessment: Sixth symp. ASTM STP 802*. Amer. Soc. for Test. and Materials, Philadelphia, pp. 389–405.

Hodson, P. V., Whittle, D. M., Wong, P. T. S., Borgmann, U., Thomas, R. L., Chau, Y. K., Nriagu, J. O., and Hallett, D. J. (1984). Lead contamination of the Great Lakes and its potential effects on aquatic biota. *Adv. Env. Sci. Technol.* In press.

Holcombe, G. W., Benoit, D. A., Leonard, E. M., and McKim, J. M. (1976). Long term effects of lead exposure on three generations of brook trout (*Salvelinus fontinalis*). *J. Fish. Res. Board Can.* **33**, 1731–1741.

Jackim, E. (1973). Influence of lead and other metals on fish δ-aminolevulinate dehydrase activity. *J. Fish. Res. Board Can.* **30**, 560–562.

Johansson-Sjöbeck, M.-L. and Larsson, A. (1978). The effect of cadmium on the hematology and on the activity of δ-aminolevulinic acid dehydratase (ALA-D) in blood and hematopoietic tissues of the flounder, *Pleuronectes flesus L. Environ. Res.* **17**, 191–204.

Johansson-Sjöbeck, M.-L. (1979). The effects of splenectomy on the hematology and on the activity of delta-aminolevulinic acid dehydratase (ALA-D) in hematopoietic tissues of the European eel *Anguilla anguilla* (L). *Comp. Biochem. Physiol.* **63**, 333–338.

Johansson-Sjöbeck, M.-L. and Larsson, A. (1979). Effects of inorganic lead on delta-aminolevulinic acid dehydratase activity and hematological variables in the rainbow trout, *Salmo gairdneri. Arch. Environ. Contam. Toxicol.* **8**, 419–431.

Krajnović-Ozretić, M. and Ozretić, B. (1980). The ALA-D activity test in lead-exposed grey mullet, *Mugil auratus. Mar. Ecol. Progr. Ser.* **3**, 187–191.

Posner, H. S. (1977). Indices of potential lead hazard. *Environ. Health Perspect.* **19**, 261–284.

Soivio, A. and Nyholm, K. (1973). Notes on hematocrit determinations on rainbow trout (*Salmo gairdneri*). *Aquaculture* **2**, 31–35.

Varanasi, U. and Gmur, D. J. (1978). Influence of water-borne and dietary calcium on uptake and retention of lead by coho salmon (*Oncorhynchus kisutch*). *Toxicol. Appl. Pharmacol.* **46**, 65–75.

Westman, I., Johansson-Sjöbeck, M.-L., and Fänge, R. (1975). The effect of PCB on the activity of delta-aminolevulinic acid dehydratase (ALA-D) and on some hematological parameters in the rainbow trout. In: *Sublethal effects of toxic chemicals on aquatic animals. Proceedings of the Swedish-Netherlands symposium*, Wageningen, the Netherlands, September 2–5, 1975. Elsevier Scientific Publishing, New York, 1975, pp. 111–119.

9

METALLOTHIONEIN
AND ACCLIMATION TO
HEAVY METALS IN FISH:
A REVIEW

*J. F. Klaverkamp, W. A. Macdonald, D. A. Duncan,
and R. Wagemann*

Department of Fisheries and Oceans
Freshwater Institute
Winnipeg, Manitoba

1. INTRODUCTION

The purpose of this paper is to provide a review of research on the ability of fish to acclimate to heavy metals, on metallothionein in fish, and on attempts to correlate these two aspects. Tables are used to summarize information in a concise manner.

Humanity's activities release massive amounts of heavy metals into the environment (Lantzy and Mackenzie, 1979; Jeffries and Snyder, 1981). Global anthropogenic emissions of cadmium (Cd), copper (Cu), lead (Pb), nickel (Ni), and zinc (Zn) into the atmosphere to date have been estimated to be 0.32×10^9 kg, 2.2×10^9 kg, 20×10^9 kg, 1.0×10^9 kg, and 14×10^9 kg, respectively (Nriagu, 1979). Anthropogenic emissions of mercury (Hg), which mostly arise from mineral processing and fossil fuel combustion (Harriss and Hohenemser, 1978; Airey, 1982), are estimated to be 1×10^7 kg per year (National Academy of Sciences, 1978). Release of these metals into the atmosphere during the past several decades has resulted in gradual contamination of aquatic ecosystems (Franzin et al., 1979; Galloway and Likens, 1979; DePinto et al., 1981; Nriagu et al., 1982).

1.1. Acclimation to Metal Toxicity

Few investigations have been conducted on the effects of gradual increases in metal concentrations on fish in spite of the fact that field, laboratory, and hatchery observations of over 30 years ago indicated that these animals could acclimate to metals. While earlier laboratory studies demonstrate that fish acclimate to Zn, more recent investigations have shown that these animals also acclimate to nonessential metals, such as Cd and Hg. Acclimation to Cu and arsenic (As) by rainbow trout has also been reported. Table 1 presents a summary of investigations pertaining to acclimation and other compensatory responses by fish to the metals Zn, Cd, Hg and Cu and to As, selenium (Se), and cyanide (CN).

1.1.1. Mammals

In mammals, acclimation to Cd is thought to be due to the sequestering of Cd by a metal-binding protein, metallothionein (MTN), thereby preventing adverse reactions of Cd with enzymes and other cellular molecules (Leber and Miya, 1976; Probst et al., 1977; Kotsonis and Klaassen, 1978). MTN is ubiquitously distributed throughout biological kingdoms and is thought to function primarily to regulate intracellular metabolism of Zn and Cu and, in higher animals, to absorb these metals from the intestine (Cherian and Goyer, 1978). In mammals, MTN is a cytosolic protein with an apparent molecular weight of approximately 10000–12000 daltons and contains about 30 residue percent cysteine with few, if any, aromatic amino acids (Banerjee et al., 1982). Synthesis of this protein is induced by exposure to metals in Groups IB and IIB, and it is thought to function as a general detoxifying mechanism for these metals (Nordberg and Kojima, 1979).

Table 1. A Summary of Investigations on Acclimation to Metals by Fish

Year	Investigator(s)	Observation(s)
1937	King	Trout hatched in water from galvanized pipe continue to survive, but others do not
1951	Goodman	Increased resistance to zinc by trout held in water from galvanized pipe
1952	Affleck	Zinc was less toxic to trout held in water from galvanized pipe
1952	Paul	Resident fish populations survived in stream polluted with Zn and Cu; nonresident could not
1960	Lloyd	MST to 10 ppm Zn = 290 min for nonacclimated trout; = 400 min for trout in 2.5 ppm Zn; = 500 min for trout in 3.5 ppm Zn
1965	Schofield	Trout from hatchery water with high zinc could survive in zinc-polluted lake, others could not
1967	Edwards and Brown	First dose-response relating "acclimatization concentration of zinc" to increases in 48 hr LC_{50} values for Zn in trout.
1973	Bouquegneau	Eels adapted to Hg-induced gill damage and imbalance
1974	Sinley et al.	Rainbow trout not exposed to Zn as eggs were 4 times more susceptible to Zn than trout exposed to Zn as eggs
1976	Lett et al.	Rainbow trout acclimated to Cu concentrations up to 0.5 of the 96 hr LC_{50} by returning to control rates of food consumption and growth
1976	Spehar	Flagfish larvae exposed to Cd or Zn as embryos were more tolerant than those not previously exposed
1978	Beattie and Pascoe	Rainbow trout alevins from hatching eggs exposed to sublethal Cd survived longer in Cd than alevins not exposed as eggs
1978	Chapman	Sockeye salmon alevins and juveniles exposed to sublethal Zn were more tolerant (>2 times in 96 hr LC_{50}) than nonexposed
1979	Bouquegneau	Eels exposed to sublethal Hg or Cd were less susceptible to Hg toxicity
1979	Dixon	Rainbow trout acclimate to Cu and As, but not to CN
1979	Pascoe and Beattie	Rainbow trout exposed to sublethal Cd were more resistant, but not more tolerant, to Cd
1980	Duncan and Klaverkamp	White suckers exposed to Cd, Hg, or Zn were less susceptible to Cd toxicity
1981a	Dixon and Sprague	Rainbow trout developed concentration and time dependent increases in tolerance to Cu; these increases were not retained after exposure to Cu ceased, and Cu-tolerant trout were more sensitive to Zn toxicity

(Table continues on pg. 102)

Table 1.   (*Continued*)

Year	Investigator(s)	Observation(s)
1981b	Dixon and Sprague	Rainbow trout exposed to As developed a gradual increase in tolerance to As, but when exposed to CN demonstrated an initial sensitization to CN
1981	Rahel	Flagfish showed increased resistance to lethal Zn concentrations but continued selection for 3 additional generations did not produce further increases in tolerance; common shiners from a Zn and Fe polluted stream were not more tolerant to Zn than common shiners from unpolluted streams
1982	Buckley et al.	Coho salmon acclimated to 0.25 and 0.50 of the 168 hr LC_{50} of Cu, but consumed less food and showed decreased rates of growth during initial phase of exposure
1982c	Kito et al.	Carp exposed to sublethal Cd or Zn were more resistant to Cd toxicity, and contained elevated Cd concentrations in MTN fraction of kidney and liver, but not gill
1983	McCarter and Roch	Coho salmon exposed to Cu for 1 week were more tolerant to Cu as observed in 2.5 fold increases in the 168 hr LC_{50}; with increased exposure duration, tolerance declined somewhat; parallel increases in liver MTN concentrations were observed with tolerance development over the exposure period
1983a	Duncan and Klaverkamp	White suckers acclimated to Cd toxicity by exposure to Cd, Zn, or Hg, but not to Se; increased exposure duration to Cd from 1 to 2 weeks did not affect development of acclimation

1.1.2. Fish

MTN has been shown to exist in fish tissues and to exhibit many of the properties of mammalian MTN. Table 2 provides a summary of species and organs analyzed and of methodology used in studies pertaining to MTN in fish.

Early investigations on fish were primarily concerned with identification and simple characterization of low molecular-weight metal binding proteins in liver, kidney, and gill. Methodology in these studies was largely limited to crude fractionation by gel filtration, although some investigators subjected these fractions to further purification by electrophoresis. Observations from these studies established that the protein is inducible (Olafson and Thompson, 1974; Yamamoto et al., 1978); it has an apparent molecular weight of 9000–12000 daltons (Coombs, 1975); it binds Zn (Brown, 1977), Cd (Marafante, 1976), Cu (Coombs, 1975), and Hg (Bouquegneau et al., 1975); it contains relatively high percentages of cysteine (Overnell et al., 1977; Noel-Lambot et al., 1978); and there are apparent differences between organs in metal binding affinities (Coombs, 1975; Marafante, 1976; Olson et al., 1978).

Table 2. A Summary of Investigations on Metallothionein in Fish

Year	Investigator(s)	Fish Species	Tissue(s)	Methodology[a]
1972	Marafante et al.	Goldfish	Liver	UC–GF–RT
1974	Olafson and Thompson	Copper rock fish	Liver	UC–GF–AA
1975	Bouquegneau et al.	Eel	Gill	UC–GF–AA
1975	Coombs	Plaice	Liver and kidney	UC–GF–AA
1976	Marafante	Goldfish	Liver and kidney	UC–GF–RT
1977	Overnell et al.	Plaice	Liver	UC–GF–AA
				IEC
				PAGE
1977	Brown	Flounder	Liver	HD–GF–AA
1978	Noel-Lambot et al.	Eel	Liver and gill	UC–GF–AA
				PAGE
				SGE
				AAA
1978	Olson et al.	Rainbow trout	Gill, liver, kidney, and spleen	UC–GF–AA
1978	Yamamoto et al.	Carp	Liver	UC–GF–AA
				SHA
1978	Brown and Parsons	Chum salmon	Liver	HD–GF–AA
1979	Overnell and Coombs	Plaice	Liver	UC–GF–AA
				IEC, UF, SHA, AAA,
				PAGE, IEF

103

Table 2. (*Continued*)

Year	Investigator(s)	Fish Species	Tissue(s)	Methodology[a]
1981c	Dixon and Sprague	Rainbow trout	Liver	UC–HD–UF–TCAppt–RT
1982	McCarter et al.	Coho salmon	Liver	UC–GF–AA
				SHA, IEC, PAGE, AAA
1982a	Takeda and Shimizu	Yellowtail	Liver from all species;	C–GF–AA
		Skipjack	muscle, kidney, and intestine	
		Eel	from yellowtail	
		Carp		
		Black seabream		
1982b	Takeda and Shimizu	Skipjack	Liver	C–GF–AA
				PAGE
				AAA
				SHA
				IEC
1982a	Kito et al.	Carp	Liver	UC–HD–C–GF–IEC
				UF–GF–AA
				PAGE
				AAA–EAEM
1982b	Kito et al.	Crucian carp	Liver, kidney, gill, gastro-	UC–GF–AA
		Japanese dace	intestine, spleen, bile, muscle	
		Japanese sea perch		
		Ayu-fish		
1982	Roch et al.	Rainbow trout	Liver	C–GF–AA
				HD–DPP

104

| 1983 | McCarter and Roch | Coho salmon | Liver | C–GF–AA
HD–DPP |
| 1983b | Duncan and Klaverkamp | White sucker | Liver and gill | UC–GF–AA and RT |

^a Abbreviations:

C	centrifugation
UC	ultracentrifugation
GF	gel filtration
AA	atomic absorption
RT	radiotracer metal
IEC	ion exchange chromatography
PAGE	polyacrylamide gel electrophoresis
HD	heat denatured
SGE	starch gel electrophoresis
AAA	amino acid analyses
SHA	sulfhydryl analyses
UF	ultrafiltration
DPP	differential pulse polarography
IEF	isoelectric focusing
TCAppt	protein precipitation by trichloroacetic acid
EAEM	elemental analyses with analytical electron microscopy

Initial MTN separation procedures are indicated by combining abbreviations. For example, UC–GF–AA indicates gel filtration of ultracentrifugation supernatant with fractions analyzed for metal(s) by atomic absorption. In some studies, these fractions were further purified (viz. IEC, PAGE) as indicated.

1.2. "Spillover" Hypothesis

The first study to implicate directly the metal-binding capacity of MTN to metal toxicity in fish was reported by Brown and Parsons (1978). Chum salmon fry exposed to Hg in controlled experimental ecosystems demonstrated pathology when liver MTN became saturated with Hg. Because this Hg-induced pathology coincided with the appearance of Hg in enzyme-containing high-molecular-weight fractions, the "spillover" hypothesis, first proposed by Winge et al. (1973), was used to explain the role of MTN in protecting these fish against the toxicity of Hg and other heavy metals. According to this hypothesis, Hg toxicity occurred when both the displacement of less toxic Zn and Cu from MTN by Hg, and the Hg-induced de novo synthesis of MTN, could no longer serve to sequester Hg, resulting in the excess Hg "spilling over" and reacting with enzymes.

Other investigators have demonstrated that concentrations of metals that produce acclimation to metal toxicity in fish also produce elevated MTN concentrations. Rainbow trout exposed to sublethal Cu concentrations exhibited increased tolerance to Cu, increased protein content in the low-molecular-weight protein fraction of liver, and increased incorporation of $[^{14}C]$ leucine into that protein fraction (Dixon and Sprague, 1981c). In another investigation on acclimation to Cu, increased levels of liver MTN (McCarter et al., 1982) were associated with tolerance to Cu (Buckley et al., 1982) in coho salmon. After 4 weeks of exposure to Cu by coho salmon, this association can be described (McCarter and Roch, 1983) mathematically as:

$$168 \text{ hr } LC_{50} = 2.423 \cdot \text{liver [MTN]} + 48.6$$
$$P < .03$$
$$r^2 = .937$$

Exposure of carp to sublethal concentrations of Cd or Zn produced increased resistance to a lethal concentration of Cd and increased Cd-binding in the MTN fraction of hepatopancreas and kidney, but not gill (Kito et al., 1982c). In white suckers, sublethal exposures to Cd, Zn, or Hg produced fish more tolerant to Cd (Duncan and Klaverkamp, 1983a), increased Cd-binding capacity in liver from Cd- and Zn-exposed fish and in gill from Hg-exposed animals, and increased MTN content in liver from Zn-exposed fish, in gill from Cd-exposed animals, and in liver and gill from Hg-exposed fish (Duncan and Klaverkamp, 1983b). This study also demonstrated that exposure to any of these three metals caused a decrease in Zn and an increase in Cu levels in gill MTN.

Contradictory and, in at least one case, controversial observations have been described in investigations on low-molecular-weight metal-binding proteins in fish. The spillover hypothesis proposed by Brown and Parsons (1978) has been described as being too simple to explain observations on coho salmon exposed to Cu (McCarter et al., 1982). These investigators reason that, if the spillover hypothesis is valid, then fish showing an increase in Cu concentrations in the high-molecular-weight protein fraction should be at a threshold of toxic effects and that slight increases in Cu should be harmful. They found, however, that such animals can withstand large increases in Cu and suggest that the rate of MTN synthesis, rather than the actual concentration of the protein, is the critical factor determining the ability of fish to acclimate to metals.

2.   DIFFERENCES BETWEEN MAMMALIAN AND FISH METALLOTHIONEIN

Because the spillover hypothesis was first proposed from observations made on mammals (Winge et al., 1973), it is important to consider differences between mammalian and piscine MTN. In spite of claims that the protein is very similar between these animal groups (Noel-Lambot et al., 1978; Overnell and Coombs, 1979), some observations indicate that there may be significant differences.

2.1.   Cysteine

For example, mammalian hepatic MTN contains approximately 30 residue percent of cysteine (Cherian and Goyer, 1978; Nordberg and Kojima, 1979). While several investigators have found similar concentrations in MTN isolated from fish livers (Overnell and Coombs, 1979; Kito et al., 1982a; McCarter et al., 1982; Takeda and Shimizu, 1982b), other values have also been reported. Noel-Lambot et al. (1978) found a cysteine content of less than 25 residue percent, Overnell et al. (1977) report it to be approximately 16, and Bouquegneau et al. (1975) report a value of less than 9.

2.2.   Metallothionein Isomers

A second example pertains to the number of MTN isomers ("isometallothioneins") that can be resolved in tissue samples from different sources using such secondary isolation techniques as ion exchange chromatography and electrophoresis. Mammalian MTN generally consists of two isometallothioneins (Buhler and Kagi, 1974; Nordberg, 1978; Garvey et al., 1982). While carp liver MTN also occurs in two isomeric forms (Kito et al., 1982a), three isomers were isolated from liver preparations of coho salmon (McCarter et al., 1982) and skipjack (Takeda and Shimizu, 1982b), and only one was found in plaice liver (Overnell and Coombs, 1979). It is not yet clear whether these differences are due to real biological differences or to variations in isolation procedures (Table 2).

2.3.   Copper and Zinc Content

A third apparent difference between mammalian and piscine MTN is that the latter generally contains a higher content of Cu than Zn (Kito et al., 1982a; Takeda and Shimizu, 1982b; Duncan and Klaverkamp, 1983b), whereas the opposite is usually true in adult mammals (Buhler and Kagi, 1974; Kagi et al., 1974; Kissling and Kagi, 1977). In mammals, a different low-molecular-weight protein, such as Cu chelatin (Winge et al., 1975; Brady et al., 1979) or a polymerized form of MTN (Nordberg and Kojima, 1979) is thought to be involved in sequestering Cu in mammals. Investigators entering this area should be aware that there are technical and operational difficulties involved in the isolation and purification of these rather unstable Cu-binding proteins (Bremner and Young, 1976; Hartmann and Weser, 1977; Webb et al., 1979).

2.4. Metal-Specific Proteins

Different metal-specific proteins may also explain recent contradictory observations on cross-acclimation between metals in fish. In rainbow trout, tolerance to Cu (Dixon and Sprague, 1981a) and synthesis of a leucine-containing Cu-binding protein (Dixon and Sprague, 1981c) are induced only by exposure to Cu, and not to Zn and Cd. In contrast, in white suckers acclimation to Cd, (Duncan and Klaverkamp, 1983a), MTN production, and Cd-binding activity (Duncan and Klaverkamp, 1983b) were not specific because they were induced not only by Cd, but also by Zn and Hg.

3. METALLOTHIONEIN AS AN INDICATOR OF METAL CONTAMINATION

The value of MTN as an exposure indicator to heavy metal contamination of natural freshwater ecosystems has recently been demonstrated by Roch et al. (1982). Rainbow trout were captured in metal-contaminated lakes in a river watershed receiving mine effluent and in a nearby uncontaminated lake. While the MTN fraction of livers from these fish primarily consisted of Cu, the correlation coefficient between waterborne Zn concentrations and liver MTN concentrations was 0.967. Relative to the uncontaminated lake, fish from the most contaminated lake had 4.6-fold increases in liver MTN concentration.

From field observations on lakes and fish populations contaminated by metals emitted to the atmosphere by a base-metal smelter in the Flin Flon, Manitoba area, Van Loon and Beamish (1977) suggested that these fish were more tolerant than would be expected from laboratory toxicity tests to the toxic effects of Zn and other heavy metals. Preliminary investigations in our laboratory demonstrated that white suckers from a contaminated Flin Flon lake had 3.3-, 3.6-, and 4.5-fold increases in gill, liver, and intestine MTN concentrations, respectively, when compared to suckers from a relatively uncontaminated lake in the area. In situ lake toxicity tests using Cd were conducted on suckers in these two lakes (Klaverkamp, unpublished observations). These tests demonstrated that suckers from the contaminated lake were up to 2.3 times more resistant to Cd, suggesting that acclimation or adaptation to atmospheric emissions of heavy metals occurs in fish in the natural environment.

4. RESEARCH NEEDS

Additional research is required to resolve issues in the areas of acclimation and MTN in fish. There is also need to determine biological costs, such as reduced growth (Dixon and Sprague, 1981a; Buckley et al., 1982), of producing metal acclimation; to understand the importance of other biochemical mechanisms, such as selenoproteins (Ganther, 1978) and glutathione (Congiu et al., 1979); and to establish the relevance of MTN and these other molecules in producing metal-resistant fish populations capable of surviving in contaminated natural ecosystems (Luoma, 1977). To clarify the role of MTN in acclimation to heavy metals by fish, studies designed to correlate rates of acclimation development and decline with rates of MTN synthesis and catabolism, respectively, are required. Expanding these studies to include other biochemical observations, such as lipid peroxidation (Tappel, 1980) and essential metal deficiencies (Coombs, 1975), on metal toxicology would help in understanding the validity of the spillover hypothesis.

These investigations would be extremely useful because fossil fuel combustion, base-metal smelting, and other activities of humanity are causing increased metal contamination of aquatic environments (National Academy of Sciences, 1981). Sound environmental management decisions concerning these metals will require a better understanding of acclimation processes.

REFERENCES

Affleck, R. J. (1952). Zinc poisoning in a trout hatchery. *Aust. J. Mar. Freshwater Res.* **3**, 142–169.

Airey, D. (1982). Contributions from coal and industrial materials to mercury in air, rainwater and snow. *Sci. Total Environ.* **25**, 19–40.

Banerjee, D., Onosaka, S., and Cherian, M. G. (1982). Immunohistochemical localization of metallothionein in cell nucleus and cytoplasm of rat liver and kidney. *Toxicology* **24**, 95–105.

Beattie, J. H. and Pascoe, D. (1978). Cadmium uptake by rainbow trout, *Salmo gairdneri*, eggs and alevins. *J. Fish Biol.* **13**, 631–637.

Bouquegneau, J. M. (1973). Etude de l'intoxication par le mercure d'un poisson teleosteen, *Anguilla anguilla*. I. Accumulation du mercure dans les organes. *Bull. Soc. R. Sci. Liege* **9–10**, 440–446.

Bouquegneau, J. M. (1979). Evidence for the protective effect of metallothioneins against inorganic mercury injuries to fish. *Bull. Environ. Contam. Toxicol.* **23**, 218–219.

Bouquegneau, J. M., Gerday, C., and Disteche, A. (1975). Fish mercury-binding thionein related to adaptation mechanisms. *FEBS Lett.* **55**, 173–177.

Brady, F. O., Panemangalore, M. and Day, F. A. (1979). *In vivo* and *ex vivo* induction of rat liver metallothionein. In: J. H. R. Kagi and M. Nordberg (Eds.), *Metallothionein: Proc. 1st int. meet. on metallothionein and other low molecular weight metal-binding proteins.* Zurich, July 17–22, 1978. Birkhauser Verlag, Boston, pp. 41–117.

Bremner, I. and Young, B. W. (1976). Isolation of (copper, zinc)-thioneins from the livers of copper-injected rats. *Biochem J.* **157**, 517–520.

Brown, D. A. (1977). Increases of Cd and the Cd:Zn ratio in the high molecular weight protein pool from apparently normal liver of tumor-bearing flounders (*Parophrys vetulus*). *Mar. Biol.* **44**, 203–209.

Brown, D. A. and Parsons, T. R. (1978). Relationship between cytoplasmic distribution of mercury and toxic effects to zooplankton and chum salmon (*Oncorhynchus keta*) exposed to mercury in a controlled ecosystem. *J. Fish. Res. Board Can.* **35**, 880–884.

Buckley, J. T., Roch, M., McCarter, J. A., Rendell, C. A., and Matheson, A. T. (1982). Chronic exposure of coho salmon to sublethal concentrations of copper. I. Effect on growth, on accumulation and distribution of copper, and on copper tolerance. *Comp. Biochem. Physiol.* **72C**, 15–19.

Buhler, R. H. O. and Kagi, J. H. R. (1974). Human hepatic metallothioneins. *FEBS Lett.* **39**, 229–234.

Chapman, G. A. (1978). Effects of continuous zinc exposure on sockeye salmon during adult-to-smolt freshwater residency. *Trans. Amer. Fish. Soc.* **107**, 828–836.

Cherian, M. G. and Goyer, R. A. (1978). Metallothioneins and their role in the metabolism and toxicity of metals. *Life Sci.* **23**, 1–10.

Congiu, L., Corongiu, F. P., Dore, M., Montaldo, C., Vargiolu, S., Casula, D., and Spiga, G. (1979). The effect of lead nitrate on the tissue distribution of mercury in rats treated with methylmercury chloride. *Toxicol. Appl. Pharmacol.* **52**, 363–366.

Coombs, T. L. (1975). The significance of multielement analyses in metal pollution studies. In: A. D. McIntrye and C. F. Mills (Eds.), *Ecological toxicology research: Effects of heavy metal and organohalogen compounds.* Plenum Press, New York, pp. 187–195.

DePinto, J. V., Young, T. C., and Marting, S. C. (1981). Aquatic sediments. *J. Wat. Poll. Cont. Fed.* **53**, 999–1007.

Dixon, D. G. (1979). Acclimation to toxicants by rainbow trout and its potential use in predicting safe levels. In: P. T. S. Wong, P. V. Hodson, A. J. Niimi, V. Cairns, and U. Borgmann (Eds.), *Proc. 5th ann. aquatic toxicology workshop.* Fish. Environ. Canada Fish. Mar. Serv. tech. rep. no. 862, pp. 162–166.

Dixon, D. G. and Sprague, J. B. (1981a). Acclimation to copper by rainbow trout (*Salmo gairdneri*)— A modifying factor in toxicity. *Can. J. Fish. Aquat. Sci.* **38**, 880–888.

Dixon, D. G. and Sprague, J. B. (1981b). Acclimation induced changes in toxicity of arsenic and cyanide to rainbow trout, *Salmo gairdneri* Richardson. *J. Fish. Biol.* **18**, 579–589.

Dixon, D. G. and Sprague, J. B. (1981c). Copper bioaccumulation and hepatoprotein synthesis during acclimation to copper by juvenile rainbow trout. *Aquat. Toxicol.* **1**, 69–81.

Duncan, D. A. and Klaverkamp, J. F. (1980). Induced tolerance to cadmium in white suckers (*Catostomus commersoni* Lacepede) by exposure to sublethal concentrations of heavy metals. In: J. F. Klaverkamp, S. L. Leonhard, and K. E. Marshall (Eds.), *Proc. 6th ann. aquat. toxicology workshop.* Fish. Oceans Canada. Can. tech rep. of fish. aquat. sci. no. 975, p. 108.

Duncan, D. A. and Klaverkamp, J. F. (1983a). Tolerance and resistance to cadmium in white suckers (*Catostomus commersoni*) previously exposed to cadmium, mercury, zinc or selenium. *Can. J. Fish. Aquat. Sci.* **40**, 128–138.

Duncan, D. A. and Klaverkamp, J. F. (1983b). Acclimation to cadmium toxicity by white sucker (*Catostomus commersoni*): Metal distribution in liver and gill cytosol. *Can. J. Fish. Aquat. Sci.* (submitted).

Edwards, R. W., and Brown, V. M. (1967). Pollution and fisheries: A progress report. *Wat. Poll. Contr.* **66**, 63–78.

Franzin, W. G., McFarlane, G. A., and Lutz, A. (1979). Atmospheric fallout in the vicinity of a base metal smelter at Flin Flon, Manitoba, Canada. *Environ. Sci. Technol.* **13**, 1513–1522.

Galloway, J. N. and Likens, G. E. (1979). Atmospheric enhancement of metal deposition in Adirondack lake sediments. *Limnol. Oceanogr.* **24**, 427–433.

Ganther, H. E. (1978). Modification of methylmercury toxicity and metabolism by selenium and vitamin E. Possible mechanisms. *Environ. Health Perspect.* **25**, 71–76.

Garvey, J. S., Vander Mallie, R. J., and Chang, C. C. (1982). Radioimmunoassay of metallothioneins. *Meth. Enzymol.* **84**, 121–138.

Goodman, J. R. (1951). Toxicity of zinc for rainbow trout (*Salmo gairdneri*). *Calif. Fish Game* **37**, 191–194.

Harriss, R. C. and Hohenemser, C. (1978). Mercury: Measuring and managing the risk. *Environment* **20**, 25–36.

Hartmann, H. J. and Weser, U. (1977). Copper-thionein from fetal bovine liver. *Biochim. Biophys. Acta* **491**, 211–222.

Jeffries, D. S. and Snyder, W. R. (1981). Atmospheric deposition of heavy metals in central Ontario. *Water, Air Soil Pollut.* **15**, 127–152.

Kagi, J. H. R., Himmelhoch, S. R., Whanger, P. D., Bethune, J. L., and Vallee, B. L. (1974). Equine hepatic and renal metallothioneins. Purification, molecular weight, amino acid composition, and metal content. *J. Biol. Chem.* **249**, 3537–3542.

King, W. (1937). Mortality of hatchery trout, Great Smoky Mountains National Park. *Prog. Fish-Cult.* No. 29, 14–17.

Kissling, M. M. and Kagi, J. H. R. (1977). Primary structure of human hepatic metallothionein. *FEBS Lett.* **82**, 247–250.

Kito, H., Ose, Y., Mizuhira, Z., Sato, T., Ishikawa, T., and Tazawa, T. (1982a). Separation and purification of (Cd, Cu, Zn)-metallothionein in carp hepato-pancreas. *Comp. Biochem. Physiol.* **73C**, 121–127.

Kito, H., Tazawa, T., Ose, Y., Sato, T., and Ishikawa, T. (1982b). Formation of metallothionein in fish. *Comp. Biochem. Physiol.* **73C**, 129–134.

Kito, H., Tazawa, T., Ose, Y., Sato, T. and Ishikawa T. (1982c). Protection by metallothionein against cadmium toxicity. *Comp. Biochem Physiol.* **73C**, 135–139.

Kotsonis, F. N. and Klaassen, C. D. (1978). The relationship of metallothionein to the toxicity of cadmium after prolonged oral administration to rats. *Toxicol. Appl. Pharmacol.* **46**, 39–54.

Lantzy, R. J. and Mackenzie, F. T. (1979). Atmospheric trace metals: Global cycles and assessment of man's impact. *Geochim. Cosmochim. Acta.* **43**, 511–525.

Leber, A. P. and Miya, T. S. (1976). A mechanism for cadmium- and zinc-induced tolerance to cadmium toxicity: Involvement of metallothionein. *Toxicol. Appl. Pharmacol.* **37**, 403–414.

Lett, P. F., Farmer, G. J., and Beamish, F. W. H. (1976). Effect of copper on some aspects of the bioenergetics of rainbow trout (*Salmo gairdneri*). *J. Fish. Res. Board Can.* **33**, 1335–1342.

Lloyd, R. (1960). The toxicity of zinc sulphate to rainbow trout. *Ann. Appl. Biol.* **48**, 84–94.

Luoma, S. N. (1977). Detection of trace contaminant effects in aquatic ecosystems. *J. Fish. Res. Board Can.* **34**, 436–439.

Marafante, E. (1976). Binding of mercury and zinc to cadmium-binding protein in liver and kidney of goldfish (*Carassius auratus* L.). *Experienta* **32**, 149–150.

Marafante, E., Pozzi, G., and Scoppa, P. (1972). Detossicazione dei metalli pesanti pesci: Isolamento della metallotioneina dal fegato di Carassius auratus. *Boll. Soc. Ital. Biol. Speriment.* **48** (20), 109.

McCarter, J. A., and Roch, M. (1983). Hepatic metallothionein and resistance to copper in juvenile coho salmon. *Comp. Biochem. Physiol.* **74C**, 133–137.

McCarter, J. A., Matheson, A. T., Roch, M., Olafson, R. W., and Buckley, J. T. (1982). Chronic exposure of coho salmon to sublethal concentrations of copper. II. Distribution of copper between high- and low-molecular-weight proteins in liver cytosol and the possible role of metallothionein in detoxification. *Comp. Biochem. Physiol.* **72C**, 21–26.

National Academy of Sciences (1978). An assessment of mercury in the environment. A report prepared by the panel on mercury of the coordinating committee for scientific and technical assessments of environmental pollutants. Washington, DC, pp. 15–29.

National Academy of Sciences (1981). Atmosphere-biosphere interactions: Toward a better understanding of the ecological consequences of fossil fuel combustion. A report prepared by the committee on the atmosphere and the biosphere. Washington, DC, 263 pp.

Noel-Lambot, F., Gerday, C., and Disteche, A. (1978). Distribution of Cd, Zn and Cu in liver and gills of the eel *Anguilla anguilla* with special reference to metallothioneins. *Comp. Biochem. Physiol.* **61C**, 177–187.

Nordberg, M. (1978). Studies on metallothionein and cadmium. *Environ. Res.* **15**, 381–404.

Nordberg, M. and Kojima, Y. (Eds.) (1979). Metallothionein and other low molecular weight metal-binding proteins. In: J. H. R. Kagi and M. Nordberg (Eds.), *Metallothionein: Proc. 1st int. meet. on metallothionein and other low molecular weight metal-binding proteins.* Zurich, July 17–22, 1978. Birkhäuser Verlag, Boston, pp. 41–117.

Nriagu, J. O. (1979). Global inventory of natural and anthropogenic emissions of trace metals to the atmosphere. *Nature* **279**, 409–411.

Nriagu, J. O., Wong, H. K. T., and Coker, R. D. (1982). Deposition and chemistry of pollutant metals in lakes around the smelters of Sudbury, Ontario. *Environ. Sci. Technol.* **16**, 551–560.

Olafson, R. W. and Thompson, J. A. J. (1974). Isolation of heavy metal binding proteins from marine invertebrates. *Mar. Biol.* **28**, 83–86.

Olson, K. R., Squibb, K. S., and Cousins, R. J. (1978). Tissue uptake, subcellular distribution, and metabolism of $^{14}CH_3HgCl$ and $Ch_3^{203}HgCl$ by rainbow trout, *Salmo gairdneri*. *J. Fish. Res. Board Can.* **35**, 381–390.

Overnell, J. and Coombs, T. L. (1979). Purification and properties of plaice metallothionein, a cadmium-binding protein from the liver of the plaice (*Pleuronectes platessa*). *Biochem. J.* **183**, 277–283.

Overnell, J., Davidson, I. A., and Coombs, T. L. (1977). A cadmium-binding glycoprotein from the liver of the plaice (*Pleuronectes platessa*). *Biochem. Soc. Trans.* **5**, 267–269.

Pascoe, D. and Beattie, J. H. (1979). Resistance to cadmium by pretreated rainbow trout alevins. *J. Fish Biol.* **14**, 303–308.

Paul, R. M. (1952). Water pollution: A factor modifying fish populations in Pacific coast streams. *Sci. Mon.* **74**, 14–17.

Probst, G. S., Bousquet, W. F., and Miya, T. S. (1977). Correlation of hepatic metallothionein concentrations with acute cadmium toxicity in the mouse. *Toxicol. Appl. Pharmacol.* **39**, 61–69.

Rahel, F. J. (1981). Selection for zinc tolerance in fish: Results from laboratory and wild populations. *Trans. Amer. Fish. Soc.* **110**, 19–28.

Roch, M., McCarter, J. A., Matheson, A. T., Clark, M. J. R., and Olafson, R. W. (1982). Hepatic metallothionein in rainbow trout (*Salmo gairneri*) as an indicator of metal pollution in the Campbell River system. *Can. J. Fish. Aquat. Sci.* **39**, 1596–1601.

Schofield, C. L., Jr. (1965). Water quality in relation to survival of brook trout *Salvelinus fontinalis* (Mitchill). *Trans. Amer. Fish. Soc.* **94**, 227–235.

Sinley, J. R., Goettl, J. P., Jr., and Davies, P. H. (1974). The effects of zinc on rainbow trout (*Salmo gairdneri*) in hard and soft water. *Bull. Environ. Contam. Toxicol.* **12**, 193–201.

Spehar, R. L. (1976). Cadmium and zinc toxicity to flagfish, *Jordanella floridae*. *J. Fish. Res. Board Can.* **33**, 1939–1945.

Takeda, H. and Shimizu, C. (1982a). Existence of the metallothionein-like protein in various fish tissues. *Bull. Jap. Soc. Sci. Fish.* **48**, 711–715.

Takeda, H. and Shimizu, C. (1982b). Purification of metallothionein from the liver of skipjack and its properties. *Bull. Jap. Soc. Sci. Fish.* **48**, 717–723.

Tappel, A. L. (1980). Vitamin E and selenium protection from *in vivo* lipid peroxidation. In: O. A. Levander and L. Cheng (Eds.), *Micronutrient interactions: Vitamins, minerals, and hazardous elements.* Ann. N.Y. Acad. Sci. **355**, 18–31.

Van Loon, J. C., and Beamish, R. J. (1977). Heavy-metal contamination of atmospheric fallout of several Flin Flon area lakes and the relation to fish populations. *J. Fish. Res. Board Can.* **34**, 899–906.

Webb, M., Plastow, S. R., and Magos, L. (1979). (Copper, zinc)-thionein in pig liver. *Life Sci.* **24**, 1901–1906.

Winge, D., Krasno, J., and Colucci, A. V. (1973). Cadmium accumulation in rat liver: Correlation between bound metal and pathology. In: W. G. Hoekstra, J. W. Suttie, H. E. Ganther, and W. Mertz (Eds.), *Trace element metabolism in animals*, Vol. 2. University Park Press, Baltimore, pp. 500–501.

Winge, D. R., Premakumar, R., Wiley, R. D., and Rajogopalan, K. V. (1975). Copper-chelatin: Purification and properties of a copper-binding protein from rat liver. *Arch. Biochem. Biophys.* **170**, 253–266.

Yamamoto, Y., Ishii, T., and Ikeda, S. (1978). Studies on copper metabolism in fishes. III. Existence of metallothionein-like protein in carp hepatopancreas. *Bull. Jap. Soc. Sci. Fish.* **44**, 149–153.

10

PATHOBIOLOGICAL RESPONSES OF FERAL TELEOSTS TO ENVIRONMENTAL STRESSORS: INTERLAKE STUDIES OF THE PHYSIOLOGY OF GREAT LAKES SALMON

J. F. Leatherland

Department of Zoology
University of Guelph
Guelph, Ontario

R. A. Sonstegard

Departments of Biology and Pathology
McMaster University
Hamilton, Ontario

1.   INTRODUCTION

There is obvious value in monitoring the effects of known stressors on the physiology of animal species. Such studies provide a standard by which the severity and significance of other putative environmental stressors can be measured. One purpose of this review is to reexamine the concept of stress in teleost fish, particularly with respect to the way in which the stress response has been monitored in a variety of species and under various experimental conditions. In light of Hans Selye's concept of a "general adaptation syndrome" one might expect to find a common expression of the stress response in a variety of species. The stressor mediated increase in plasma cortisol might represent such a common denominator in teleosts. While this appears to be the case in certain situations, other physiological responses to stressors, particularly chronic stressors, might be better indicators. In the second part of the review a "case study" of a chronic environmental stressor situation, that of the oncorhynchid populations in the Great Lakes of North America, is examined. The considerable interlake differences in the physiology of these salmon suggest that the fish are adapting to a unique set of environmental factors (stressors) within a given lake. Since the fish are exposed to these environmental factors (stressors) for the major part of their life cycle, the stressor effect is generally multifactorial and chronic. For reasons discussed below, the "primary" stress indicators (such as blood cortisol elevation) are inappropriate in this type of situation. The interlake physiological differences are considered with regard to their value as stress indicators.

2.   THE CONCEPT AND MONITORING OF STRESS IN FISH

Hans Selye's studies of stress physiology in mammals led him to propose the "general adaptation syndrome" hypothesis (Selye, 1956). He observed that a variety of environmental stressors, such as food-deprivation, temperature extremes, acute physical exertion, social interaction, and psychological pressure all elicited a similar adaptive physiological change, which he termed stress. The various adverse physical or psychological stimuli induced changes in the neuroendocrine system, activating corticotropin release from the pituitary and an elevation of adrenocorticosteroid secretion, particularly the glucocorticosteroids. The glucocorticosteroids stimulated gluconeogenesis from protein and lipid reserves and reduced antibody production. Thus environmental stressors, if applied for any period of time, may affect protein catabolism, suppress growth, and suppress the immune system, thereby increasing the animal's susceptibility to disease. Christian (1975) extended the concept by proposing that high population density acts as a potent stressor, eliciting changes in sex steroid production by the adrenal cortex and consequently lowering the reproductive success of the population.

Despite a relatively large amount of published data pertaining to the physiology of stress in fish, the concept of stress is enigmatic, generally poorly defined by most authors, and seemingly equally poorly understood. The term "stress" in fish, and other animals, is all-too-often applied to physiological responses that cannot be explained in another way. Recent reviews, specifically addressing the hypothesis and interpretation of stress in fish (Peters, 1979; Pickering, 1981), have done much to clarify the concept, and form the foundation on which a better understanding of the stress phenomenon in fish can be built.

Although it is likely that such a sequence of responses is exhibited by teleosts, the available data concerning the physiological responses to adaptive stressors, applied at levels that might be considered within a normal environmental range, are few. Much of the literature of stress in fish pertains to stressors such as capture, acute overcrowding, asphyxiation, handling, acute temperature shock, acute osmotic shock, or exposure to acute levels of toxicants. It is questionable whether any of these types of stressors can be considered as components of a normal range of environmental stimuli. Many of these situations are more appropriately defined as "distressful" rather than "stressful," and the physiological responses may be more akin to "fright and flight" responses than to the general adaptation responses hypothesized by Selye (1956).

In addition, the parameters of an "acceptable" stress level in fish and other animals has yet to be determined. Selye (1956) differentiates between the physiological responses to stressful as opposed to distressful situations. There is increasing evidence that for all animals there is a range of stress comparable to the tolerance range for any other homeostatic system; above and below the tolerance range are the distressful (or resistance) ranges. As with any other homeostatic process, the tolerance and resistance ranges of stress can be altered by acclimation, and there is considerable individual variation as regards susceptibility to stressful situations (Rose and Sachar, 1981). McLeay (1970) found that interrenal nuclear size (a sensitive indicator of the activity of the pituitary-interrenal axis) in coho salmon was lowest when fish were maintained individually in aquaria in quiet surroundings and isolated from visual contact with other fish or the experimentors. Clearly, this represents an atypical situation, one that would not be experienced by most fish in a feral state, and certainly not by fish maintained in a hatchery. Moreover, there is increasing evidence of differences in the activity of the pituitary-interrenal axis associated with social hierarchy in fish (Erickson, 1967; Noakes and Leatherland, 1977; Klinger et al., 1979; Ejike and Schreck, 1980; Peters et al., 1980; Scott and Currie, 1980), suggesting that in a given physical environment, social interactions evoke a variable stress within the same population.

A variety of physiological changes in response to diverse stressors have been reported in fish (Tables 1–5). There is some consensus that the stress response can be separated into two or three stages (Mazeaud et al., 1977; Wedemeyer and McLeay, 1981; Pickering et al., 1982). The primary stage involves neuroendocrine responses such as catecholamine release and activation of the corticotropin-interrenal axis. The secondary stage involves physiological responses such as hematological, osmoregulatory, blood enzymatic, and metabolic changes. The tertiary stage due to chronic exposure of fish to a stressor(s) includes growth inhibition, reduced fecundity, increased susceptibility to pathogens, and behavioral changes. It is generally assumed that the primary stress responses bring about the secondary and tertiary responses, although the progression is not as yet confirmed.

Table 1.   Studies Examining the Effect of Various Environmental Stressors on the Activity of the Pituitary-Interrenal Axis

Stressors	Genera	Authors
Various (review)	Various	Donaldson, 1981
Handling, hypoxia, transportation, confinement, electroshocking, injection, or anaesthetization	*Salmo*	Leloup-Hatey, 1964; Hane et al., 1966; Donaldson and McBride, 1967; Wedemeyer, 1969a,b, 1970a, 1972; Simpson, 1975/1976; Schreck et al., 1976; Strange et al., 1977; Barton et al., 1980; Swift, 1981, 1982
	Oncorhynchus	Hane et al., 1966; Fagerlund, 1967; Donaldson and Fagerlund, 1969, 1970; Fagerlund and McBride 1969; Wedemeyer, 1969b, 1972; Fagerlund and Donaldson, 1970; Mazeaud et al., 1977; Strange et al., 1977, 1978; Strange and Schreck, 1978; Specker and Schreck, 1980
	Carassius	Spieler, 1974; Fryer, 1975, Singley and Chavin, 1975a
	Poecilia	Mezhnin, 1978
	Coregonus	Fuller et al., 1974

Table 1.   (*Continued*)

Stressors	Genera	Authors
Temperature shock	*Salmo*	Wedemeyer, 1969a,b, 1973; Strange et al., 1977
	Oncorhynchus	Wedemeyer, 1973; Strange et al., 1977
	Ictalurus	Strange, 1980
	Poecilia	Mezhnin, 1978
	Chrysophrys	Ishioka, 1980a,b
Osmotic shock	*Oncorhynchus*	Donaldson and Dye, 1975; Specker and Schreck, 1980; Strange and Schreck, 1980
	Chrysophrys	Ishioka, 1980a,b
Social hierarchy	*Salmo*	Noakes and Leatherland, 1977
	Oncorhynchus	Ejike and Schreck, 1980
	Lepomis	Erikson, 1967
	Xiphophorus	Scott and Currie, 1980
	Anguilla	Klinger et al., 1979; Peters, 1979; Peters et al., 1980
Exposure to infectious agents	*Salmo*	Simpson, 1975/1976
	Oncorhynchus	Fagerlund, 1967; Fagerlund and McBride, 1969
Exposure to toxicants, acid stress, or antibiotics	*Salmo*	Hill and Fromm, 1968; Wedemeyer, 1969a,b, 1971; Grant and Mehrle, 1973; McBride et al., 1975, 1979, 1981; Mazeaud et al., 1977; Simpson, 1975/1976; Pearce and McBride, 1978
	Oncorhynchus	Wedemeyer, 1971; McLeay, 1973; Donaldson and Dye, 1975; Lorz et al., 1978; Schreck and Lorz, 1978
	Salvelinus	Ashcom et al., 1977
	Fundulus	DiMichele and Taylor, 1978

Of the three stages of stress, the primary and secondary stages are perhaps the easiest to monitor in a laboratory because measurement of hematocrit and blood cortisol and glucose levels is now routine. However, the reliance on these parameters as the principle indicators of stress is questionable since they are notoriously sensitive to handling stress. For example, Pickford (1973) noted large changes in plasma cortisol levels in *Fundulus heteroclitus* within minutes of commencing blood sampling,

Table 2.   Studies Examining the Effect of Various Environmental Stressors
on Blood Glucose Levels

Stressors	Genera	Authors
Handling, hypoxia, transportation, confinement, electroshocking, anaesthetization	*Salmo*	Wedemeyer, 1972, 1976; Wendt and Saunders, 1973; Schreck et al., 1976; Wydoski et al., 1976; Swift, 1981, 1982; Pickering et al., 1982
	Oncorhynchus	Wedemeyer, 1972, 1976; Mazeaud et al., 1977; Specker and Schreck, 1980
	Salvelinus	Houston et al., 1971a,b
	Cyprinus	Smit et al., 1979a
	Labeo	Hattingh and Van Pletzen, 1974; Hattingh, 1977; Soivio and Oikari, 1976
	Esox	Soivio and Oikari, 1976
	Ictalurus	Strange, 1980
Osmotic and temperature shock	*Chrysophrys*	Ishioka, 1980a,b
Exposure to toxicants	*Salmo*	Swift, 1981, 1982
	Fundulus	DiMichele and Taylor, 1978
	Etheostoma	Silbergeld, 1974

and Chavin and co-workers found that *Carassius auratus* developed hyperglycemia and hypercortisolism within seconds of capture. To minimize the problem, it was necessary to maintain the fish in containers of suitable size to permit their capture with one pass of the net (Chavin and Young, 1970; Singley and Chavin, 1975a,b), or to subject the fish to training regimes that lower the stress response to capture and handling (Slicher et al., 1966; Rush and Umminger, 1978). The sensitivity of the cortisol and glucose responses gives rise to discrepancies in cortisol and glucose levels between the first and last fish sampled in an experimental group; capture of the first animal will inevitably affect cortisol and glucose levels in the remainder of the group, as will the activity of the experimentor in the vicinity of the experimental aquaria (DiMichele and Taylor, 1978).

The difficulties with using primary stress responses as the principle indicator of the effects of environmental stressors were illustrated by two studies of the effect of environmental pollutants on blood cortisol levels in fish. Schreck and Lorz (1978) found that cortisol was elevated in coho salmon exposed to copper, but not in fish exposed to cadmium, despite the obvious toxicity of both heavy metals. DiMichele and Taylor (1978) found that low-level naphthalene exposure effected a dose-related increase in cortisol levels in the mummichog; at high doses, however, blood cortisol levels were lowered because of necrosis of the interrenal tissue. In both cases, despite the apparent absence of a primary stress response, the animals exhibited

Table 3. Studies Examining the Effect of Various Environmental Stressors on Hematological Parameters (HP), Blood Ion (BI), or Blood Metabolite (BM) Levels Other than Glucose

Stressors	Genera	Authors
Handling, transportation, confinement, electro-shocking, hemorrhagia, or anaesthetization	Salmo	Wedemeyer, 1971, 1972 (BI, BM), 1976 (BI); Schreck et al., 1976 (HP, BI, BM); Cairns and Christian, 1978 (HP); Smit et al., 1979a,b (BI, HP)
	Oncorhynchus	Wedemeyer, 1971, 1972 (BI, BM), 1976 (BI); Mazeaud et al., 1977 (BM)
	Salvelnius	Houston et al., 1971a,b (HP, BI, BM)
	Sarotherodon	Smit et al., 1979a,b (BI)
	Cyprinus	Smit et al., 1979a,b (BI)
	Label	Hattingh and Van Pletzen, 1974 (BM, HP)
	Esox	Soivio and Oikari, 1976 (BI, HP)
Temperature or osmotic shock	Chrysophrys	Ishioka, 1980a,b (BI, HP)
	Colisa	Srivastava anad Agrawal, 1977 (HP)
Exposure to infectious agents	Salmo	Richards and Pickering, 1979 (BI, BM)
	Carassius	Munkittrick and Leatherland, 1983 (HP)
Exposure to toxicants or acid stress	Various (review)	Fromm, 1980 (HP, BI, BM)
	Salmo	Wedemeyer, 1971, 1972 (BI, BM); McLeay and Gordon, 1977 (HP); Zeitoun, 1977 (HP, BM); Hlavek and Bulkley, 1980 (HP); Swift, 1981, 1982 (BI, BM)
	Oncorhynchus	Wedemeyer, 1971, 1972 (BI, BM); McLeay and Gordon, 1977 (HP); Schreck and Lorz, 1978 (BI)
	Fundulus	DiMichele and Taylor, 1978 (BM)
	Aphanius	Hilmy et al., 1980 (HP)
	Clarias	Tandon and Chandra, 1978 (BM)

clear signs of secondary and teritary responses. These findings argue against a cause-effect relationship between primary and secondary and tertiary responses postulated by some authors (Donaldson, 1981; Schreck, 1981). The results of a recent study by Swift (1982) similarly question the progression of stress responses from primary to tertiary. Swift used betamethasone to block the stressor mediated elevation in cortisol. The secondary stress effects in response to phenol exposure or hypoxia were found in betamethasone-treated and control rainbow trout alike, despite an elevation of cortisol in the former.

Table 4. Studies Examining the Effect of Various Environmental Stressors on Endocrine Systems Other than the Pituitary-Interrenal Axis

Stressors	Genera	Endocrine System[a]	Authors
Several (reviews)	Several (reviews)	PGA	Billard et al., 1981; Bresch, 1982
Capture, handling, salinity change	*Salmo*	PTA	Osborn and Simpson, 1974; Leatherland et al., 1977; Leatherland, 1982a; Brown et al., 1978; Milne et al., 1979
	Oncorhynchus	AMH	Mazeaud et al., 1977
	Pseudopleuronectus	PTA	Eals and Fletcher, 1982
	Platichthys	PTA	Osborn and Simpson, 1974
Background, color change, salinity change, temperature shock injection, decapitation, electroshocking	*Anguilla*	H	Leatherland and Dodd, 1969
Exposure to infectious agents	*Carassius*	PGA	Munkittrick and Leatherland, 1984
Exposure to toxicants or acid	*Salmo*	PGA	Burdick et al., 1972
	Salvelinus	PGA	Freeman and Idler, 1975
	Jordanella	PGA	Nebeker et al., 1974; Ruby et al., 1977
	Pimephales	PGA	Nebeker et al., 1974; McCormick et al., 1980
	Lepomis	PGA	Benoit, 1975
	Phoxinus	PGA	Bengtsson, 1980
	Saratherodon	PGA	Kling, 1981
	Cyprinus	PGA	Kapur et al., 1978
	Heteropneustes	PGA	Singh and Singh, 1980

[a] PGA, pituitary-gonad axis; PTA, pituitary-thyroid axis; AMH, adrenal medulla homologue; H, hypothalamus.

Table 5.   Studies Examining the Effect of Various Environmental Stressors on Skins and Gills and Enzymes of Blood and Liver

Stressor	Genera	Physiological Parameter[a]	Authors
Capture and/or hypoxia	Various (*Mugil, Chanos, Caranx, Albula, Chaetodon*)	SMH	Smith and Ramos, 1976
	Salmo	PE	Bouck et al., 1978
	Ophisurus	SMS	Zaccone, 1980
Exposure to toxicants or acid stress	*Salmo*	GSA,LE	Arillo et al., 1979a,b
	Notopterus	LE	Dalela et al., 1980

[a] SMH, skin mucus hemoglobin; PE, plasma enzymes; SMS, skin mucus secretion; GSA, gill sialic acid content; LE, liver enzymes.

A further problem with cortisol as an indicator of stress in fish is the influence of the life cycle, particularly smoltification (Specker and Schreck, 1982) and late stages of gonadal maturation (Chester-Jones et al., 1969; Idler and Truscott, 1972; Fontaine, 1976; Pickering and Christie, 1981). It is arguable whether hypercortisolism under these conditions, presumably with its subsequent secondary effects, represents a stress response in the sense that the term stress is commonly used. Similarly, elevation of plasma cortisol associated with the adaptation of some fish to seawater (the process associated with the osmotic adaptation, rather than with osmotic shock) (Butler, 1973; Johnson, 1973; Hirano and Mayer-Gostan, 1978) has an equally questionable stress-associated etiology, unless the concept of stress is expanded to include adaptive stress. It is as yet impossible to determine "normal" values for the activity of the pituitary-interrenal axis; the lowest blood cortisol values probably do not represent the least stressed or the most desirable ("optimal") level.

Plasma or serum cortisol levels are, for the most part, an unreliable indicator of environmental stress in feral fish. For example, if fish are trapped live it is not always feasible to sample tissues immediately. Moreover, in some studies, such as those made in our laboratory, which are concerned with sexually mature and maturing oncorhynchids in the last stages of their life cycle, cortisol and other blood measurements are probably not particularly appropriate as stress indicators. In such situations, well-defined secondary or tertiary stress responses, providing they can be well defined, are probably better indicators of the effects of environmental stressors.

3. INTERLAKE STUDIES OF THE PHYSIOLOGY OF GREAT LAKES COHO AND CHINOOK SALMON

3.1. Background

Oncorhynchid species were introduced into several of the Great Lakes of North America during the late 1950s and early 1960s as part of the Great Lakes restoration program. The most successful of these introductions have been coho salmon (*On-corhynchus kisutch*) and chinook salmon (*Oncorhynchus tshawytscha*), which form the basis of a multimillion dollar salmon sportsfishing industry in the Great Lakes basin.

By the early 1970s thyroid lesions were reported in coho salmon from Lakes Erie (Black and Simpson, 1974), Michigan (Drongowski et al., 1975), and Ontario (Sonstegard and Leatherland, 1976). Extensive studies undertaken from 1975 to the present have attempted to determine the etiology and physiological consequences of the apparent thyroid dysfunction. In part, the studies have been concerned with the etiology of the thyroid lesion in terms of the implications to human health of possible waterborne or tissue associated goitrogens in the Great Lakes ecosystem. In addition, it was hoped that interlake studies of the physiology of the presumably clinically hypothyroid coho salmon would provide an insight into the physiological roles of thyroid hormones in adult salmonid fish.

3.2. Etiology of Thyroid Lesions in Great Lakes Oncorhynchids

3.2.1. Iodide Deficiency

Early reports of apparent thyroid hyperplasia in Great Lakes fishes were considered initially to indicate an iodide-deficient environment. The thyroid in all vertebrates requires sufficient amounts of dietary or waterborne iodide in order to synthesize the iodinated thyronine compounds (predominantly L-thyroxine (T4) and some 3,3',5 triiodo-L-thyronine (T3)), which represent the principal thyroid hormones. In iodide-deficiency situations, the amount of T4 that can be synthesized by the thyroid is reduced, resulting in a lowered plasma T4 level. Pituitary secretion of the hormone thyrotropin (TSH), the major pituitary hormone involved in stimulating thyroid hormone synthesis and secretion, is regulated by negative feedback and is dependant upon blood T4 levels. Thus any reduction in blood T4 levels brings about an increased secretion of TSH and a subsequent increase in blood TSH levels. The TSH acts on the thyroid to stimulate iodide uptake, thyroid hormone synthesis, and T4 (and T3) release. If the level of dietary and environmental iodide is insufficient to meet the demands for T4 synthesis or release, the secretion of TSH escalates and blood TSH levels rise. The resultant chronic blood TSH levels act to deplete the reserve of thyroid hormones in the colloid of the thyroid follicles, and bring about hypertrophy and hyperplasia of the thyroid epithelial cells.

The thyroid tissue of Great Lakes oncorhynchids has all the cytological characteristics indicative of high TSH stimulation. It is composed of masses of thyroid follicle cells with only small amounts of follicular colloid. In light micrographs the goiters show marked regional variations in appearance with peripheral areas containing follicles with rather more colloid than those in the central mass of tissue. The central area is, for the most part, composed of tall columnar epithelial cells that often lack the follicular arrangement typical of normal thyroid tissue. Even when present, the follicle lumina contain little or no colloid material (Sonstegard and Leatherland,

1976; Leatherland et al., 1978; Leatherland and Sonstegard, 1980a; Moccia et al., 1981).

The central areas of North America have long been recognized as iodide-deficient regions. Consequently, it would not be surprising to find iodide-deficiency thyroid hyperplasia in teleosts in the area, particularly in marine teleosts introduced into the region. Hence, several authors proposed an iodide-deficiency etiology to explain the thyroid hyperplasia observed in several freshwater species in the Great Lakes area (Marine and Lenhart, 1910; Marine, 1914; Hoar, 1952; LaRoche, 1952; Robertson and Chaney, 1953; Black and Simpson, 1974; Drongowski et al., 1975).

Although an iodide-deficiency etiology for the thyroid hyperplasia in Great Lakes oncorhynchids cannot be excluded, field epizootiological studies suggest that it is not the critical factor. If iodide-deficiency were the major etiologic agent, the lake level of iodide, and/or the accumulated level of iodide should be correlated with the degree of thyroid hyperplasia. As seen in Table 6, although there is good correlation between lake iodide levels and tissue-accumulated iodide levels in coho salmon from Lakes Ontario, Erie, and Michigan, there is no evidence of a correlation between iodide levels and the degree of thyroid hyperplasia measured by means of goiter frequency and thyroid goiter index. In fact, the tissue iodide level of Lake Ontario salmon was comparable to that of salmon collected as they migrated from the Pacific Ocean (Sonstegard and Leatherland, 1978).

Several studies suggest that the ovaries in fish compete with the thyroid for available iodide (Tarrant, 1971). Thus, if the thyroid hyperplasia was simply iodide-related, a sex difference might be expected in the degree of thyroid hyperplasia within a given lake. No such sex differences were found (Moccia, 1978; Leatherland and Sonstegard, 1980a; Moccia et al., 1981).

A further argument against an iodide-deficiency etiology was the observation that Lake Ontario and Lake Erie coho salmon, collected during the spring, had high

Table 6. Interlake Differences in Thyroid Goiter Frequency, Thyroid Goiter Index, and Tissue Iodide Content of Sexually-Mature Coho Salmon (*Oncorhynchus kisutch*) from Lakes Ontario, Erie, and Michigan, Compared with Lake Iodide Levels[a]

Lake	Lakewater Iodide Content[a] (μg/l)	Goiter Frequency[b] % (95% Confidence Limit) [n]	TGI[c]	Tissue Iodide[d] (mg/kg wet weight)
Ontario	2.9	35 (24, 47) [81]	3.3 ± 1.1 [122]	2.1
Erie	1.7	89 (94, 83) [224]	4.8 ± 0.1 [122]	1.7
Michigan	0.9	16 (12, 21) [384]	2.2 ± 0.1 [217]	0.8

[a] From Winchester, 1970.

[b] Frequency of nodular lesions < 1 cm in diameter at the base of the gill arches (data taken from pooled 1979 and 1980 collections.

[c] Value derived from arbitrary scaling of the size of the lesions based on external examination (1 = no overt hyperplasia; 2 = hyperplasia; 3 = marked hyperplasia or small nodular lesion; 4 = nodular lesion > 1 cm in diameter; 5 = multinodular, bilateral lesions).

[d] From Sonstegard and Leatherland, 1978.

serum T4 and T3 levels. In fact, the serum T3 levels at that time are the highest in all of the thyroid literature (Moccia, 1978; Leatherland and Sonstegard, 1980a, 1981a,b), despite the marked thyroid hyperplasia exhibited by these fish. Clearly, the animals were obtaining sufficient iodide to maintain relatively high levels of thyroid hormones in their blood.

Since oncorhynchids are predatory fish high on the food web of the Great Lakes ecosystem, it is perhaps not surprising that they can obtain sufficient iodide. The iodide turnover of salmonid fish is low (as is their dietary iodide requirement) (Woodall and LaRoche, 1964), so that prey species would likely provide sufficient iodide for the salmon's metabolic needs, even if waterborne iodide levels were insufficient.

3.2.2. Organochlorines

Great Lakes salmon are known to be contaminated with complex mixtures of chlorinated organic compounds (Moccia et al., 1978; Sonstegard and Leatherland, 1978; Leatherland and Sonstegard, 1982a). Some of these, for example polychlorinated biphenyls (PCBs) are known to be goitrogenic in tetrapods (Bastomsky, 1974, 1977a,b; Bastomsky and Wyse, 1975; Bastomsky et al., 1976; Collins et al., 1977) and are therefore suspect etiologic agents. However, in an extensive series of feeding trials, in which both *Salmo gairdneri* and *Oncorhynchus kisutch* were fed PCBs and/or Mirex at levels considerably in excess of those found in the Great Lakes environment, there was no evidence of induced thyroid hyperplasia (Leatherland and Sonstegard, 1978a,b, 1980b, 1982a; Leatherland et al., 1979). In addition, there appeared to be no correlation between bioaccumulated levels of organochlorines in salmon from Lakes Ontario, Erie, or Michigan and the degree of thyroid hyperplasia exhibited by the fish from the respective lakes (Moccia et al., 1978). However, when Great Lakes salmon were fed to rats the rats became hypothyroid and there appeared to be a correlation between the flesh organochlorine content of the salmon and the degree of thyroid hyperplasia (Sonstegard and Leatherland, 1979a; Leatherland and Sonstegard, 1980c, 1982b).

3.2.3. Waterborne Goitrogens

Epidemiological studies have recently shown evidence of endemic goiters (thyroid hyperplasia) that do not have an iodide-deficient etiology (Vought et al., 1967, 1974; Gaitán, 1973; Gaitán et al., 1980a,b). In some instances the hyperplasia appears to be related to the type of underlying rock in an area; rocks containing high levels of organic material were particularly well-correlated with the degree of thyroid hyperplasia (Gaitán, 1973; Delange and Ermans, 1976; Meyer et al., 1978; Gaitán et al., 1978, 1980a,b). In other cases the hyperplasia was related to bacterial contamination of the drinking water (Vought et al., 1967, 1974; Gaitán et al., 1978, 1980a,b). Yet other studies found evidence to suggest that certain minerals, heavy metals, and industrial pollutants such as phenol, sulphide, and ammonia affect thyroid function in various vertebrates (Bloomfield et al., 1960; Stolc and Podoba, 1960; Demole, 1970; Day and Powell-Jackson, 1972; Bobek and Kahl, 1973; Kahl and Ewy, 1975; Bobek et al., 1976; Nath De and Bhattacharya, 1976; Buthieau and Autissier, 1977; Triantaphyllidis et al., 1977; Der et al., 1977; Bech and Madsen, 1978).

Published data on the concentration of minerals and heavy metals in the Great Lakes (e.g., Weiler and Chawla, 1969; Dobson et al., 1974) do not support the hypothesis that they are directly involved in the thyroid hyperplasia. However, there is a good correlation with the degree of eutrophication of the Great Lakes and the degree of thyroid proliferation exhibited by the salmon in the lakes. In view of the increasing evidence of the goitrogenic effects of bacterial (and algal) metabolic products (see above), the possible involvement of the Great Lakes microflora (bacteria and/or algae) in the etiology of the thyroid lesions exhibited by coho and chinook salmon is suspected. It may be that the presumptive waterborne agent is the bacterial or algal organism itself, or a toxic by-product of these organisms.

3.3. Blood Thyroid Hormone Levels

Measurements of serum L-thyroxine (T4) and serum triiodo-L-thyronine (T3) levels, in combination with gross and histological examination of the thyroid tissue in coho salmon collected during their spawning run in 1975, led us to conclude that the fish were hypothyroid (Sonstegard and Leatherland, 1976). The diagnosis was based on the apparent low values of serum T4 (0.4 ± 0.1 µg/dl (n = 16)) and T3 (80 ± 15 ng/dl (n = 17)) levels in spawning fish. However, subsequent studies showed marked interlake differences in plasma thyroid hormone levels both within and between seasons (Table 7), none of which appear to correlate with the degree of thyroid hyperplasia. In fact, in spawning coho salmon collected from British Columbia (which exhibited no thyroid hyperplasia) serum T4 and T3 levels were somewhat lower than in some Great Lakes fish (Table 7). Moreover, seasonal collections of Lake Ontario coho salmon made between May and November, during the last year of their life cycle, showed that the fish maintained high serum thyroid hormone levels from May to September, whereafter the levels of both hormones fell precipitously to the low values seen in spawning and prespawning fish (Leatherland and Sonstegard, 1980a, 1981a,b). Thus it appears that the activity of the thyroid in migratory potamodromous coho salmon decreases shortly after the gonad reaches its maximum size. There are few published data pertaining to thyroid changes in salmonid fish during their upstream migration, but the available evidence suggests that this decreased thyroid activity is not restricted to the Great Lakes salmon (Table 8), but is a feature common to both anadromous and potamodromous species of both *Oncorhynchus* and *Salmo* genera. Consequently, the interlake differences in thyroid hormone levels in Great Lakes salmon probably reflect different degrees of hormone clearance from the blood rather than a meaningful difference in the levels of thyroid function.

The role of the thyroid hormones in adult salmon remains an enigma (Leatherland, 1982b). However, the observation of apparently high thyroid activity, as evidenced by elevated blood titres of thyroid hormones, during the spring and summer months when the fish are growing, storing food material, and at the end of the summer, incorporating food into the developing gonads, suggests a role of the hormones in

Table 7. Interlake Differences in Serum L-Thyroxine (T4) and Triiodo-L-thyronine (T3) Levels in Spawning Coho Salmon from Lakes Ontario, Michigan, and Erie, and the Fraser River, British Columbia

Collection Year	Fraser River, British Columbia		Lake Ontario		Lake Michigan		Lake Erie	
	T4[a]	T3[b]	T4[a]	T3[b]	T4[a]	T3[b]	T4[a]	T3[b]
1975[c]	—	—	0.4 ± 0.1 (16)[f]	80 ± 15 (17)	—	—	—	—
1976[d]	0.5 ± 0.1 (12)	47 ± 22 (12)	1.7 ± 0.2 (20)	592 ± 43 (20)	0.6 ± 0.1 (20)	151 ± 33 (20)	0.2 ± 0.1 (20)	76 ± 17 (20)
1978[e]	—	—	0.4 ± 0.1 (18)	272 ± 23 (18)	0.5 ± 0.1 (10)	67 ± 20 (10)	0.4 ± 0.1 (10)	89 ± 31 (10)

[a] μg/dl.
[b] ng/dl.
[c] From Sonstegard and Leatherland, 1976.
[d] From Leatherland and Sonstegard, 1980a.
[e] From Moccia et al., 1981.
[f] Mean ± s.e. (n).

Table 8.   Changes in Thyroid Activity in Salmonid Fish During Their Upstream Migration Toward the Spawning Area

Species	Observation	Authors
Oncorhynchus nerka	Progressive decrease in thyroid activity during spawning	Robertson and Wexler, 1960
Oncorhynchus tshawytscha	Decrease in thyroid epithelial cell size between entry into river and arrival at spawning areas	Robertson and Wexler, 1960
Salmo salar	Thyroid activity, estimated by plasma protein-bound iodide, thyroid iodide concentrating capability, and hormone secretion rate, was high at the onset of migration and had decreased in spawning fish	Fontaine and Leloup, 1962
Oncorhynchus kisutch	Progressive fall in thyroid epithelial cell height between entry into river and arrival at spawning area	Chestnut, 1970
Oncorhynchus nerka	Decrease in thyroid epithelial cell height between premigratory period at the mouth of the river and spawning and postspawning period	McBride, 1967
Oncorhynchus nerka	Decrease in thyroid epithelial cell height between entry into river and arrival at the spawning area	Ichikawa et al., 1974
Oncorhynchus kisutch	Progressive decrease in thyroid hormone levels during migratory and prespawning period	Leatherland and Sonstegard, 1980a, 1981b; Sower and Schreck, 1982

the regulation of food incorporation and/or mobilization. There are several studies to suggest a role of the thyroid in gonad growth and maturation in fish (Eales, 1979; Pickering and Christie, 1981; Leatherland, 1982b), but it is still unclear whether this role is related to the metabolite incorporation discussed above.

3.4.   Body Weight

The body weights of sexually mature coho salmon collected during their potamodromous migration from Lakes Ontario, Michigan, and Erie in 1976, 1978, and 1980 are shown in Table 9. The values obtained for three collection years were reasonably consistent. There was a clear interlake difference in the size of the fish, with the Lake Ontario coho salmon the largest and the Lake Erie salmon generally the smallest.

Table 9. Interlake Comparison of Body Weight of Spawning Coho Salmon from Lakes Ontario, Michigan, and Erie

Year of Collection	Sex	Source of Salmon		
		Lake Ontario	Lake Michigan	Lake Erie
1976[a]	Male	4.44 ± 0.24 (8)[b]	2.94 ± 0.23 (10)	2.12 ± 0.15 (10)
	Female	4.51 ± 0.24 (12)	2.86 ± 0.25 (10)	2.71 ± 0.22 (10)
1978[c]	Male	4.37 ± 0.28 (19)	3.24 ± 0.29 (7)	2.01 ± 0.12 (14)
	Female	4.68 ± 0.11 (27)	2.50 ± 0.27 (9)	2.27 ± 0.10 (31)
1980[c]	Male	4.15 ± 0.24 (10)	2.91 ± 0.20 (12)	2.11 ± 0.11 (20)
	Female	3.69 ± 0.14 (16)	2.48 ± 0.13 (12)	2.66 ± 0.15 (8)

[a] Based on data from Moccia, 1978.

[b] Mean ± s.e. (n).

[c] Based on data from J. F. Leatherland and R. A. Sonstegard (unpublished data).

130

3.5. Blood Lipid and Ion Levels

Lake Erie coho salmon had significantly lower levels of serum free fatty acids, triglyceride, and cholesterol than Lake Ontario coho salmon. Similarly, Lake Michigan coho salmon had significantly lower levels of serum free fatty acids and triglyceride than Lake Ontario coho salmon (Table 10). Serum calcium and magnesium levels in Lake Erie coho salmon were significantly lower, and serum potassium levels were significantly higher than in coho salmon from the other two lakes.

It is still unclear whether any of the interlake differences in blood lipid and ions are related to the thyroid hyperplasia and presumed thyroid dysfunction.

3.6. Reproductive Physiology

Coho salmon collected during their potamodromous spawning migration from Lake Erie exhibit little or no evidence of sexual dimorphism. The secondary sexual characteristics of red flank and kype in males, and marked silvering of the females are either absent, or poorly defined in the Lake Erie salmon compared with fish collected at a comparable stage of sexual development from Lakes Michigan or Ontario. Conversely, in Lakes Ontario and Michigan the salmon develop well-marked secondary sexual characteristics. Moreover, data obtained from various hatcheries indicate that the survival of eggs from zygote to 5 cm-long fingerling stage is 40–60% in stock collected from Lake Ontario and Lake Michigan coho salmon, but less than 15% in eggs collected from Lake Erie salmon (Leatherland et al., 1982). These figures are supported by pilot studies of egg survival carried out in our laboratory under controlled water temperature and water quality conditions. Survival to hatch was 80% in Lake Michigan, 80% in Lake Ontario (U.S.A. stock), 90% in Lake Ontario (Canadian stock), and 25% in Lake Erie coho salmon (P. Morrison, J. F. Leatherland, and R. A. Sonstegard, unpublished data). The poorly defined secondary sexual development of the adults and the high egg mortality in Lake Erie salmon are possibly related phenomena, indicative of endocrine dysfunction.

Lake Erie fish have apparently inactive pituitary gonadotrop cells compared with fish from Lakes Ontario or Michigan (Leatherland and Sonstegard, 1980a, 1981a), although the pituitary content of immunoreactive gonadotropin was similar in salmon from all three study lakes. Plasma gonadotropin, testosterone, and 11-ketotestosterone levels in both males and females were lower in Lake Erie fish than in comparable salmon from Lakes Ontario and Michigan (Table 11). The gonadosomatic indices revealed no evidence of reduced gonad size in Lake Erie fish, although the eggs were smaller in Lake Erie salmon than in fish from the other two lakes (Table 10). Despite the smaller egg size of the Lake Erie salmon the lipid content of the eggs (mg/egg) was comparable in eggs of salmon taken from all three lakes (Table 11).

3.7. Liver Pathobiology

The liver in teleosts is the site of production of vitellogenin, which is synthesized under the influence of steroids, particularly estrogens (Campbell and Idler, 1976;

Table 10. Interlake Comparison of Serum Lipid and Ion Levels in Spawning Coho Salmon from Lakes Ontario, Michigan, and Erie

Blood Parameter	Sex	Source of Salmon		
		Lake Ontario	Lake Michigan	Lake Erie
Free fatty acids (μEq/l)	Male	582 ± 48 (5)[a]	208 ± 27 (5)	250 ± 101 (5)
	Female	356 ± 49 (5)	166 ± 28 (5)	142 ± 15 (5)
Triglycerides (mg/dl)	Male	136 ± 19 (5)	68 ± 8 (5)	81 ± 21 (5)
	Female	435 ± 58 (5)	205 ± 65 (5)	136 ± 80 (5)
Cholesterol (mg/ml)	Male	6.2 ± 0.6 (5)	5.9 ± 0.2 (5)	4.0 ± 0.7 (5)
	Female	4.8 ± 0.5 (5)	3.7 ± 0.4 (5)	2.6 ± 0.6 (5)
Sodium (mEq/l)	NDS[b]	137 ± 4 (10)	142 ± 5 (10)	143 ± 4 (10)
Potassium (mEq/l)	NDS	1.1 ± 0.3 (10)	1.2 ± 0.2 (10)	2.8 ± 0.5 (10)
Chloride (mEq/l)	NDS	132 ± 2 (10)	126 ± 2 (10)	132 ± 2 (10)
Magnesium (mg/dl)	NDS	3.6 ± 0.4 (10)	3.4 ± 0.1 (10)	2.2 ± 0.1 (10)
Calcium (mg/dl)	Male	12 ± 1 (6)	14 ± 1 (7)	10 ± 1 (11)
	Female	26 ± 3 (10)	26 ± 4 (9)	11 ± 1 (8)
Osmotic pressure (mOsmol/l)	NDS	359 ± 5 (10)	343 ± 6 (10)	342 ± 3 (10)

[a] Mean ± s.e. (n).

[b] NDS = no differences between sexes, data combined. Data taken from Leatherland and Sonstegard, 1980d, and Leatherland et al., 1981, derived from sexually mature spawning salmon collected in 1976 and 1978.

Table 11. Interlake Differences in Gonad Size, Pituitary and Blood Gonadotropin Content, and Blood Androgen Content in Spawning Coho Salmon from Lake Ontario, Michigan, and Erie

Parameter	Sex	Source of Salmon		
		Lake Ontario	Lake Michigan	Lake Erie
Gonadosomatic index[a]	Male	5.0 ± 0.3 (22)[b]	4.2 ± 0.1 (20)	5.7 ± 0.3 (23)
	Female	19.8 ± 1.4 (18)	32.5 ± 1.3 (17)	29.7 ± 0.7 (30)
Egg volume[a] (ml)		0.188 ± 1.4 (23)	0.168 ± 4 (20)	0.154 ± 6 (30)
Egg diameter[a] (mm)		7.1 ± 0.1 (23)	6.8 ± 0.1 (20)	6.6 ± 0.1 (30)
Egg density		1.29 ± 0.06 (23)	1.25 ± 0.01 (20)	1.17 ± 0.06 (30)
Egg lipid content[a] (mg/egg)		175 ± 4 (23)	159 ± 4 (20)	67 ± 6 (30)
Egg lipid content[a] (% wet weight)		41.7 ± 0.7 (23)	38.2 ± 0.5 (20)	43.9 ± 0.7 (30)
Pituitary gonadotropin[c] content (ng/pituitary)	Male	1487 ± 157 (6)	806 ± 103 (14)	987 ± 75 (6)
	Female	1321 ± 282 (6)	1075 ± 117 (17)	1113 ± 182 (6)
Plasma gonadotropin[c] content (ng/ml)	Male	97 ± 15 (7)	91 ± 17 (17)	54 ± 8 (10)
	Female	117 ± 12 (5)	97 ± 8 (15)	68 ± 10 (9)
Plasma testosterone[c] (ng/ml)	Male	99 ± 9 (20)	137 ± 10 (20)	176 ± 30 (20)
	Female	367 ± 24 (20)	435 ± 34 (20)	155 ± 18 (30)
Plasma 11-ketotestosterone (ng/ml)	Male	705 ± 78 (7)	180 ± 19 (18)	132 ± 23 (10)
	Female	14 ± 2 (10)	12 ± 1 (27)	5 ± 1 (12)

[a] Unpublished data (P. Morrison, J. F. Leatherland, and R. A. Sonstegard) derived from sexually mature salmon collected in 1981.

[b] Mean ± s.e. (n).

[c] From Leatherland et al., 1982, derived from sexually mature salmon collected in 1978.

Table 12. Interlake Differences in Liver Morphology in Spawning Coho Salmon From Lakes Ontario, Michigan, Erie, and Superior

Parameter	Sex	Source of Salmon			
		Lake Superior	Lake Ontario	Lake Michigan	Lake Erie
HSI[a]	Male	1.7 ± 0.1 (17)[b]	1.6 ± 0.1 (10)	1.4 ± 0.1 (12)	1.8 ± 0.1 (20)
	Female	2.4 ± 0.3 (11)	1.6 ± 0.1 (18)	1.6 ± 0.2 (10)	1.3 ± 0.1 (20)
HNA[c]	Male	20.6 ± 0.4 (80)	23.6 ± 0.3 (80)	18.1 ± 0.3 (80)	22.4 ± 0.3 (80)
	Female	19.6 ± 0.4 (80)	20.4 ± 0.2 (80)	18.4 ± 0.4 (80)	20.2 ± 0.3 (80)
HVI[d]	Male	41.3 ± 1.0 (40)	43.1 ± 0.8 (40)	53.4 ± 1.6 (40)	51.5 ± 0.8 (40)
	Female	54.9 ± 1.8 (40)	61.8 ± 1.7 (40)	66.9 ± 2.2 (40)	75.1 ± 1.7 (40)

[a] Hepatosomatic index = liver weight/carcass weight × 100.

[b] Mean ± s.e. (n).

[c] Hepatocytic nuclear area (μm^2).

[d] Hepatocytic volume index = the number of nuclei per unit area.

Hori et al., 1979; de Vlaming et al., 1980). In addition, the liver in some teleosts catabolizes steroids such as cortisol to produce 11 β-hydroxandrostenedione and androstentrione (Kime, 1978). The former is considered to be a precursor of 11 β-hydroxytestosterone and 11-ketotestosterone (Idler and MacNab, 1967). Such a metabolic pathway has implications in the control of reproduction since the liver, in addition to the gonad, may be an important source of androgenic steroids, thus removing androgen steroidogenesis from the restrictive negative-feedback regulation of the hypothalamus-pituitary-gonad axis (Kime, 1978). Because the liver is involved in steroid catabolism, and in addition is the site of vitellogenin production, the sex differences in hepatic morphology of Great Lakes coho salmon are not surprising. Significantly smaller hepatocytes were found in female spawning coho salmon from Lakes Superior, Ontario, Michigan, or Erie than in males from the same lakes based on the hepatocytic volume index (HVI) (Table 12). Similarly, the hepatocytic nucleus size was larger in males than in female salmon of Lakes Ontario and Erie; no sex differences were evident in salmon from Lakes Superior or Michigan.

The liver also plays an important intermediate role in the incorporation, mobilization, and distribution of nutrient reserves, as well as acting as the principal organ for detoxification. Thus, alterations in the physiology or morphology of the liver reflect responses to nutrient or toxic stressors. Consequently, the liver is a potentially important stress indicator. Relatively little is known of the structure and function of the liver of feral teleostean species since most studies have dealt with hatchery-reared fish, usually salmonids propagated on formulated diets (Leatherland, 1982c). However, there were considerable interlake differences in hepatocyte morphology of coho salmon from Lakes Superior, Ontario, Michigan, or Erie. Although the significance of many of these differences remains unresolved, they undoubtedly reflect the different conditions of the four lake ecosystems (Leatherland and Sonstegard, 1983).

4. INTERLAKE DIFFERENCES IN THE PHYSIOLOGY OF GREAT LAKES COHO SALMON AS INDICATORS OF INTERLAKE STRESS?

As discussed above, in field studies of feral populations it is not always possible to use hypercortisolism as the principal indicator of the degree of response to various stressors. Environmental stressors are probably multifactorial, more complex and more subtle than the acute, distressful stressors applied in most laboratory studies of teleostean stress response. Little is known of the additive, synergistic, and/or permissive* interaction of environmental stressors, although there are suggestions

---

* *Permissive interaction*: exposure of the fish to a low intensity of one stressor (e.g., a shift in ambient pH, or temperature) that may not itself evoke a physiological stress response, but that predisposes the animal to the effects of a second stressor (e.g., heavy metal exposure or disease challenge). The phenomenon differs from *additive* or *synergistic interactions* in that the first stressor elicits the physiological (adaptive) response that makes the animal susceptible to the second stressor. *Additive* interactions are the cumulative effects of several above-threshold stressors, and *synergistic interactions* are evident when two (or more) stressors act together to effect a response greater than can be accounted for by the additive interactions of the two stressors.

of such interactions as evidenced by increased disease-susceptibility of "stressed" fish (Wedemeyer, 1970b; Roth, 1972; Snieszko, 1974; Hunter et al., 1980; Ellis, 1981). Indeed, epizootic data have been used as a means of monitoring the relative "health of ecosystems" (Sonstegard, 1977, 1978; Sonstegard and Leatherland, 1979b, 1982), in the same way that epidemiological studies have been used to identify sources of environmental concern for human populations (e.g., cigarette smoke, asbestos fibers, toxic chemicals, and radiation sources).

Such a sentinel system provides a means of monitoring the effects of complex and subtle interactions of numerous diverse biotic and abiotic ecological factors without necessarily knowing fully the identity of the stressors. In this regard, coho salmon are ideal sentinel organisms for evaluating the "health" of the Great Lakes ecosystem. Fish from common genetic stock, originating from the State of Michigan salmon program, are introduced annually into several of the Great Lakes. They feed primarily on the same forage species in all the lakes (smelts, *Osmerus mordax*, and alewives, *Alosa pseudoharengus*). As one of the top predators in the aquatic ecosystem they are exposed to the full extent of xenobiotic bioaccumulation by diet, in addition to uptake of waterborne factors.

Eighteen months after their stocking as smolts they become sexually mature and congregate in the rivers at the original stocking areas. Consequently, annual synchronized collections can be made from the lakes using simple collection techniques of electroshocking, weir traps, snagging, or dipnetting. A disadvantage of the species is that, at this time in its life cycle, it is at an advanced stage of sexual maturation and much has yet to be learned of the normal physiological changes associated with this phase before abnormal events can be recognized.

Perennial diverse empiric interlake differences in the physiology of Great Lakes coho salmon are suggestive of interlake differences in the degree of exposure to stressors within the respective lakes. For example, the different growth rates of coho salmon within the three study lakes reflect markedly different environmental conditions. The size differences might be attributed to several factors; they reflect differences in genetic stock, food availability, or food conversion. With regard to the differences being due to genetic differences, similar genetic stocks were introduced into Lakes Ontario and Michigan where they exhibit markedly different growth rates (Moccia, 1978; Leatherland et al., 1981). Moreover, fish of two different genetic stocks (i.e., eggs taken from the State of Michigan stock (originating from the State of Oregon) and from the Province of Ontario stock (originating from the Province of British Columbia) planted in Lake Ontario both achieve a similar final body weight. This suggests that the interlake differences in body weights are not due to genetic variation. Differential food availability in the different Great Lakes may be a factor. However, since the biomass of prey species in Lake Erie is considered to be among the highest of the three study lakes, it is difficult to reconcile the small size of the sexually mature adult coho salmon in Lake Erie with a food-deprived hypothesis. Nevertheless, without a better understanding of the competition for the prey species within the three lakes, it is impossible to eliminate food availability as a factor. Food conversion differences could explain some of the variability although these would likely not be due to thermal differences, since in all three lakes the salmon appear to exploit the thermocline. Since the thyroid hormones are thought to be involved in some aspects of teleostean growth (Eales, 1979; Leatherland, 1982b), it is possible that food conversion by the Lake Erie salmon is deleteriously affected by the severe thyroid dysfunction exhibited by the Lake Erie salmon. Conversely, bioaccumulated xenobiotics may interfere with metabolism, possibly by the liver, and impair food incorporation by salmon in Lakes Erie or Michigan. This is supported by the lower lipid content of the flesh of spawning salmon from Lakes Erie or Michigan (1.52 and 1.05% wet weight, respectively) compared with the flesh of spawning Lake Ontario (5.82% wet weight) or Pacific Ocean salmon (7.68% wet weight) (Sonstegard and Leatherland, 1978).

Whether or not the low reproductive success of Lake Erie salmon is related to a metabolic dysfunction is not yet clear. The egg lipid content of Lake Erie salmon is comparable with that of the eggs from Lake Ontario and Lake Michigan salmon (Table 11), but the quality of lipid and of other nutrients is not yet known. This forms the subject of a study currently underway in our laboratory. One of the most likely causes of the low egg survival of Lake Erie salmon is the apparently low steroid secreting capabilities of these fish (Table 11). Plasma testosterone and 11-ketotestosterone levels in female Lake Erie fish are significantly lower than in comparable fish from Lakes Ontario and Michigan. Testosterone is thought to be an important precursor for estrogen synthesis by the developing eggs (Lambert and von Bohemen, 1979); consequently the low androgen levels in the females may signal an impaired ability to synthesize estrogens, which are necessary for vitellogenin synthesis by the liver (Campbell and Idler, 1976; Fontaine, 1976; De Vlaming et al., 1980; Idler and Campbell, 1980). If, as Kime (1978) suggests, the liver is implicated in steroid catabolism, and is an important site of androgen synthesis from cortisol precursor in teleosts, the interlake differences in liver morphology (Table 12; Leatherland and Sonstegard, 1983) might be indicative of diffrent degrees of steroid catabolic activity, rather than, or in addition to, interlake differences in nutrient metabolism.

The mode of interaction of environmental and endocrine factors affecting gonadal function in teleosts is poorly understood but very complex (Fontaine, 1976; Hirose, 1976; Peter and Crim, 1979; Billard et al., 1981; Suzuki et al., 1981a,b; Sower et al., 1982; Peter, 1982). As discussed by Billard et al. (1981), environmental factors (temperature, thermoperiod, photoperiod, water quality, pollution, food availability, social pressures, etc.) affecting one or more of the regulatory systems can adversely affect reproductive viability. It is not yet clear how these factors affect gonad development and/or maturation in Great Lakes coho salmon. However, irrespective of the physiological basis, the low reproductive success of the Lake Erie salmon compared with that of the Lake Ontario and Lake Michigan fish suggests that environmental stressors in Lake Erie effect a greater response in the salmon reared in that lake compared with salmon from the other two study lakes.

Interlake differences in serum ions similarly reflect the dynamic equilibrium established between the fish (by virtue of the various osmo- and ionoregulatory processes acting at gill, kidney, skin, intestinal sites, etc.) and the ambient environment. The differences in ion level between lakes (Table 13) are relatively insignificant,

Table 13. Interlake Comparison of the Concentrations of Major Ions in Lakes Ontario, Michigan, and Erie[a]

Ion[b]	Lake Ontario	Lake Michigan	Lake Erie
Calcium	39	32	37
Magnesium	8.2	10.0	8.0
Sodium	9.5	3.4	10.0
Potassium	1.8	0.9	1.4
Sulphate	32	16	24
Chloride	24	6	21
Bicarbonate	113	130	113
pH	8.1	8.0	8.1

[a] Data from Kramer, 1964, Dobson, 1967, Weiler and Chawla, 1969, and Dobson et al., 1974.
[b] Expressed as mg/l.

and, in themselves, cannot account for the differences in serum ions of salmon in the three study lakes. In particular, the low calcium and magnesium and high potassium of the Lake Erie coho salmon cannot be explained by differences in ambient ion levels. The ionic differences may be related to endocrine dysfunction (e.g., thyroid) but undoubtedly reflect real interlake differences in the adaptation of fish to different Great Lakes, and therefore reflect different degrees of adaptive stress response.

More work is necessary before the full significance of the interlake pathobiological differences can be assessed. In particular, more extensive information is needed pertaining to normal cyclic physiological changes in teleostean species. In addition, if such phenomena as growth and reproductive success are to be used as monitors, a detailed understanding of the normal processes of growth and reproduction is essential before the effect of environmental stressors can be fully elucidated.

In order to understand and appreciate the stress response to environmental stressors (i.e., nonexperimental stressors), further studies are needed of the effects of interactions of stressors on the primary, secondary, and tertiary stress responses of fish. The nature of stressor effects on feral fish is presently an unknown quantity; consequently, it is impossible to extrapolate from laboratory situations involving chronic distressful stimulation to the situation experienced by free-living feral fish. Inasmuch as the health of feral teleosts reflects, in part, the health of the ecosystem, comparative epidemiological (epizootiological) studies have a relevance both to human health and to fisheries management programs.

5. CONCLUSIONS

The use of the so-called primary and secondary stress responses as indicators of stressor effects, while of some value in well-controlled laboratory experiments, is not always applicable to studies of feral teleostean populations. Under these conditions the so-called tertiary responses may be more meaningful indicators. Disease susceptibility and tumor frequencies may be useful monitors of aquatic systems, particularly in studies of the effects of environmental pollution. Similarly, reproductive success and growth are also ideal criteria for assessing the relative quality of different ecosystems. Examples are presented of interlake comparisons of the growth, reproductive success, and physiology of coho salmon taken from Lakes Ontario, Michigan, and Erie. However, the interpretation of physiological data obtained in this type of study relies upon a firm understanding of the seasonal physiology of the species under investigation; such a data base is restricted to few teleostean species.

REFERENCES

Arillo, A., Margiocco, C., and Melodia, F. (1979a). The gill sialic acid content as an index of environmental stress in rainbow trout, *Salmo gairdneri* Richardson. *J. Fish. Biol.* **15**, 405–410.

Arillo, A., Maniscalco, N., and Margiocco, C. (1979b). Fructose 1, 6-biphosphatase and total proteolytic activity in the liver of *Salmo gairdneri*: Effects of pH and ammonia. *Comp. Biochem. Physiol.* **63C**, 325–331.

Ashcom, T. L., Markle, J. T., Neff, W. H., and Anthony, A. (1977). Serum cortisol and interrenal ribonucleic acid changes in brook trout subjected to sublethal acid stresses. *Amer. Zool.* **17**, 873.

Barton, B. A., Peter, R. E., and Paulencu, C. R. (1980). Plasma cortisol levels of fingerling rainbow trout (*Salmo gairdneri*) at rest, and subjected to handling, confinement, transport, and stocking. *Can. J. Fish. Aquat. Sci.* **37**, 805–811.

Bastomsky, C. H. (1974). Effects of a polychlorinated biphenyl mixture (Aroclor 1254) and DDT on biliary thyroxine excretion in rats. *Endocrinology* **95**, 1150–1155.

Bastomsky, C. H. (1977a). Enhanced thyroxine metabolism and high uptake goiters in rats after a single dose of 2,3,7,8-tetrachlorodibenzo-*p*-dioxin. *Endocrinology* **101**, 292–296.

Bastomsky, C. H. (1977b). Goiters in rats fed polychlorinated biphenyls. *Can. J. Physiol. Pharmacol.* **55**, 288–292.

Bastomsky, C. H. and Wyse, J. M. (1975). Enhanced thyroxine metabolism following cutaneous application of microscope immersion oil. *Res. Comm. Chem. Path. Pharmacol.* **10**, 725–733.

Bastomsky, C. H., Murthy, P. V. N., and Banovac, K. (1976). Alterations in thyroxine metabolism produced by cutaneous application of microscope immersion oil: Effects due to polychlorinated biphenyls. *Endocrinology* **98**, 1309–1314.

Bech, K. and Madsen, S. N. (1978). Human thyroid adenylate cyclase in non-toxic goitre: Sensitivity to TSH, fluoride and thyroid stimulating immunoglobulins. *Clin. Endocrinol.* **8**, 457–466.

Bengtsson, B. E. (1980). Long term effects of PCB (Clophen A50) on growth, reproduction and swimming performance in the minnow, *Phoxinus phoxinus. Water Res.* **14**, 681–687.

Benoit, D. A. (1975). Chronic effects of copper on survival, growth and reproduction of the bluegill (*Lepomis macrochirus*). *Trans. Amer. Fish. Soc.* **104**, 353–358.

Billard, R., Bry, C., and Gillet, C. (1981). Stress, environment and reproduction in teleost fish. In: A. D. Pickering (Ed.), *Stress and fish.* Academic Press, London, pp. 185–208.

Black, J. J. and Simpson, C. L. (1974). Thyroid enlargement in Lake Erie coho salmon. *J. Nat. Cancer Inst.* **54**, 725–730.

Bloomfield, R. A., Welsh, C. W., Garner, G. B., and Muhrer, M. E. (1960). Effect of dietary nitrate on thyroid function. *Science* **134**, 1690.

Bobek, S. and Kahl, S. (1973). The effect of perchlorate and thiocyanate anions on the PBI level and exogenic I-thyroxine binding to the blood plasma proteins and diaphragm muscle of the rat. The experiments *in vivo* and *in vitro. Endokrin. Pol.* **24**, 21–31.

Bobek, S., Kahl, S. and Ewy, Z. (1976). Effect of long-term fluoride administration on thyroid hormone levels in blood in rats. *Endocrinol. Exp.* **10**, 289–295.

Bouck, G. R., Cairns, M. A., and Christian, A. R. (1978). Effect of capture stress on plasma enzyme activities in rainbow trout (*Salmo gairdneri*). *J. Fish. Res. Board Can.* **325**, 1495–1498.

Bresch, M. (1982). Investigation of the long term action of xenobiotics on fish with special regard to reproduction. *Ecotoxicol. Environ. Safety* **6**, 102–112.

Brown, J., Fedoruk, K., and Eales, J. G. (1978). Physical injury due to injections or blood removal causes transitory elevations of plasma thyroxine in rainbow trout, *Salmo gairdneri. Can. J. Zool.* **56**, 1998–2003.

Burdick, G. E., Deem, H. J., Harris, E. J., Skea, J., Karcher, R., and Frisca, C. (1972). Effect of rate and duration of feeding DDT on the reproduction of salmonid fishes reared and bred under controlled conditions. *N.Y. Fish Game J.* **19**, 97–115.

Buthieau, A.-M. and Autissier, N. (1977). Action des ions Mn^{2+} sur le métabolisme iode thyroiden du rat. *C.R. Sean. Soc. Biol.* **171**, 1024–1028.

Butler, D. E. (1973). Structure and function of the adrenal gland of fishes. *Amer. Zool.* **13**, 839–897.

Cairns, M. A. and Christian, A. R. (1978). Effects of hemorrhagic stress on several blood parameters in adult rainbow trout (*Salmo gairdneri*). *Trans. Amer. Fish. Soc.* **107**, 334–340.

Campbell, C. M. and Idler, D. R. (1976). Hormonal control of vitellogenesis in hypophysectomized winter flounder (*Pseudopleuronectes americanus* Walbaum). *Gen. Comp. Endocrinol.* **28**, 143–150.

Chavin, W. and Young, J. E. (1970). Factors in the determination of normal serum glucose levels of goldfish, *Carassius auratus* L. *Comp. Biochem. Physiol.* **33**, 629–653.

Chester-Jones, I., Chan, D. K. O., Henderson, I. W., and Ball, J. N. (1969). The adrenocortical steroids, adrenocorticotropin and the corpuscles of Stannius. In: W. S. Hoar and D. J. Randall (Eds.), *Fish physiology*, Vol. II. Academic Press, New York, pp. 321–376.

Chestnut, C. W. (1970). The pituitary gland of coho salmon (*Oncorhynchus kisutch* Walbaum) and its function in gonad maturation and thyroid activity. Ph.D. dissertation, Simon Fraser Univ., B.C.

Christian, J. J. (1975). Hormonal control of population growth. In: B. E. Eleftheriou and R. Sprott (Eds.), *Hormonal correlates of behavior*, Vol. I. Plenum Press, New York, pp. 205–274.

Collins, W. T., Jr., Capen, G. C., Kasza, L., Carter, C., and Dailey, R. E. (1977). Effect of polychlorinated biphenyl (PCB) on the thyroid gland of rats. *Amer. J. Pathol.* **89**, 119–136.

Dalela, R. C., Rani, S., and Verma, S. R. (1980). Physiological stress induced by sublethal concentrations of phenol and pentachlorophenol in *Notopterus notopterus*: Hepatic acid and alkaline phosphatases and succinic dehydrogenase. *Environ. Pollut.* **21A**, 3–8.

Day, T.K., and Powell-Jackson, P. R. (1972). Fluoride, water hardness, and endemic goitre. *Lancet* **i**, 1135–1138.

Delange, F. M. and Ermans, A. M. (1976). Endemic goiter and cretinism. Naturally occurring goitrogens. *Pharmacol. Ther. C* **1**, 57–93.

Demole, V. (1970). Toxic effects on the thyroid. In: *Fluorides and human health*. W.H.O. Monograph, Ser. 59, Geneva, pp. 255–262.

Der, R., Yousef, M., Fahim, Z., and Fahim, M. (1977). Effects of lead and cadmium on adrenal and thyroid functions in rats. *Res. Comm. Chem. Pathol. Pharmacol.* **17**, 237–253.

DeVlaming, V. L., Wiley, H. S., Delahunty, G., and Wallace, R. A. (1980). Goldfish (*Carassius auratus*) vitellogenin: Induction, isolation, properties and relationship to yolk proteins. *Comp. Biochem. Physiol.* **67**, 613–624.

DiMichele, L. and Taylor, M. H. (1978). Histopathological and physiological responses of *Fundulus heteroclitus* to naphthalene exposure. *J. Fish. Res. Board Can.* **35**, 1060–1066.

Dobson, H. F. H. (1967). Principal ions and dissolved oxygen in Lake Ontario. In: *Proc. of the 10th conf. Great Lakes res.*, pp. 337–356.

Dobson, H. F. H., Gilbertson, M., and Sly, P. G. (1974). A summary and comparison of nutrients and related water quality in Lakes Erie, Ontario, Huron, and Superior. *J. Fish. Res. Board Can.* **31**, 731–738.

Donaldson, E. M. (1981). The pituitary-interrenal axis as an indicator of stress in fish. In: A. D. Pickering (Ed.), *Stress and fish*. Academic Press, London, pp. 11–47.

Donaldson, E. M. and Dye, H. M. (1975). Corticosteroid concentrations in sockeye salmon (*Oncorhynchus nerka*) exposed to low concentrations of copper. *J. Fish. Res. Board Can.* **32**, 533–539.

Donaldson, E. M. and Fagerlund, U. H. M. (1969). Cortisol secretion rate in gonadectomized female sockeye salmon (*Oncorhynchus nerka*): Effects of estrogen and cortisol treatment. *J. Fish. Res. Board Can.* **26**, 1789–1799.

Donaldson, E. M. and Fagerlund, U. H. M. (1970). Effect of sexual maturation and gonadectomy at sexual maturity on cortisol secretion rate in sockeye salmon (*Oncorhynchus nerka*). *J. Fish. Res. Board Can.* **27**, 2287–2296.

Donaldson, E. M. and McBride, J. R. (1967). The effects of hypophysectomy in the rainbow trout, *Salmo gairdneri* (Rich.) with special reference to the pituitary-interrenal axis. *Gen. Comp. Endocrinol.* **9**, 93–101.

Drongowski, R. A., Wood, J. S., and Bouck, G. R. (1975). Thyroid activity in coho salmon from Oregon and Lake Michigan. *Trans. Amer. Fish. Soc.* **2**, 349–352.

Eales, J. G. (1979). Thyroid functions in cyclostomes and fishes. In: E. J. W. Barrington (Ed.), *Hormones and evolution*, Vol. I. Academic Press, New York, pp. 341–346.

Eales, J. G. and Fletcher, G. L. (1982). Circannual cycles of thyroid hormones in plasma of winter flounder (*Pseudopleuronectes americanus* Walbaum). *Can. J. Zool.* **60**, 304–309.

Ejike, C. and Schreck, C. B. (1980). Stress and social hierarchy rank in coho salmon. *Trans. Amer. Fish. Soc.* **109**, 423–426.

Ellis, A. E. (1981). Stress and the modulation of defence mechanisms in fish physiological systems. In: A. D. Pickering (Ed.), *Fish and stress*. Academic Press, London, pp. 147–169.

Erickson, J. G. (1967). Social hierarchy, territoriality, and stress reactions in sunfish. *Physiol. Zool.* **40**, 40–80.

Fagerlund, U. H. M. (1967). Plasma cortisol concentration in relation to stress in adult sockeye salmon during the freshwater stage of their life cycle. *Gen. Comp. Endocrinol.* **8**, 197–207.

Fagerlund, U. H. M. and Donaldson, E. M. (1970). Dynamics of cortisone secretion in sockeye salmon (*Oncorhynchus nerka*) during sexual maturation and after gonadectomy. *J. Fish. Res. Board Can.* **27**, 2323–2331.

Fagerlund, U. H. M. and McBride, J. R. (1969). Suppression by dexamethasone of interrenal activity in adult sockeye salmon (*Oncorhynchus nerka*) *Gen. Comp. Endocrinol.* **12**, 651–657.

Fontaine, M. (1976). Hormones and the control of reproduction in aquaculture. *J. Fish. Res. Board Can.* **33**, 922–939.

Fontaine, M. and Leloup, J. (1962). Le fonctionnement thyroïdien du saumon adulte (*Salmo salar* L.) a quelques étapes du son cycle migratoire. *Gen. Comp. Endocrinol.* **2**, 317–322.

Freeman, H. C. and Idler, D. R. (1975). The effects of polychlorinated biphenyls on steroidogenesis and reproduction in the brook trout (*Salvelinus fontinalis*). *Can. J. Biochem.* **53**, 666–670.

Fromm, P. O. (1980). A review of some physiological and toxicological responses of freshwater fish to acid stress. *Environ. Biol. Fish.* **5**, 79–93.

Fryer, J. N. (1975). Stress and adrenocorticosteroid dynamics in the goldfish, *Carassius auratus. Can. J. Zool.* **53**, 1012–1020.

Fuller, J. D., Scott, D. B. C. and Fraser, R. (1974). Effect of catching techniques, captivity and reproductive cycle on plasma cortisol concentration in powan (*Coregonus lavaretus*), a freshwater teleost from Loch Lomond. *J. Endocrinol.* **63**, 24.

Gaitán, E. (1973). Water-borne goitrogens and their role in the etiology of endemic goiter. *World Rev. Nutr. Diet* **17**, 53–90.

Gaitán, E., Merino, H., Rodriguez, G., Medina, P., Meyer, J. D., DeRoue, T. A., and MacLennan, R. (1978). Epidemiology of endemic goitre in western Colombia. *Bull. W.H.O.* **56**, 403–416.

Gaitán, E., Medina, P. and DeRouen, R. A. (1980a). Goiter prevalence and bacterial contamination of water supplies. In: J. R. Stocking and S. Nagataki (Eds.), *Thyroid research VIII*. Australian Academy of Sciences, Canberra, pp. 210–214.

Gaitán, E., Medina, P., DeRouen, R. A., and Zia, M. S. (1980b). Goiter prevalence and bacterial contamination of water supplies. *J. Clin. Endocrinol. Metab.* **53** 957–961.

Grant, B. F. and Mehrle, P. M. (1973). Endrin toxicosis in rainbow trout (*Salmo gairdneri*). *J. Fish. Res. Board Can.* **30**, 31–40.

Hane, S., Robertson, O. H., Wexler, B. C., and Krupp, A. (1966). Adrenocortical response to stress and ACTH in Pacific salmon (*Oncorhynchus tshawytscha*) and steelhead trout (*Salmo gairdneri*) at successive stages in the sexual cycle. *Endocrinology* **78**, 791–800.

Hattingh, J. (1977). Blood sugar as an indicator of stress in the freshwater fish, *Labeo capensis* (Smith). *J. Fish. Biol.* **10**, 191–195.

Hattingh, J. and Van Pletzen, A. J. J. (1974). The influence of capture and transportation on some blood parameters of freshwater fish. *Comp. Biochem. Physiol.* **49A**, 607–609.

Hill, C. W. and Fromm, P. O. (1968). Response of the interrenal gland of rainbow trout (*Salmo gairdneri*) to stress. *Gen. Comp. Endocrinol.* **11**, 69–77.

Hilmy, A. M., Shabbana, M. B., and Said, M. M. (1980). Haematological responses to mercury toxicity in the marine teleost, *Aphanius dispar* (Ruepp). *Comp. Biochem. Physiol.* **67C**, 147–158.

Hirano, T. and Mayer-Gostan, N. (1978). Endocrine control of osmoregulation in fish. In: P. J. Gaillard and H. H. Boer (Eds.), *Comparative endocrinology*. Elsevier-North Holland Press, New York, pp. 209–212.

Hirose, K. (1976). Endocrine control of ovulation in medaka (*Oryzias latipes*) and ayu (*Plecoglossue altivelis*). *J. Fish. Res. Board Can.* **33**, 989–994.

Hlavek, R. R. and Bulkley, R. V. (1980). Effect of malachite green on leucocyte abundance in rainbow trout, *Salmo gairdneri* (Richardson). *J. Fish. Biol.* **17**, 431–444.

Hoar, W. S. (1952). Thyroid function in some anadromous and landlocked teleosts. *Trans. Royal Soc. Can.* **66**, 39–53.

Hori, S. H., Kodama, T., and Tanahoshi, K. (1979). Induction of vitellogenin synthesis in goldfish by massive doses of androgens. *Gen. Comp. Endocrinol.* **17**, 306–320.

Houston, A. H., Madden, J. A., Woods, R. J., and Miles, H. M. (1971a). Some physiological effects of handling and tricaine methanesulphonate anesthetization upon the brook trout, *Salvelinus fontinalis*. *J. Fish. Res. Board Can.* **28**, 625–631.

Houston, A. H., Madden, J. A., Woods, R. J., and Miles, H. M. (1971b). Variations in the blood and tissue chemistry of brook trout, *Salvelinus fontinalis*, subsequent to handling, anaesthesia, and surgery. *J. Fish. Res. Board Can.* **28**, 635–642.

Hunter, V. A., Knittel, M. D., and Fryer, J. L. (1980). Stress-induced transmission of *Yersinia ruckeri* infection from carriers to recipient steelhead trout, *Salmo gairdneri* Richardson. *J. Fish. Dis.* **3**, 467–472.

Ichikawa, M., Mori, T., Kawashima, S., Ueda, K., and Shirahata, S. (1974). Histological changes in the thyroid and the interrenal tissue of the kokanee (*Oncorhynchus nerka*) during sexual maturation and spawning. *J. Fac. Sci. Univ. Tokyo, Sect. 4, Zool.* **13**, 175–182.

Idler, D. R. and Campbell, C. M. (1980). Gonadotropin stimulation of estrogen and yolk precursor synthesis in juvenile rainbow trout. *Gen. Comp. Endocrinol.* **41**, 384–391.

Idler, D. R. and MacNab, H. C. (1967). The biosynthesis of 11-ketotestosterone and 11 β-hydroxy-testosterone by Atlantic salmon tissue *in vitro*. *J. Biochem. Physiol.* **45**, 581–589.

Idler, D. R. and Truscott, B. (1972). Corticosteroids in fish. In: D. R. Idler (Ed.), *Steroids in nonmammalian vertebrates*. Academic Press, New York, pp. 127–252.

Ishioka, H. (1980a). Stress reactions induced by environmental salinity changes in red sea bream. *Bull. Jap. Soc. Sci. Fish.* **46**, 1323–1331.

Ishioka, H. (1980b). Stress reactions in marine fish. I. Stress reactions induced by temperature change. *Bull. Jap. Soc. Sci. Fish.* **46**, 523–531.

Johnson, D. W. (1973). Endocrine control of hydromineral balance in teleosts. *Amer. Zool.* **13**, 799–818.

Kahl, S. and Ewy, Z. (1975). Effect of single and long-term sodium fluoride administration on thyroid hormone biosynthesis in rats. *Fluoride* **8**, 191–198.

Kapur, K., Kamaldeep, K., and Toor, H. S. (1978). The effect of fenitrothion on reproduction of a teleost fish, *Cyprinus carpio communis* Linn: A biochemical study. *Bull. Environ. Contam. Toxicol.* **20**, 438–442.

Kime, D. E. (1978). The hepatic catabolism of cortisol in teleost fish—Adrenal origin of 11-oxotestosterone precursors. *Gen. Comp. Endocrinol.* **35**, 322–328.

Kling, D. (1981). Total atresia of the ovaries of *Tilapia leucostica* (Cichlidae) after intoxication with the insecticide Lebaycid. *Experentia* **37**, 73–74.

Klinger, H., Peters, G., and Delventhal, H. (1979). Physiologische und morphologische Effects von sozialem Stress bien Aal, *Anguilla anguilla* L. *Verh. Dt. Zool. Ges.*, **1979**, 246.

Kramer, J. R. (1964). Theoretical model for the chemical composition of fresh water with application to the Great Lakes. In: *Proc. of the 7th conf. on Great Lakes res.*, pp. 147–160.

Lambert, J. G. D. and Van Boheman, C. G. (1979). Steroidogenesis in the ovary of the rainbow trout, *Salmo gairdneri*, during the reproductive cycle. *Proc. Ind. Nat. Sci. Acad.* **B45**, 414–420.

LaRoche, G. (1952). Effets de preparations thyroidiennes et d'iodures sur le goitre ("pseudo-cancer") des salmonides. *Rev. Can. Biol.* **11**, 439–445.

Leatherland, J. F. (1982a). Effect of ambient salinity, food deprivation and prolactin on the thyroidal response to TSH, and *in vitro* hepatic T4 to T3 conversion in yearling coho salmon, *Oncorhynchus kisutch*. *Acta Zool., Stockh.* **63**, 55–64.

Leatherland, J. F. (1982b). Environmental physiology of the teleostean thyroid gland: A review. *Environ. Biol. Fish.* **7**, 83–110.

Leatherland, J. F. (1982c). Effect of a commercial diet on liver ultrastructure of fed and fasted yearling coho salmon, *Oncorhynchus kisutch* Walbaum. *J. Fish.Biol.* **21**, 311–319.

Leatherland, J. F. and Dodd, J. M. (1969). Activity of the hypothalamo-neurohypophysial complex of the European eel (*Anguilla anguilla* L.) assessed by the use of an *in situ* staining technique and by autoradiography. *Gen. Comp. Endocrinol.* **13**, 45–59.

Leatherland, J. F. and Sonstegard, R. A. (1978a). Lowering of serum thyroxine and triiodothyronine levels in yearling coho salmon, *Oncorhynchus kisutch*, by dietary Mirex or PCB's. *J. Fish. Res. Board Can.* **35**, 1285–1289.

Leatherland, J. F. and Sonstegard, R. A. (1978b). Effect of dietary PCB, Mirex and a combination of Mirex and PCB on thyroid activity in rainbow trout (*Salmo gairdneri*) and coho salmon (*Oncorhynchus kisutch*). In: *Int.Symp. on Analysis hydrocarbons and halogenated hydrocarbons*, vol. 54, pp. 1–5.

Leatherland, J. F. and Sonstegard, R. A. (1980a). Seasonal changes in thyroid hyperplasia, serum thyroid hormone and lipid concentrations and pituitary gland structure in Lake Ontario coho salmon (*Oncorhynchus kisutch*) and a comparison with coho salmon from Lakes Michigan and Erie. *J. Fish Biol.* **16**, 539–562.

Leatherland, J. F. and Sonstegard, R. A. (1980b). Effect of dietary Mirex and PCB's in combination with food deprivation and testosterone administration on thyroid activity and bioaccumulation of organochlorines in rainbow trout, *Salmo gairdneri* Richardson. *J. Fish Dis.* **3**, 115–124.

Leatherland, J. F. and Sonstegard, R. A. (1980c). Structure of thyroid and adrenal glands in rats fed diets of Great Lakes coho salmon (*Oncorhynchus kisutch*). *Environ. Res.* **23**, 77–86.

Leatherland, J. F. and Sonstegard, R. A. (1980d). Interlake differences in growth, thyroid physiology and blood chemistry of coho salmon (*Oncorhynchus kisutch*) in the Great Lakes of North America:

Effects of environmental contamination. In: R. Gilles (Ed.), *Animals and environmental fitness*. Pergamon Press, Oxford, pp. 89–90.

Leatherland, J. F. and Sonstegard, R. A. (1981a). Thyroid function, pituitary structure and serum lipids in Great Lakes coho salmon, *Oncorhynchus kisutch* Walbaum, "Jacks" compared with sexually immature spring salmon. *J. Fish Biol.* **18**, 643–653.

Leatherland, J. F. and Sonstegard, R. A. (1981b). Thyroid dysfunction in Great Lakes coho salmon, *Oncorhynchus kisutch* (Walbaum): Seasonal and interlake differences in serum T3 uptake and serum total and free T4 and T3 levels. *J. Fish Dis.* **4**, 413–423.

Leatherland, J. F. and Sonstegard, R. A. (1982a). Bioaccumulation of organochlorines by yearling coho salmon (*Oncorhynchus kisutch* Walbaum) fed diets containing Great Lakes' coho salmon, and the pathophysiological responses of the recipients. *Comp. Biochem. Physiol.* **72C**, 91–99.

Leatherland, J. F. and Sonstegard, R. A. (1982b). Thyroid responses in rats fed diets formulated with Great Lakes' coho salmon. *Bull. Environ. Contam. Toxicol.* **29**, 341–346.

Leatherland, J. F. and Sonstegard, R. A. (1983). Interlake comparison of liver morphology and *in vitro* monodeiodination of L-thyroxine in sexually mature coho salmon, *Oncorhynchus kisutch* Walbaum, from Lakes Erie, Ontario, Michigan, and Superior. *J. Fish Biol.,* **22**, 519–536.

Leatherland, J. F., Cho, C. Y., and Slinger, S. J. (1977). Effects of diet, ambient temperature, and holding conditions of plasma thyroxine levels in rainbow trout (*Salmo gairdneri*). *J. Fish. Res. Board Can.* **34**, 677–682.

Leatherland, J. F., Moccia, R., and Sonstegard, R. (1978). Ultrastructure of the thyroid gland in goitered coho salmon (*Oncorhynchus kisutch*). *Cancer Res.* **38**, 149–158.

Leatherland, J. F., Sonstegard, R. A., and Holdrinet, M. V. (1979). Effect of dietary Mirex and PCB's on hepatosomatic index, liver lipid, carcass lipid and PCB and Mirex accumulation in yearling coho salmon, *Oncorhynchus kisutch*. *Comp. Biochem. Physiol.* **63C**, 243–246.

Leatherland, J. F., Sonstegard, R. A., and Moccia, R. D. (1981). Interlake differences in body and gonad weights and serum constituents of Great Lakes coho salmon (*Oncorhynchus kisutch*). *Comp. Biochem. Physiol.* **69A**, 701–704.

Leatherland, J. F., Copeland, P., Sumpter, J. P., and Sonstegard, R. A. (1982). Hormonal control of gonadal maturation and development of secondary sexual characteristics in coho salmon, *Oncorhynchus kisutch*, from Lakes Ontario, Erie, and Michigan. *Gen. Comp. Endocrinol.* **48**, 196–204.

Leloup-Hatey, J. (1964). Functionnement de l'interrenal anterieur de deux teleosteens: Le saumon Atlantique et l'anguille Europeene. *Ann. Inst. Oceanogr., Monaco* **42**, 221–338.

Lorz, H. W., Williams, R. H., and Fustish, C. A. (1978). Effects of several metals on smolting of coho salmon, *Oncorhynchus kisutch*. U.S. Environ. Prot. Agency, Rep. EPA-6—/3-78-090, 85 pp.

Marine, D. (1914). Further observations and experiments on goiter (so-called thyroid carcinoma) in brook trout (*Salvelinus fontinalis*). III. Its prevention and cure. *J. Exp. Med.* **19**, 70–88.

Marine, D. and Lenhart, C. (1910). On the occurrence of goiter (active thyroid hyperplasia) in fish. *Bull. Johns Hopkins Hosp.* **21**, 95–98.

Mazeaud, M. M., Mazeaud, F., and Donaldson, E. M. (1977). Primary and secondary effects of stress in fish: Some new data with a general review. *Trans. Amer. Fish. Soc.* **106**, 201–212.

McBride, J. R. (1967). Effects of feeding on the thyroid, kidney, and pancreas in sexually ripening adult sockeye salmon (*Oncorhynchus nerka*). *J. Fish. Board Can.* **24**, 67–76.

McBride, J. R., Strasdine, G., and Fagerlund, U. H. M. (1975). Acute toxicity of kanamycin to steelhead trout (*Salmo gairdneri*). *J. Fish. Res. Board Can.* **32**, 554–558.

McBride, J. R., Donaldson, E. M., and Derksen, G. (1979). Toxicity of land-fill leachates to underyearling rainbow trout (*Salmo gairdneri*). *Bull. Environ. Contam. Toxicol.* **23**, 806–813.

McBride, J. R., Dye, H. M. and Donaldson, E. M. (1981). Stress response of juvenile sockeye salmon (*Oncorhynchus nerka*) to the butoxyethanol ester of 2,4-dichlorophenoxyacetic acid. *Bull. Environ. Contam. Toxicol.* **27**, 877–884.

McCormick, J. H., Stokes, G. N., and Portele, G. J. (1980). Environmental acidification impact detection by examination of mature fish ovaries. *Can. Tech. Rep. Fish. Aquat. Sci.* **975**, 41–48.

McLeay, D. J. (1970). A histometric investigation of the activity of the pituitary-interrenal axis in juvenile coho salmon (*Oncorhynchus kisutch*) Walbaum. Ph.D. dissertation, Univ. of British Columbia.

McLeay, D. J. (1973). Effects of a 12-hr and 25-day exposure to kraft pulp mill effluent on the blood and tissues of juvenile coho salmon (*Oncorhynchus kisutch*). *J. Fish. Res. Board Can.* **30**, 395–400.

McLeay, D. J. and Gordon, M. R. (1977). Leucocrit: A simple haematological technique for measuring acute stress in salmonid fish including stressful concentrations of pulpmill effluent. *J. Fish. Res. Board Can.* **34**, 2164–2175.

Meyer, J. D., Gaitán, E., Merino, H., and DeRouen, T. (1978). Geologic implications in the distribution of endemic goiter in Colombia, South America. *Int. J. Epidem.* **7**, 25–30.

Mezhnin, F. I. (1978). The interrenal and suprarenal glands and Stannius' bodies of the guppy, *Lebistes reticulatus*, under conditions of extreme stress. *J. Ichthyol.* **18**, 611–632.

Milne, R. S., Leatherland, J. F., and Holub, B. J. (1979). Changes in plasma thyroxine, triiodothyronine and cortisol associated with starvation in rainbow trout. (*Salmo gairdneri*). *Environ. Biol. Fish.* **4**, 185–190.

Moccia, R. D. (1978). Environmental pathobiology of thyroid hyperplasia in coho (*Oncorhynchus kisutch*) and chinook (*Oncorhynchus tshawytscha*) salmon in the Great Lakes. M.Sc. dissertation, Univ. of Guelp, Ont.

Moccia, R. D., Leatherland, J. F., Sonstegard, R. A., and Holdrinet, M. V. H. (1978). Are goiter frequencies in Great Lakes salmon correlated with organochlorine residues? *Chemosphere* **8**, 649–652.

Moccia, R. D., Leatherland, J. F., and Sonstegard, R. A. (1981). Quantitative interlake comparison of thyroid pathology in Great Lakes coho (*Oncorhynchus kisutch*) and chinook (*Oncorhynchus tshawytscha*) salmon. *Cancer Res.* **41**, 2200–2210.

Munkittrick, K. R. and Leatherland, J. F. (1983). Hematocrit values in feral goldfish (*Carassius auratus* L.) as indicators of the health of the population. *J. Fish Biol.* **23**, 153–161.

Munkittrick, K. R. and Leatherland, J. F. (1984). Abnormal pituitary-gonad function in two feral populations of goldfish (*Carassius auratus* L.) suffering epizootics of an ulcerative disease. *J. Fish Dis.* (in press).

Nath De, S. and Bhattacharya, S. (1976). Effect of some industrial pollutants on fish thyroid peroxidase activity and role of cytochrome C thereon. *Ind. J. Exp. Biol.* **14**, 561–563.

Nebeker, A. V., Puglisi, F. A., and Defoe, D. L. (1974). Effect of polychlorinated biphenyl compounds on survival and reproduction of the fathead minnow and flagfish. *Trans. Amer. Fish. Soc.* **103**, 562–568.

Noakes, D. L. G. and Leatherland, J. F. (1977). Social dominance and interrenal cell activity in rainbow trout, *Salmo gairdneri* (Pisces, Salmonidae). *Environ. Biol. Fish.* **2**, 131–136.

Osborn, R. H. and Simpson, T. H. (1974). The effect of stress and food deprivation on the thyroidal status of plaice and of rainbow trout. Cons. Perm. Int. Explor. Mer. CM 1974 Gear and Behaviour Comm. B9. Cited by Pickering et al., 1982.

Pearce, B. C. and McBride, J. R. (1978). A preliminary investigation of the effect of 2,4-dichlorophenoxyacetic acid (butoxy ethanol ester) on juvenile rainbow trout. Fish. Mar. Serv. Ms. report no. 1478. Cited by Donaldson, 1981.

Peter, R. E. (1982). Neuroendocrine control of reproduction in teleosts. *Can. J. Fish. Aquat. Sci.* **39**, 48–55.

Peter, R. E. and Crim, L. W. (1979). Reproductive endocrinology of fishes: Gonadal cycles and gonadotropin in teleosts. *Ann. Rev. Physiol.* **41**, 323–335.

Peters, G. (1979). Zur Interpretation des Begriffs "Stress" bien Fisch. *Fisch Tiershutz* **7**, 25–32.

Peters, G., Delventhal, H., and Kluiger, H. (1980). Physiological and morphological effects of social stress on the eel (*Anguilla anguilla* L.). *Arch. Fisch Weiss* **30**, 157–180.

Pickering, A. D. (1981). Introduction: The concept of biological stress. In: A. D. Pickering (Ed.), *Stress and fish*. Academic Press, London, pp. 1–9.

Pickering, A. D. and Christie, P. (1981). Changes in the concentrations of plasma cortisol and thyroxine during sexual maturation of the hatchery-reared brown trout, *Salmo trutta*. *Gen. Comp. Endocrinol.* **44**, 487–496.

Pickering, A. D., Pottinger, T. G., and Christie, R. (1982). Recovery of the brown trout, *Salmo trutta* L., from acute handling stress: A time course study. *J. Fish. Biol.* **20**, 229–244.

Pickford, G. E. (1973). Current status of fish endocrine systems. Introductory remarks. *Zoology* **13**, 711–717.

Richards, R. H. and Pickering, A. D. (1979). Changes in serum parameters of *Saprolegnia*-infected brown trout, *Salmo trutta* L. *J. Fish Dis.* **2**, 197–206.

Robertson, O. H. and Chaney, A. L. (1953). Thyroid hyperplasia and tissue iodine content in spawning rainbow trout: A comparative study of Lake Michigan and California sea-run trout. *Physiol. Zool.* **26**, 328–340.

Robertson, O. H. and Wexler, R. C. (1960). Histological changes in the organs and tissues of migrating and spawning Pacific salmon (Genus *Oncorhynchus*). *Endocrinology* **66**, 222–239.

Rose, R. M. and Sachar, E. (1981). Psychoendocrinology. In: R. H. Williams (Ed.), *Textbook of endocrinology*, W. B. Saunders Co., Philadelphia, pp. 645–671.

Roth, R. R. (1972). Some factors contributing to the development of fungus infection in freshwater fish. *J. Wildl. Dis.* **8**, 24–28.

Ruby, S. M., Aczel, J., and Craig, G. R. (1977). The effects of depressed pH on oogenesis in flagfish *Jordanella floridae*. *Water Res.* **11**, 757–762.

Rush, S. B. and Umminger, B. L. (1978). Elimination of stress-induced changes in carbohydrate metabolism of goldfish (*Carassius auratus*) by training. *Comp. Biochem. Physiol.* **60A**, 69–73.

Schreck, C. B. (1981). Stress and compensation in teleostean fishes: Responses to social and physical factors. In: A. D. Pickering (Ed.), *Stress and fish*. Academic Press, London, pp. 295–321.

Schreck, C. B. and Lorz, H. W. (1978). Stress response of coho salmon (*Oncorhynchus kisutch*) elicited by cadmium and copper and potential use of cortisol as an indicator of stress. *J. Fish. Res. Board Can.* **35**, 1124–1129.

Schreck, C. B., Whaley, R. A., Bass, M. L., Maughan, O. E., and Solazzi, M. (1976). Physiological responses of rainbow trout (*Salmo gairdneri*) to electroshock. *J. Fish. Res. Board Can.* **33**, 76–84.

Scott, D. B. and Currie, C. E. (1980). Social hierarchy in relation to adrenocortical activity in *Xiphophorus helleri* Heckel. *J. Fish Biol.* **16**, 265–277.

Selye, H. (1956). *The stress of life*. McGraw-Hill Book Co., New York, 324 pp.

Silbergeld, E. K. (1974). Blood glucose: A sensitive indicator of environmental stress. *Bull. Environ. Contam. Toxicol.* **11**, 20–25.

Simpson, T. H. (1975/1976). Endocrine aspects of salmonid culture. *Proc. Roy. Soc. Edinburgh Ser. B.* **75**, 241–252.

Singh, H. and Singh, T. P. (1980). Effect of two pesticides on ovarian ^{32}P uptake and gonadotrophin concentration during different phases of the annual reproductive cycle in the freshwater catfish, *Heteropneustes fossilis* (Bloch). *Environ. Res.* **22**, 190–200.

Singley, J. A. and Chavin, W. (1975a). The adrenocortical-hypophyseal response to saline stress in the goldfish, *Carassius auratus* L. *Comp. Biochem. Physiol.* **51A**, 749–756.

Singley, J. A. and Chavin, W. (1975b). Serum cortisol in normal goldfish (*Carassius auratus* L.). *Comp. Biochem. Physiol.* **50A**, 77–82.

Slicher, A. M., Pickford, G. E., and Pang, P. K. T. (1966). Effects of "training" and of volume and composition of the injection fluid on stress-induced leukopenia in the mummichog. *Prog. Fish Cult.* **28**, 216–219.

Smit, G. L., Hattingh, J., and Burger, A. P. (1979a). Haematological assessment of the effects of the anaesthetic MS 222 in natural and neutralized form in three freshwater fish species: Interspecies differences. *J. Fish Biol.* **15**, 633–643.

Smit, G. L., Hattingh, J., and Burger, A. P. (1979b). Haematological assessment of the effects of the anaesthetic MS 222 in natural and neutralized form in three freshwater fish species: Haemoglobin electrophoresis, ATP levels and corpuscular fragility curves. *J. Fish Biol.* **15**, 655–663.

Smith, A. C. and Ramos, F. (1976). Occult haemoglobin in fish skin mucus as an indicator of early stress. *J. Fish Biol.* **9**, 537–541.

Snieszko, S. F. (1974). Effects of environmental stress on outbreaks of infectious diseases of fish. *J. Fish. Biol* **6**, 197–208.

Soivio, A. and Oikari A. (1976). Haematological effects of stress on a teleost, *Esox lucius* L. *J. Fish Biol.* **8**, 397–411.

Sonstegard, R. A. (1977). Environmental carcinogenesis studies in fishes of the Great Lakes of North America. *Ann. N.Y. Acad. Sci.* **298**, 261–269.

Sonstegard, R. A. (1978). Feral aquatic organisms as indicators of waterborne environmental carcinogens. In: O. Hutzinger, L. H. van Lelyveld, and B. C. J. Zoeteman (Eds.), *Aquatic pollutants: Transformation and biological effects*. Pergamon Press, New York, pp. 349–358.

Sonstegard, R. A. and Leatherland, J. F. (1976). The epizootiology and pathogenesis of thyroid hyperplasia in coho salmon (*Oncorhynchus kisutch*) in Lake Ontario. *Cancer Res.* **36**, 4467–4475.

Sonstegard, R. A. and Leatherland, J. F. (1978). Growth retardation in rats fed coho salmon collected from the Great Lakes of North America. *Chemosphere* **11**, 903–910.

Sonstegard, R. A. and Leatherland, J. F. (1979a). Hypothyroidism in rats fed Great Lakes coho salmon. *Bull. Environ. Contam. Toxicol.* **22**, 779–784.

Sonstegard, R. A. and Leatherland, J. F. (1979b). Aquatic organism pathobiology as a sentinel system to monitor environmental carcinogens. In: B. K. Afghan and D. Mackay (Eds.), *Analysis of hydrocarbons and halogenated hydrocarbons*. Plenum Press, New York, pp. 513–520.

Sonstegard, R. A. and Leatherland, J. F. (1982). Comparative epidemiology: The use of fishes in assessing carcinogenic contaminants. In this volume, Chapter 15.

Sower, S. A. and Schreck, C. B. (1982). Steroid and thyroid hormones during sexual maturation of coho salmon (*Oncorhynchus kisutch*) in sea water and freshwater. *Gen. Comp. Endocrinol.* **47**, 42–53.

Sower, S. A., Schreck, C. B., and Donaldson, E. M. (1982). Hormone-induced ovulation of coho salmon (*Oncorhynchus kisutch*) held in sea water and freshwater. *Can. J. Fish. Aquat. Sci.* **39**, 627–632.

Specker, J. L. and Schreck, C. B. (1980). Stress response to transportation and fitness for marine survival in coho salmon (*Oncorhynchus kisutch*) smolts. *Can. J. Fish. Aquat. Sci.* **37**, 765–769.

Specker, J. L. and Schreck, C. B. (1982). Change in plasma corticosteroids during smoltifaction of coho salmon, *Oncorhynchus kisutch*. *Gen. Comp. Endocrinol.* **46**, 53–58.

Spieler, R. E. (1974). Short-term serum cortisol concentrations in goldfish (*Carassius auratus*) subjected to serial sampling and restraint. *J. Fish. Res. Board Can.* **31**, 1240–1243.

Srivastava, A. K. and Agrawal, U. (1977). Involvement of pituitary-interrenal axis and cholinergic mechanisms during the cold-shock leucocyte sequence in a freshwater tropical teleost, *Colisa fasciatus*. *Arch. Anat. Microsc. Morphol. Exp.* **66**, 97–108.

Stolc, V. and Podoba, J. (1960). Effect of fluoride on the biogenesis of thyroid hormones. *Nature, Lond.* **188**, 855–856.

Strange, R. J. (1980). Acclimation temperature influences cortisol and glucose concentrations in stressed channel catfish. *Trans. Amer. Fish. Soc.* **109**, 298–303.

Strange, R. J. and Schreck, C. B. (1978). Anesthetic and handling stress on survival and cortisol concentration in yearling chinook salmon (*Oncorhynchus kisutch*). *J. Fish. Res. Board. Can.* **35**, 345–349.

Strange, R. J. and Schreck, C. B. (1980). Seawater and confinement alters survival and cortisol concentration in juvenile chinook salmon. *Copeia*, **1980**, 351–353.

Strange, R. J., Schreck, C. B., and Golden, J. T. (1977). Corticoid stress responses to handling and temperature in salmonids. *Trans. Amer. Fish. Soc.* **106**, 213–218.

Strange, R. J., Schreck, C. B., and Ewing, R. D. (1978). Cortisol concentrations in confined juvenile chinook salmon (*Oncorhynchus tshawytscha*). *Trans. Amer. Fish. Soc.* **107**, 812–819.

Suzuki, K., Tamaoki, B. I., and Hirose, K. (1981a). *In vitro* metabolism of 4-pregnenes in ovaries of a freshwater teleost ayu (*Plecoglassus astivelis*): Production of 17α-, 20β-dihydroxy-4-pregnene-3-one and its 5-reduced metabolites, and activation of 3- and 20-hydroxysteroid dehydrogenases by treatment with a fish gonadotropin. *Gen. Comp. Endocrinol.* **45**, 473–481.

Suzuki, K., Tamaoki, B. I., and Nagahama, Y. (1981b). *In vitro* synthesis of an inducer for germinal vesicle breakdown of fish oocytes, 17α-,20β-dihydroxy-4-pregnene-3-one by ovarian tissue preparations of amago salmon (*Oncorhynchus rhodurus*). *Gen. Comp. Endocrinol.* **45**, 533–535.

Swift, D. J. (1981). Changes in selected blood component concentrations of rainbow trout (*Salmo gairdneri* Richardson) exposed to hypoxia or sublethal concentrations of phenol or ammonia. *J. Fish Biol.* **19**, 45–61.

Swift, D. J. (1982). Changes in selected blood component values of rainbow trout, *Salmo gairdneri* Richardson, following the blocking of the cortisol stress response with betamethasone and subsequent exposure to phenol or hypoxia. *J. Fish Biol.* **21**, 269–277.

Tandon, R. S. and Chandra, S. (1978). Effect of asphyxiation stress on serum transaminases (GOT and GPT) levels of freshwater catfish, *Clarius batrachus*. *Z. Tierphysiol. Tierernahr. Futtermittelkd.* **40**, 34–38.

Tarrant, R. M., Jr. (1971). Seasonal variation in the accumulation and loss of ^{125}I by tissues of adult female channel catfish, *Ictalurus punctatus* (Rafinesque). *Trans. Amer. Fish. Soc.* **100**, 237–246.

Triantaphyllidis, H., Dugas du Villard, J. A., and Guichard, C. (1977). Regulation, *in vivo*, par le calcium du métabolisme thyroïdien de l'iode, chez le rat. *C.R. Sean. Soc. Biol.* **171**, 1167–1172.

Vought, R. L., London, W. T., and Stebbing, G. T. (1967). Endemic goiter in Northern Virginia. *J. Clin. Endocrinol.* **27**, 1381–1389.

Vought, R. L., Brown, F. A., and Sibinovic, K. A. (1974). Antithyroid compound(s) produced by *Escherichia coli*: Preliminary report. *J. Clin. Endocrinol. Metab.* **38**, 861–865.

Wedemeyer, G. (1969a). Pituitary activation by bacterial endotoxins in rainbow trout (*Salmo gairdneri*). *J. Bact.* **100**, 532–543.

Wedemeyer, G. (1969b). Stress-induced ascorbic acid depletion and cortisol production in two salmonid fishes. *Comp. Biochem. Physiol.* **29**, 1247–1251.

Wedemeyer, G. (1970a). Stress of anesthesia with MS222 and benzocaine in rainbow trout (*Salmo gairdneri*). *J. Fish. Res. Board Can.* **27**, 909–914.

Wedemeyer, G. (1970b). The role of stress in the disease resistance of fishes. In: S. F. Snieszko (Ed.), *Symp. on diseases of fish and shellfishes*. Amer. Fish. Soc., Washington, DC, pp. 30–35.

Wedemeyer, G. (1971). The stress of formalin treatments in rainbow trout (*Salmo gairdneri*) and coho salmon (*Oncorhynchus kisutch*). *J. Fish. Res. Board Can.* **28**, 1899–1904.

Wedemeyer, G. (1972). Some physiological consequences of handling stress in the juvenile coho salmon (*Oncorhynchus kisutch*) and steelhead trout (*Salmo gairdneri*). *J. Fish. Res. Board Can.* **29**, 1780–1783.

Wedemeyer, G. (1973). Some physiological aspects of sublethal heat stress in the juvenile steelhead trout (*Salmo gairdneri*) and coho salmon (*Oncorhynchus kisutch*). *J. Fish. Res. Board Can.* **30**, 831–834.

Wedemeyer, G. A. (1976). Physiological response of juvenile coho salmon (*Oncorhynchus kisutch*) and rainbow trout (*Salmo gairdneri*) to handling and crowding stress in intensive fish culture. *J. Fish. Res. Board Can.* **33**, 2699–2702.

Wedemeyer, G. and McLeay, D. J. (1981). Methods for determining the tolerance of fishes to environmental stressors. In: A. D. Pickering (Ed.), *Stress and fish*. Academic Press, New York, pp. 247–275.

Weiler, R. R. and Chawla, V. K. (1969). Dissolved mineral quality of Great Lakes waters. In: *Proc. of the 12th conf. on Great Lakes res*, pp. 801–818.

Wendt, C. A. G. and Saunders, R. L. (1973). Changes in carbohydrate metabolism in young Atlantic salmon in response to various forms of stress. Spec. Publ. Int. Atlantic Salmon Foundation, vol. 4, pp. 55–82.

Winchester, J. W. (1970). Chemical equilibria of iodine in Great Lakes waters. In: *Proc. 13th conf. on Great Lakes res.*, vol. 11, pp. 137–140.

Woodall, A. N. and LaRoche, G. (1964). Nutrition of salmonid fishes: Iodide requirement of chinook salmon. *J. Nutr.* **82**, 475–481.

Wydoski, R. S., Wedemeyer, G. A., and Nelson, N. C. (1976). Physiological response to hooking stress in hatchery and wild rainbow trout (*Salmo gairdneri*). *Trans. Amer. Fish. Soc.* **105**, 601–606.

Zaccone, G. (1980). Structure, histochemistry and effect of stress on the epidermis of *Ophisurus serpens* (L.) (Teleostei: Ophichthidae). *Cell. Mol. Biol.* **26**, 663–674.

Zeitoun, I. H. (1977). The effect of chlorine toxicity on certain blood parameters of adult rainbow trout (*Salmo gairdneri*). *Environ. Biol. Fish.* **1**, 189–195.

11

ADENYLATE ENERGY CHARGE AS A MEASURE OF STRESS IN FISH

Robert E. Reinert and David W. Hohreiter

School of Forest Resources
University of Georgia
Athens, Georgia

1. INTRODUCTION

Adenylate energy charge (AEC) is an index of metabolic energy available to an organism from the adenylate pool (Atkinson, 1968, 1977) and is calculated from tissue concentrations of the three adenine nucleotides—adenosine triphosphate (ATP), adenosine diphosphate (ADP), and adenosine monophosphate (AMP). The AEC is calculated by the formula

$$AEC = \frac{(ATP) + \frac{1}{2}(ADP)}{(ATP) + (ADP) + (AMP)}$$

By definition, AEC can range from 0 to 1. For a wide variety of organisms (bacteria, Chapman et al., 1971; rats, Ridge, 1972; soybean nodules, Ching et al., 1975; and bivalve molluscs, Wijsman, 1976), high values (0.9–0.8) have typically been found for organisms under optimal conditions. These values typify animals having high rates of growth and the ability to reproduce. Lower AECs (0.70–0.50) are generally associated with a limiting or partial stress and are characterized by slow or zero growth and reduced or no reproduction (Chapman et al., 1971; Wiebe and Bancroft, 1975). In studies that primarily involve bacteria, AEC values less than 0.50 indicate severe stress, and organisms are characterized by no growth, no reproduction, and no recovery even after the stress has been removed (Chapman et al., 1971; Ridge, 1972; Montague and Dawes, 1974).

1.1. Adenylate Energy Charge in Invertebrates

Although the majority of stress studies concerning AEC have involved noncontaminant stressors such as dissolved oxygen (DO), temperature, restricted diet, and exercise, there have been some AEC studies with invertebrates that have dealt with the effect of contaminants. For example, Giesy et al. (1981) found that a concentration of 10 ppb of cadmium, which caused mortality of crayfish after several months of exposure, caused a significant drop in AEC after only 7 days of exposure. They also found that a similar concentration of cadmium caused a significant drop in the AEC of freshwater shrimp after 2 days of exposure. Ivanovici (1979) reported that the AEC of molluscs moved from a noncontaminated river to one that was contaminated with hydrocarbons was lower than the AEC of those left in the noncontaminated river, with the lowest AECs recorded for molluscs from the most contaminated sites. Controls moved to adjacent noncontaminated sites did not differ from those of molluscs from the original noncontaminated sites. AEC values of molluscs taken from contaminated water and returned to the original noncontaminated site increased to control levels, and of 30 molluscs that were left at each site, no live ones were found at any of the contaminated sites after 9 months. It should be noted, however, that along with the contaminants, which were a variety of heavy oils, greases, waxes, and tars, sections of the polluted river had sufficiently low DO concentrations for anaerobic conditions to exist (Furzer, 1975). Therefore, in this instance it cannot be determined if the contaminants, the low DO, or a combination of the two were responsible for the low AEC values in the molluscs.

1.2.   Adenylate Energy Charge in Fish

1.2.1.   Temperature Effects

At present the only published information concerning AEC and fish has dealt with the effects of changes in water temperature, DO, and pH. Bashamohideen and Kunnemann (1979) found that ATP content of dorsal muscle and brain of *Idus idus*, a cyprinid, was dependent on the acclimation temperature. They also found that when the water temperature of fish acclimated to 10°C was increased by 1.0°C per hour to 20°C, that during the 25 days of the study the AEC of these fish did not reach that of fish that had been held at 20°C. When the fish were exposed to a temperature change of 0.5°C per day, however, the AEC levels did reach those of fish held at 20°C. They concluded that rapid temperature changes act as stressors and can block the setting of new metabolic states for a long time.

1.2.2.   Dissolved Oxygen Effects

Two studies of AEC in fish concern the effect of oxygen. Van Den Thillart et al. (1976) reported that the AEC of white muscle in goldfish (*Carassius auratus*) decreased from 0.98 to 0.81 during the time (about 3 hours) it took for the DO concentration to drop from 60% saturation to zero. The AEC stabilized at about 0.81 for the remaining 13 hours of exposure to anoxic conditions. In the other study, which concerned the effects of hypoxia on adenylates and AEC of flounder (*Platichthys flesus* L.), Jorgensen and Mustafa (1980) found the total adenylates and AEC in heart, kidney, and muscle were relatively unaffected. In the liver, however, hypoxia caused a 90% drop in ATP, a rise in ADP and AMP, and a drop in total adenylates and in AEC. Jorgensen and Mustafa suggested that the ATP concentration in muscle is stabilized by a large creatine phosphate pool.

1.2.3.   pH Effects

In another study, changes in pH were found to affect AEC of the Gulf killifish *Fundulus grandis* (MacFarlane, 1981). Fish were acclimated to a pH of 7.8 for 2 weeks and then three groups of fish were exposed for 96 hr to pHs of 6.5, 5.0, and 4.0, respectively, and a control group was maintained at a pH of 7.8. The AEC in brain, gill, muscle, and liver was determined. In each of these tissues the AEC was lower in fish exposed to each of the experimental pHs than in the control and was lowest in the brain and gill tissue.

In our study we examined the effects of three possible sources of stress on AEC and adenylate concentrations in fish. In the laboratory we examined the effects of handling and electroshocking on the adenylate system in fathead minnows (*Pimephales promelas*) because we wanted to determine if the capture method (electrofishing) that we planned to use in the field and handling would affect a fish's adenylate system. Our field study concerned the effects of stressors associated with low dissolved oxygen (DO) concentrations on the adenylate system in rainbow trout (*Salmo gairdneri*) collected from the Chattahoochee River in Georgia immediately below the Lake Lanier Dam. This area of the river has large seasonal variations in DO concentrations.

2. ADENYLATE SAMPLING AND ANALYTICAL PROCEDURE

The procedure we used for measuring AEC was developed by Chapman et al. (1971) and modified by John Giesy, Savannah River Ecology Laboratory, University of Georgia. The procedure is generally done in three stages: (1) sample collection and freezing; (2) adenylate extraction; and (3) adenylate analysis.

2.1. Sampling

Sample collection and freezing involved rapidly removing a small amount of tissue ($\leq$ 150 mg) from the animal and then quickly freeze clamping it between two aluminum blocks. The time between dissecting and freezing the samples was about 10 sec. The frozen samples were then stored in a plastic envelope in liquid nitrogen at $-196°C$.

2.2. Extraction

Adenylates are extracted by grinding the sample to a fine powder in a steel grinder at $-196°C$ with a 10:1 volume to tissue weight of 6% (V:V) perchloric acid. The mixed powdered sample and powdered perchloric acid were thawed in a preweighed centrifuge tube in an ice water bath for about 1 hr to inactivate enzymes that would break down the adenylates. The sample was then centrifuged in a high-speed refrigerated centrifuge at 5°C for 20 min at 40,000 $\times$ g. The supernatant was poured into a clean centrifuge tube. The tissue was rinsed with an additional 0.5 ml of 6% perchloric acid. This was then added to the supernatant and neutralized to pH 7.4 with K_2CO_3 and centrifuged again for 20 min. The supernatant was then diluted five- to tenfold with 0.02 M Tris-HCL (7.4) buffer, frozen in a plastic scintillation bottle, and stored frozen for later analysis. The preweighed centrifuge tube containing the tissue was oven dried at about 50°C for 24 hr and then weighed to obtain the tissue weight.

2.3. Analysis

The frozen supernatant was thawed and diluted five- to tenfold with Tris-HCL buffer (pH 7.4). One ml of diluted sample was then added to each of three tubes. The sample in one tube was used to measure the amount of ATP; the ADP was determined by adding pyruvate kinase to one of the remaining tubes to convert ADP to ATP; the AMP was determined by adding pyruvate kinase and myokinase to the remaining tube to convert AMP and ADP to ATP. The three tubes were incubated at 38°C for 10 min and then cooled in an ice bath.

The amount of ATP is determined by adding 250 μl of sample to 2 ml of luciferin in a scintillation vial placed in a biophotometer. The reaction that takes place,

$$\text{luciferin} + \text{ATP} + O_2 \underset{\text{luciferase}}{\overset{Mg^{2+}}{\rightleftharpoons}} \text{oxyluciferin} + \text{PPi} + \text{AMP} + H_2O + h\nu$$

produces light (hν), which is measured by the biophotometer. One molecule of ATP produces one photon of light.

Standard curves are prepared by running three–four known concentrations of ATP. Because myokinase causes some inhibition of the formation of ATP, a separate standard curve is prepared for AMP by incubating three–four known concentrations of ATP with myokinase.

3. FACTORS AFFECTING ADENYLATE ENERGY CHARGE

3.1. Laboratory Studies

3.1.1. Handling

In the handling studies, groups of five fathead minnows 25–35 mm in length were placed in each of three 33-l aquaria. Fish from the first group were netted and muscle samples were immediately taken. Fish from the second group werre netted and held in the net for 30 sec before muscle samples were taken. During the 30 sec the fish were in the net they were struggling. Fish in the third group were netted, held 30 sec in the net, replaced in their aquarium, and 2 hr later they were netted and immediately dissected and processed. Water quality parameters were: DO 7.3– 7.5 mg/l; temperature 19.5–20°C; and pH 7.0–7.1.

Handling had no apparent effect on AEC, ATP, or total adenylates in the muscle of fathead minnows. There was no significant difference in any of these parameters among the three treatments in the handling experiment as determined by Duncan's Multiple Range Test (Table 1). The AEC values were 0.77–0.81, which are in the lower part of the range for healthy animals (Ivanovici, 1979).

Although the ATP concentration of vertebrate muscle is sufficient to sustain only a few muscle twitches, phosphocreatine (creatine phosphate) in the muscle forms an energy reservoir (Prosser, 1973; Carlson and Wilkie, 1974; Jorgensen and Mustafa, 1980). As ATP is hydrolized to supply energy for muscle contraction, it is regenerated by the "Lohman reaction":

$$\text{ADP} + \text{PCr} \overset{CPT}{\rightleftharpoons} \text{ATP} + \text{Cr}$$

Table 1.   Adenylate Concentrations and Adenylate Energy Charge in Dorsal Muscle of Fathead Minnows During Handling Experiment; For the Same Parameter, Means That Have Different Letters in Their Superscripts Are Significantly Different (α = .05)

Group[1]	AEC[2]	Mean (Standard Deviation) ATP (nM/mg dry tissue wt)	Total[3]
Group 1 (n = 5)	.770[a]	3.80[a]	8.23[a]
	(1.73)	(1.14)	(2.68)
Group 2 (n = 5)	.807[a]	3.40[a]	5.20[a]
	(.221)	(0.96)	(1.77)
Group 3 (n = 5)	.799[a]	3.72[a]	5.90[a]
	(.070)	(1.82)	(3.11)

[1] Group 1—dissected immediately. Group 2—handled 30 seconds, then dissected. Group 3—handled 30 seconds and dissected 2 hours later.
[2] AEC = (ATP + ½ ADP)/(ATP + ADP + AMP).
[3] Total = ATP + ADP + AMP.

(Carlson and Wilkie, 1974). After the phosphocreatine reserve is depleted, ATP and AEC decrease, and this stimulates glycolysis, the citric acid cycle, and oxidative phosphorylation to generate ATP (Stryer, 1975).

In our experiment, netting and handling fish for 30 sec apparently did not exhaust the phosphocreatine reserve in the muscle. Driedzic and Hochachka (1976) also found that extreme exertion did not affect AEC in white muscle of carp; however, ATP and total adenylates were depleted. In their experiment, however, the fish were exercised for at least 1 hr, and in contrast to our study, the fish were completely exhausted when they were sampled. In another study of an active muscle, Sahlin et al. (1978) found that, immediately after exercise, values for AEC and ATP in the human quadriceps muscle were similar to those of resting muscle.

In contrast to the results obtained for fish and human muscle, AEC and adenylates in crayfish (Dickson, 1980) and scallops (Grieshaber, 1978) declined rapidly after muscular activity. A possible explanation for this difference is that the dorsal muscle of fishes and human muscle are used frequently and often for prolonged periods, whereas the tail muscle of crayfish and the abductor muscle of scallops are used much less frequently. Consequently, crayfish and scallop muscle may not require an adenylate system capable of sustaining prolonged activity.

3.1.2. Electroshocking

In the electroshocking experiment, groups of 10 fathead minnows were placed in each of 2 glass containers. Fish in the control group were netted individually and dissected. A frayed electrical cord was placed in the other container and then plugged into a 120 V AC outlet for 15 sec. All fish in the container were stunned. After muscle samples were taken from six fish (about 7 min from the initial shock) the remaining four had recovered and were swimming normally. They were shocked again for 15 sec, and muscle samples were taken. The entire procedure took about 12 min. Water quality parameters were the same as those in the handling study.

Regression analysis indicated that the time between shocking and dissection had no significant (α = .10) effect on the adenylates, so simple ANOVA was used on the pooled data. Although treatment group values were slightly lower than the control group, there were no significant differences in AEC between control and treatment groups. There were, however, significant differences in ATP concentration and total adenylates, with shocked fish having lower values (Table 2).

Stress induced by electroshock resembles that caused by hypoxia and severe muscular activity (Schreck et al., 1976). After shocking, breathing is disrupted or stopped altogether for about 60 sec, and breathing amplitude and blood lactate are increased, which indicates an oxygen debt (Schreck et al., 1976). The difference in ATP concentration and total adenylates between control and shocked fish may be because the ATP is utilized but not regenerated quickly by oxidative phosphorylation because of that temporary lack of oxygen in the shocked fish. Therefore, when the phosphocreatine reserve is used up, the ATP concentration and total adenylates decline. The AEC, however, appears to be more stable than the individual adenylates because, in this study, AEC was maintained in the shocked fish despite a 50% decrease in ATP and total adenylates.

3.2. Field Studies

Rainbow trout were sampled by electrofishing about once every 1 ½ months from August 1980 through July 1981. The area of the Chattahoochee River sampled was from the Lake Lanier Dam to about 1 km downstream. All sampling was done during low flow when water was not being released for generation of electricity. Muscle samples were taken from four trout 200–250 mm in length, quick frozen

Table 2.   Adenylate Concentrations and Energy Charge in Dorsal Muscle of Shocked and Unshocked Fathead Minnows; For the Same Parameter, Means with Different Letters in Their Superscripts Are Significantly Different (α = .05)

| | | Mean (Standard Deviation) | |
| | | ATP | |
Group[1]	AEC[2]	(nM/mg dry tissue wt.)	Total[3]
Group 1 (n = 10)	.839[a]	11.81[a]	16.96[a]
	(.039)	(5.82)	(6.70)
Group 2 (n = 10)	.801[a]	5.96[b]	9.45[b]
	(.045)	(4.06)	(5.36)

[1] Group 1—control, unshocked. Group 2—shocked.
[2] AEC = (ATP + ½ ADP)/(ATP + ADP + AMP).
[3] Total = ATP + ADP + AMP.

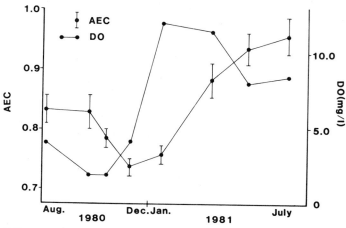

Figure 1. AEC ($\pm$ 1 S.E.) of rainbow trout and DO concentration in the Chattahoochee River, August 1980–July 1981.

in liquid nitrogen, and taken back to the laboratory. Dissolved oxygen concentrations at low flow varied from a low of 1.5 mg/l in October and November to a high of 12 mg/l in January. Water temperatures were 8–11°C.

In this study there appeared to be a relation between AEC in the muscle of rainbow trout and the oxygen concentration in the water (Fig. 1). The AEC at the beginning of the study in August 1980 was 0.85 and the DO concentration at low flow was 4.2 mg/l. The AEC values declined during the fall as the DO decreased and AEC reached a low of 0.73 in December. The DO concentration at low flow in October and November preceding turnover in Lake Lanier was less than 2 mg/ l and this was possibly a stress to the fish. When Lake Lanier turned over in December, DO concentration levels increased rapidly to a high of 12 mg/l in early January. The AEC also increased during January and continued to increase throughout the rest of the study to a high of 0.95 in July. The water temperature throughout the study was 8–12°C. The differences in AEC between periods of low DO concentration (October, November, and December 1980) and periods of high DO concentration (March, May, and July 1981) were significant (α = .05) as determined by analysis of variance and the "least significant difference" method for multiple comparisons.

Although there appears to be a relation between a decline in AEC and decline in DO concentration, there was a lag in the response of AEC to both a decline in DO concentration and an increase in DO concentration. This lag in the response of AEC in the muscle to a stress is consistent with the way AEC in the muscle seems to be rigidly controlled. We believe this lag is due to the "buffering" of the adenylate system in muscle by the creatine phosphate reserve. Other tissues that contain less creatine phosphate may respond more rapidly to stress than muscle. For example, in a laboratory study Jorgensen and Mustafa (1980) found that exposure for 29 hr to a low DO concentration had little effect on AEC or total adenylates in the dorsal muscle of flounder. In the liver, however, hypoxia caused a 90% decrease in ATP, a rise in ADP and AMP, and a drop in total adenylates and in AEC. They suggested that ATP concentration in muscle was stabilized by a large creatine phosphate pool.

Although the work of Jorgensen and Mustafa (1980) and Van Den Thillart et al. (1976) shows that low DO concentration can affect AEC in fish, we cannot be sure DO concentration was the only factor affecting AEC during our field study. For example, elevated concentrations of metals during the periods of low DO concentration may also have stressed the fish. It is also possible that the changes in AEC we observed were normal physiological changes caused by seasonal factors such as reproduction. However, we did not observe such changes in fish kept in the laboratory. Also in a study of the effect of season on AEC of two species of clams, Giesy et al. (1981) found that *Corbicula fluminea*, which spawn during the spring and fall, had their highest AEC values during the winter. They also found that AEC in *Anodonta imbecillis* decreased during the spawning period. At present we are trying to determine if other sources of stress that are associated with low DO concentration can affect the adenylate system of rainbow trout.

4. CONCLUSIONS

Information concerning AEC in fish and other organisms suggests that changes in AEC can prove to be early and sensitive indicators of certain types of stress-induced metabolic changes. However, the value of this technique as a general measure of stress in the field depends on the number of stressors that affect the adenylate system. If AEC only responds to a few, it will only have limited application as a measure of the well-being of fish and other aquatic organisms in field situations. If, however, AEC is sensitive to a wide variety of stressors, it could be a very useful tool for determining stress in field situations.

One phase of the AEC technique that needs further study is the method of dissecting samples. Although it is relatively easy to get homogenous tissue samples from a large fish, it is extremely difficult to dissect small fish quickly without having the muscle samples contaminated with small pieces of other tissue. For example, in our study we could not rapidly dissect muscle samples from fathead minnows without having them contaminated with small pieces of skin, scales, bone, or viscera. These other tissues could have affected the values for the adenylate concentration. The concentration of total adenylates in our fathead minnow muscle samples were lower (1.64–3.39 μM/g wet weight) than concentrations (2.90–10.49 μM/g wet weight) reported by other authors for 14 species of fishes (Beis and Newsholme, 1975; Van Den Thillart et al., 1976; Driedzic et al., 1981; Vetter and Hodson, 1982). In contrast to the values for fathead minnows, our data for rainbow trout and bluegill (*Lepomis macrochirus*) were 2.31–7.32 μM/g, which is in the same range as other studies. It is likely that the inclusion of low ATP-containing material (e.g., bone, scales) in our "muscle" samples from fathead minnows resulted in calculation of lower adenylate concentrations than were reported elsewhere. In large fish where it was easy to dissect homogeneous muscle samples, our technique has produced results comparable to other studies.

We feel an area of great potential for the use of AEC in the laboratory would be in chronic bioassay studies where the effects of known stressors such as contaminants are tested under controlled conditions. However, to fully assess the applicability of AEC as a bioassay tool, it will be necessary to determine the sensitivity of the adenylate system in different species of fishes and in different fish tissues to a variety of stressors. These should include water quality parameters such as DO concentration, temperature, and pH and a variety of xenobiotics such as pesticides, industrial components, and heavy metals. Of special importance would be the relation of AEC values determined for fish over short exposure periods to long-term metabolic changes that affect the fish's growth, reproduction, behavior, and mortality.

ACKNOWLEDGMENTS

We thank Barry Moser and the students and faculty of the Georgia Cooperative Fishery Research Unit who helped with many of the collections in the field, and the staff of the Georgia Game and Fish Division who supplied us with trout from their Buford Trout Hatchery. We would also like to thank Sue Anthony for her valuable assistance in the preparation of the manuscript. This project was funded in part through Dr. Ed Chin by monies from the Georgia Sea Grant Program (Project No. R/EE-10).

REFERENCES

Atkinson, D. E. (1968). The energy charge of the adenylate pool as a regulatory parameter. Interaction with feedback modifiers. *Biochemistry* **7**, 4030–4034.

Atkinson, D. E. (1977). *Cellular energy metabolism and its regulation.* Academic Press, New York.

Bashamohideen, M. and Kunneman, H. (1979). Responses of the fish *Idus idus* L. (Cyprinidae) to environmental load. II. ATP, ADP, AMP, and energy charge in muscle and brain. *Zool. Anz. Jena.* **202**, 163–171.

Beis, I. and Newsholme, E. A. (1975). The contents of adenine nucleotides, phosphagens and glycolytic intermediates in resting muscles from vertebrates and invertebrates. *Biochem J.* **152**, 23–32.

Carlson, F. D. and Wilkie, D. R. (1974). *Muscle physiology.* Prentice-Hall, Englewood Cliffs, NJ.

Chapman, A. G., Fall, L., and Atkinson, D. E. 1971. Adenylate energy charge in *Escherichia coli* during growth and starvation. *J. Bacteriol.* **108**, 1072–1086.

Ching, T. M., Hedtke, S., Russel, S. A., and Evans, H. J. (1975). Energy state and dinitrogen fixation in soybean nodules of dark-grown plants. *Plant Physiol.* **55**, 796–798.

Dickson, G. W. (1980). Phosphoadenylate concentration and adenylate energy charge in freshwater crayfish (Decapoda:Astacidae). Natural and stress induced variation. Ph.D. dissertation, Univ. of Georgia, Athens, GA.

Driedzic, W. R. and Hochachka, P. W. (1976). Control of energy metabolism in fish white muscle. *Amer. J. Physiol.* **230**, 579–582.

Driedzic, W. R., McGuire, G., and Hatheway, M. (1981). Metabolic alterations associated with increased energy demand in fish white muscle. *J. Comp. Physiol.* **141**, 525–532.

Furzer, I. A. (1975). Polluted mud of the Parramatta River. *Search* **6**, 39–40.

Giesy, J. P., Denzer, S. R., Duke, C. S., and Dickson, G. W. (1981). Phosphoadenylate concentrations and energy charge in two freshwater crustaceans: Response to physical and chemical stressors. *Verh. int. Verein. Limnol.* **21**, 205–220.

Grieshaber, M. (1978). Breakdown and formation of high energy phosphates and octopine in the abductor muscle of the scallop, *Chlampys opercularis* (L), during escape swimming and recovery. *J. Comp. Physiol.* **126**, 269–276.

Ivanovici, A. M. (1979). Adenylate energy charge: Potential value as a tool for rapid determination of toxicity effects. In: *Proc. of the 5th annu. aquatic toxicology workshop*, Hamilton, Ont., Nov. 7–9, 1978. Fish. Mar. Ser. tech. rep. 862, pp. 241–255.

Jorgensen, J. B. and Mustafa, T. (1980). The effect of hypoxia on carbohydrate metabolism in flounder (*Platichthys flesus* L.)—II. High energy phosphate compounds and the role of glycolytic and gluconeogenetic enzymes. *Comp. Biochem. Physiol.* **67B**, 249–256.

MacFarlane, R. B. (1981). Alterations in adenine nucleotide metabolism in the Gulf killifish (*Fundulus grandis*) induced by low pH water. *Comp. Biochem. Physiol.* **66B**, 193–202.

Montague, M. D. and Dawes, E. A. (1974). The survival of *Peptococcus prevotii* in relation to the adenylate energy charge. *J. Gen. Microbiol.* **80**, 291–299.

Prosser, C. L. (1973). Muscles. In C. L. Prosser (Ed.), *Comparative animal physiology*. W. B. Saunders, Philadelphia, PA, pp. 719–788.

Ridge, W. (1972). Hypoxia and the energy charge of the cerebral adenylate pool. *Biochem. J.* **127**, 351–355.

Sahlin, K., Palmskog, G., and Hultman, E. (1978). Adenine nucleotide and IMP contents of the quadriceps muscle in man after exercise. *Pflugers Arch.* **374**, 193–198.

Schreck, C. B., Whaley, R. A., Bass, M. L., Maughan, O. E., and Solazzi, M. (1976). Physiological responses of rainbow trout (*Salmo gairdneri*) to electroshock. *J. Fish. Res. Board Can.* **33**, 76–84.

Stryer, L. (1975). *Biochemistry*. W. H. Freeman and Co., San Francisco, CA.

Van Den Thillart, G., Kesbeke, F., and van Waarde, A. (1976). Influence of anoxia on the energy metabolism of goldfish, *Carrassius auratus* (L). *Comp. Biochem. Physiol.* **55A**, 329–336.

Vetter, R. D. and Hodson, R. E. (1982). Use of adenylate concentrations and adenylate energy charge as indicators of hypoxic stress in estuarine fish. *Can. J. Fish. Aquat. Sci.* **39**, 535–541.

Wiebe, W. J. and Bancroft, K. (1975). Use of the adenylate energy charge ratio to measure growth state of natural microbial communities. *Proc. Nat. Acad. Sci.* **72**, 2112–2115.

Wijsman, T. C. M. (1976). Adenosine phosphates and energy charge in different tissues of *Mytilus edulis* L. under aerobic and anaerobic conditions. *J. Comp. Physiol.* **107**, 129–140.

12

ASSESSING THE TOLERANCE OF FISH AND FISH POPULATIONS TO ENVIRONMENTAL STRESS: THE PROBLEMS AND METHODS OF MONITORING

Gary A. Wedemeyer

U.S. Fish and Wildlife Service
National Fishery Research Center
Seattle, Washington

Donald J. McLeay

D. McLeay and Associates Ltd.
West Vancouver, British Columbia

C. Phillip Goodyear

U.S. Fish and Wildlife Service
National Fishery Center
Leetown, West Virginia

---

1.   ENVIRONMENTAL STRESS: CONCEPTS AND DEFINITIONS

Environmental stress is an inescapable aspect of life in the aquatic environment. The chemical and physical demands of life underwater impose somewhat rigorous constraints on aquatic species (Smith, 1982a). Superimposed on such demands may be the additional physiological constraints of particular ecological niches. It is true that aquatic species are adapted to these conditions, but this does not imply the absence of energy drains (Lugo, 1978). For example, thermophilic fishes must still cope physiologically with the demands of high temperatures even though they are adapted to high temperatures *per se*.

Thus the homeostatic systems of fishes are continuously challenged or "stressed" by the normal demands of the aquatic environment. For populations in natural waters, additional stress may result from anthropogenic habitat alterations, such as pollution, or energy, mineral, and water development projects (Chavin, 1973). For fish in aquacultural systems, the constraints of production requirements and operating procedures such as handling, crowding, transporting, and disease treatments result in additional stress. Unfortunately, tolerance limits are well-understood for only a few habitat alterations, and then usually only when they occur singly. More commonly, habitat changes are multiple in nature and fish populations may be subjected to some or all of the following: chronic exposure to otherwise sublethal concentrations of contaminants; unfavorable or fluctuating temperatures; adverse light levels, pH, water velocity, or sediment loads; hypoxia; and physical trauma such as entrainment or impingement at power plants. Each of these factors, singly or together, can impose a considerable load, or stress, on physiological systems. Stress that greatly exceeds tolerance limits tends to be self-evident because it is normally acutely lethal. Sublethal stress is more common but more insidious because it tends to result in adverse effects at the individual fish level that are eventually manifested at succeedingly higher levels of biological organization (Rosenthall and Alderdice,

1976). Depending on its severity, sublethal stress may load or limit physiological systems, reduce growth, impair reproduction, predispose to infectious diseases if fish pathogens are present, and reduce ability to tolerate subsequent stress. At the population level, the effects of stress will be manifested as reduced recruitment to succeeding life stages, and reduced compensatory reserve (Ryan and Harvey, 1977; McFarlane and Franzin, 1978).

The term "stress" is presently used inconsistently. Originally, stress was defined by Selye (1950) as "The sum of all the physiological responses by which an animal tries to maintain or re-establish a normal metabolism in the face of a physical or chemical force." Today, "stress" is sometimes taken to mean the environmental alteration (stressor) itself and sometimes the response of the fish, population, or ecosystem. However, although terms such as "handling stress" are still frequently encountered, the consensus is that "stressor" should be used to mean the environmental factor itself and "stress" to mean the compensating response (Pickering, 1981). More precisely, stress can be defined as the effect of any environmental alteration or force that extends homeostatic or stabilizing processes beyond their normal limits, at any level of biological organization (Esch and Hazen, 1978). That is, an environmental change (stressor) that is severe enough to cause stress is one that requires a compensating response by a fish, population, or ecosystem. If the stress response can reestablish a satisfactory relationship between the changed environment and the fish (population or ecosystem), adaptation to the stressor occurs. Although the outcome of stress may be clearly negative for an individual, the outcome at the population level may be difficult to predict. For example, mortality of individual fish due to hypoxia may actually enhance survival of a population if dissolved oxygen is limiting. In addition, because perturbations in energy flow between trophic levels may occur and result in species shifts, the original conditions may not necessarily be restored after the stressor is removed.

Thus tolerance to environmental change depends, at least in part, on the individual fish's ability to regulate stabilizing processes that will effect the required physiological or behavioral adaptation. An understanding of the stress response, and the environmental alterations to which fishes can adapt through this response, is important to a definition of the limits of change in the aquatic environment that can be tolerated. Although most species can tolerate relatively severe stress for limited periods because of their homeostatic capabilities, these should not be used as an excuse for allowing marginal conditions to occur in the aquatic environment. Instead, these capabilities should be used to set priorities and limits for species-habitat relationships so that environmental alterations can be correctly evaluated in terms of their biological costs or benefits. This monograph discusses concepts and methods for developing the improved species-habitat data base required to correctly evaluate the tolerance limits of fish and fish populations to single and multiple environmental stress factors.

2. EFFECTS OF STRESS ON FISH AND FISH POPULATIONS: THE PRIMARY, SECONDARY AND TERTIARY RESPONSES

It is widely accepted that the stress response in fishes is similar in many ways to that of the higher vertebrates (Peters, 1979). It is characterized by physiological changes that are similar for stressors as varied as handling, disease treatments, fright, forced swimming, anesthesia, rapid temperature changes, or scale loss. At the individual animal level, a series of compensatory physiological changes occur in response to stressful challenges, which, in humans, has been termed the general adaptation syndrome (Selye, 1973). As usually described, it consists of three stages:

1. An alarm reaction in which catecholamine and coricosteroid "stress hormones" are released, followed by a complex series of physiological events as the organism begins to compensate.
2. A stage of resistance if compensation is successful.
3. A stage of exhaustion if compensation fails because the stress factor was too severe or long lasting and thus exceeded biological tolerance limits.

As yet, only a general understanding exists of the compensatory alterations that occur as fishes (or fish populations) attempt to maintain themselves in the face of acute or chronic stressors. However, the following outline, developed from the work of Mazeaud et al. (1977), Donaldson (1981), Schreck (1981), Wedemeyer and McLeay (1981), Hunn (1982), and Smith (1982a) can be used as a basis for discussion.

1. *Primary effects (the endocrine system).* Adrenocorticotrophic hormone (ACTH) is released from the adenohypophysis following stimulation of the neurohypophysis via the hypothalmus. Catecholamines and corticosteroids ("stress hormones") are released from the chromaffin and interrenal cells within the anterior kidney.

2. *Secondary effects (blood and tissue alterations).* A series of hormone-induced blood, tissue, and hematological changes occur, including hyperglycemia, hyperlacticemia, hypochloremia, leukopenia, and reduced blood clotting time. Diuresis begins, with resultant blood electrolyte loss; and finally, tissue changes such as depletion of liver glycogen and interrenal Vitamin C occur, followed by interrenal hypertrophy.

3. *Tertiary effects (whole-animal, and populations).* If changes in growth, survival, and reproduction occur, succeedingly higher levels of biological organization may be affected. At the species level, reduced recruitment to succeeding life stages may exceed the compensatory reserve, leading to population declines. At the community or ecosystem level, perturbations in energy flow through trophic levels may eventually alter species composition.

3.   ASSESSING TOLERANCE TO STRESS: THE INDIVIDUAL FISH LEVEL

At present, the species-habitat relationships data base needed to correctly assess the point at which single and multiple stress factors will exceed tolerance limits is largely incomplete. Thus by necessity, current habitat-based predictive methodologies for assessing the point at which environmental alterations will begin to overstress populations are founded on inadequate data. Consequently, a posteriori fish kills or population declines still must be used, in many cases, as an indicator that some particular stress factor, or combination of stress factors, has exceeded tolerance limits. Nevertheless, certain of the physiological, whole-animal, and population changes resulting from the stress response can be quantified, and have potential as predictive indices in the needed species-habitat relationships data base.

3.1.   Effects on Physiology: The Primary and Secondary Responses

A variety of clinical tests are available to measure changes at the primary and secondary levels, few of which require complex equipment or facilities. Some, such as the occult hemoglobin test to detect acute stress, require only simple prepackaged reagents, and others are readily automated (Smith and Ramos, 1976, 1980). Several of these tests are also useful in studies of pathogenesis, in providing biological end points for bioassays, and in the determination of nutritional requirements. Example of tests having such potential are the leukocrit and C-reactive protein determinations for the detection of incipient disease, and the ketone-mucus test as an index of nutritional status (McLeay and Gordon, 1977; Ramos and Smith, 1978a,b).

Much recent research has been conducted on determining the severity of stressors by means of clinical chemistry procedures (Harman et al., 1980; Ishioka, 1980a,b). Blood chemistry tests that experience has shown to be practical include measurements of circulating levels of the catecholamine and corticosteroid stress hormones themselves (epinephrine and cortisol), or the secondary physiological changes that result from their actions such as diuresis, hyperlacticemia, hypochloremia, or hyperglycemia (Hattingh, 1976; Mazeaud et al., 1977; McLeay, 1977; Strange et al., 1978; Barton et al., 1980; Peters et al., 1981; Hunn, 1982).

Corticosteroid and catecholamine determinations provide a direct estimate of the severity and duration of the primary stress response (Donaldson, 1981; Mazeaud and Mazeaud, 1981; Schreck, 1981). However, the analytical procedures required are complex and thus are most appropriate to specialized research facilities. Thus the secondary blood chemistry changes, such as hyperglycemia, which are the result of stress hormone action, are frequently substituted as simple, albeit indirect, methods for evaluating the effects of those stressors that elicit a neuroendocrine response (McLeay and Brown, 1975; Silbergeld, 1975; Hattingh, 1976; Wedemeyer and Yasutake, 1977). In addition, measuring the extent and duration of secondary effects such as hyperglycemia integrates those aspects of the stress response mediated through both direct sympathetic (chromaffin tissue) and humoral (interrenal tissue) pathways.

Plasma glucose can be measured quickly, accurately, and precisely, with only 5–10 µl of sample, using both manual and automated methods. The small sample size required permits evaluating the acute stress response of fish weighing only a few grams as well as the repetitive blood sampling of larger individual fish. In addition, the technical procedures and equipment required for glucose determinations are simple enough that studies under field conditions are practical. To minimize variability, diet composition, time since last feeding, developmental stage, and season of the year should be standardized because these affect liver glycogen stores, and thus the magnitude of the hyperglycemic response (Nakano and Tomlinson, 1967; McLeay, 1977; Gordon and McLeay, 1978).

Blood-sugar determinations can also be used in conjunction with stressor challenge tests as a means for assessing performance in terms of ability to mount a stress response and achieve compensation within a given time period. For example, a procedure has been standardized for evaluating sublethal effects of contaminants using the hyperglycemic stress response as the biological end point (McLeay, 1977; McLeay and Gordon, 1978a, 1980). In this method, groups of fish are acclimated to test aquaria for 48 hr, followed by a 24 hr exposure to the desired dilutions of the contaminant. The hyperglycemia that results is usually dose dependent.

Blood-sugar determinations can also be used to evaluate tolerance limits to single or multiple stressors, such as handling, or handling accompanied by hypoxia, temperature, or salinity alterations—and to determine the recovery time needed. Performance tests to differentiate genetic strains physiologically best able to compensate for acute or chronic environmental stress are also feasible. These should include not only measurements of ability to mount a compensatory stress response, but also the variability of the resulting hyperglycemia (F test of variance ratios), and ability to show homeostatic recovery (regulation of plasma glucose to basal values) after the challenge is removed.

Hematological determinations can also be used to provide useful information about the severity of the stress response. Changes in the erythrocyte count (as approximated by the hematocrit) or in hemoglobin values following acute stress are useful as indicators that blood volume changes (hemodilution or hemoconcentration) have occurred. However, since anemias, polycythemia, or erythrocyte swelling may occur in certain situations (Soivio and Nikinmaa, 1981), the plasma water concentration should also be determined. Changes in the normal blood clotting time and in the differential leukocyte count are probably the most sensitive indices of acute stress of all the hematological measurements (McLeay, 1975b; Casillas and Smith, 1977). Eosinophilia may occur under certain conditions, but low sensitivity and high variability limit its usefulness in quantifying the stress response in fishes. Leukocytosis is not normally pathognomonic, but leukopenia commonly occurs during the physiological response to acute stressors (McLeay and Gordon, 1977).

In general, the moderate to severe leukopenia that accompanies stress is a result of increased pituitary-interrenal activity (Srivastava and Agrawal, 1977; Peters et al., 1980). Specifically, stress, perhaps through the mediation of ACTH and corticosteroids, results in lymphocytopenia, monocytopenia, and neutrophilia (McLeay, 1975a; Johansson-Sjobeck et al., 1978). In mammals, this result is probably due to the sequestration of circulating T-lymphocytes into the bone marrow and lymph nodes (Pearson et al., 1978). The mechanism in fishes is only poorly understood; nevertheless, the practical effect is suppression of the immune response and increased susceptibility to infectious diseases (Ellis, 1981).

The most accurate and precise determination of leukopenia or leukocytosis is the classical differential blood cell count. However, for some purposes the rapid, approximate leukocrit measurement will suffice (McLeay and Gordon, 1977). This substitute for the white blood cell count is analogous to the use of the hematocrit as a substitute for the more tedious red blood cell count. To determine the leukocrit value, blood is collected in heparinized microhematocrit tubes, centrifuged, and the layer of white blood cells separating the packed erythrocytes from the plasma is measured using a low-power microscope equipped with an ocular micrometer. The volume of the leukocytes is expressed as a percentage of the whole blood volume. The calculation is exactly analogous to that of the familiar hematocrit.

As in blood glucose determinations, leukocrit values can be obtained from (salmonid) fish as small as 1 g. However, young fish have inherently low leukocrits, and the use of fish 2 g or larger is recommended (McLeay and Gordon, 1977). This also allows somewhat larger blood samples to be taken, which eliminates the need for pooled samples when both leukocrit and plasma glucose determinations are required.

For salmonids and other fishes in which lymphocytes make up the majority of the circulating leukocytes, the leukocrit can be used as a simple, rapid test for determining the point at which single or multiple environmental alterations result in leukopenia. To calibrate the degree of severity, the response to standardized challenge tests is used as a reference point. In interpreting leukocrit data, it should be kept in mind that the significance of this measure of the secondary stress response may differ from that of other tests. For example, although hyperglycemia can be elicited by both catecholamines and glucocorticosteroids, leukopenia appears not to be associated with catecholamine production (Srivastava and Agrawal, 1977). Thus the leukocrit and plasma glucose values, which can be measured from the same centrifuged blood sample, provide indices of two aspects of the stress response for each environmental factor examined.

The leukocrit test has been used, in conjunction with the blood-sugar bioassay, to determine the threshold concentrations at which environmental contaminants become acutely stressful to salmonid fishes (McLeay and Gordon, 1977, 1979, 1980). This test has also proved useful in estimating the toxic zone of influence of waste discharges within receiving waters (McLeay and Gordon, 1978a).

Further development of the leukocrit as a stress assessment method is merited. As with most other tests, the species developmental stage, and prior history of the fish are all important. In additon, stress may in some instances result in leukocytosis rather than leukopenia and thus increase the leukocrit value (Pickford et al., 1971; McLeay, 1973; McLeay and Brown, 1974). Furthermore, in species in which lymphocytes do not constitute a preponderance of the total leukocyte volume, lymphopenia may occur while the leukocrit value remains unchanged, or is even elevated, due to a concomitant granulocytosis (Peters et al., 1980).

Tissue changes that have good potential as indices of the severity of environmental stress and merit further development include measurements of adenylate energy charge, muscle and liver glycogen, interrenal Vitamin C, uptake of ^{14}C-glycine by scales, and RNA-DNA ratios (Bulow, 1970; McLeay and Brown, 1975; Wedemeyer and Yasutake, 1977; Adelman, 1980; Ivanovici, 1980; Bulow et al., 1981; Reinert and Hohreiter, 1984). Decreased spleen weight, atrophy of the gastric mucosa, and interrenal hypertrophy have also been used as indices, especially of chronic stress (McLeay, 1975a; Peters et al., 1981). In particular, interrenal hypertrophy is well-accepted as pathognomonic, and a semiquantitative histological method has been developed in which nuclear diameters or cell size are measured in stained sections of anterior kidney tissue (Fagerlund et al., 1981). Changes in the gastric mucosa, including decreased mucous cell diameter, have been used as an index of stress in eels (*Anguilla anguilla*) in intensive culture (Peters et al., 1981). Subordinate animals in the social hierarchies that develop prove to be under measurable stress within 5–10 days.

The duration of stress has an important bearing on the severity of the tissue changes that develop. At first, stress results in a reversible depletion of vitamin C, corticosteroids, cholesterol, and other lipids, in both the interrenal tissue and the anterior pituitary. If the stress becomes chronic, it results in hypertrophy of these tissues. Eventually, atrophy and degeneration will occur with irreversible loss of cellular and tissue function. However, if the stress factor is removed soon enough, interrenal and adenohypophyseal function may gradually be recovered (Donaldson and McBride, 1974; McLeay, 1975a; Noakes and Leatherland, 1977).

A listing of selected physiological tests that can be used to evaluate the severity of the stress response at the primary and secondary levels is presented in Table 1, together with interpretive guidelines. Because of limited information on expected normal ranges, controls must be run for comparison in most casses.

Table 1.   Interpretive Guidelines for Physiological Tests to Assess Tolerance to Environmental Stress[a]

Physiological Test	Diagnostic Significance if Results Are:	
	Low	High
Adenylate energy charge (muscle, liver)	Chronic stress	Normal bioenergetic demands
Blood cell counts		
Erythrocytes	Anemias, hemodilution due to impaired osmoregulation	Stress polycythemia, dehydration, hemoconcentration due to gill damage
Leucocytes	Leucopenia due to acute stress	Leucocytosis due to bacterial infection
Thrombocytes	Abnormal blood clotting time	Thrombocytosis due to acute or chronic stress
Chloride (plasma)	Gill chloride cell damage, compromised osmoregulation	Hemoconcentration, compromised osmoregulation
Cholesterol (plasma)	Impaired lipid metabolism	Fish under chronic stress, dietary lipid imbalance
Clotting time (blood)	Fish under acute stress, thrombocytopenia	Sulfonamides or antibiotic disease treatments affecting the intestinal microflora
Cortisol (plasma)	Normal conditions	Fish under chronic or acute stress

Table 1. (*Continued*)

| Physiological Test | Diagnostic Significance if Results Are: | |
	Low	High
Gastric atrophy	Normal conditions	Chronic stress
Glucose (plasma)	Inanition	Acute or chronic stress
Glycine incorporation (scales)	Reduced growth	Normal conditions, good growth
Glycogen (liver or muscle)	Chronic stress, inanition	Liver damage due to excessive vacuolation; diet too high in carbohydrate
Hematocrit (blood)	Anemias, hemodilution	Hemoconcentration due to gill damage, dehydration; stress polycythemia
Hemoglobin (blood)	Anemias, hemodilution, nutritional disease	Hemoconcentration due to gill damage, dehydration; stress polycythemia
Hemoglobin (mucus)	Normal conditions	Acute stress
Interrenal cell size or nuclear diameter (tissue)	No recognized significance	Chronic stress
Lactic acid (blood)	Normal conditions	Acute or chronic stress; swimming fatigue
Osmolality (plasma)	External parasite infestation, heavy metal exposure, hemodilution	Dehydration, salinity increases in excess of osmoregulatory capacity, diuresis, acidosis
RNA/DNA (muscle)	Impaired growth, chronic stress	Good growth
Total protein (plasma)	Infectious disease, kidney damage, nutritional imbalance, inanition	Hemoconcentration, impaired water balance

[a] Controls are usually needed for comparison because normal range estimates are unavailable in most cases. Sample size is normally at least 10 fish per group depending on the coefficient of variation of the test employed. Compiled from Wedemeyer and Yasutake, 1977; Peters, 1979; Adelman, 1980; Ivanovici, 1980.

3.2. Effects on Physiology: The Tertiary Response

Experience gained from studies of physiology at the whole-animal level has shown that several aspects of the tertiary stress response can be used as indicators that unfavorable environmental conditions are exceeding tolerance limits (Bayne, 1980). These include: changes in metabolic processes such as oxygen consumption, efficiency of osmoregulation and energy balance; changes in behavior such as inhibited reproduction, feeding, or predator avoidance; and changes in physical condition such as relative weight, growth, or longevity (Goodyear, 1972; Anderson, 1980; Esch and Hazen, 1980; Billard et al., 1981). The increased morbidity and mortality rates include that due to increased susceptibility to fish diseases (Snieszko, 1974; Hodgins et al., 1977). At the population level, the significant effects of environmental stress on fish condition and performance are ultimately manifested as reduced recruitment rates to succeeding life stages (Ryan and Harvey, 1977; McFarlane and Franzin, 1978; Esch and Hazen, 1980). Unfortunately, population-level measurements are a posteriori, and are difficult to use in establishing cause and effect relationships because the nature, intensity, and timing of the mechanisms that regulate population size are almost always unknown.

Among the tertiary effects of stress that have significant potential as indices, effects on reproduction are of particular importance (Billard et al., 1981). There seem to be two main pathways by which stress can exert an adverse effect. First, gametogenesis (spermatogenesis or oogenesis), which is under the influence of longer term, circannual, environmental factors such as photoperiod, can be affected. Thus otherwise sublethal alterations in the normal fluctuations of environmental factors such as water temperature or salinity can have adverse effects on the onset of spermiation or ovulation. The second pathway through which effects can occur involves the fact that certain of the hormones involved in the stress response are also components of the normal endocrine pattern controlling reproduction. Thus habitat alterations can affect reproduction either directly, by physically altering normal environmental cues, or indirectly, through perturbations in the endocrine system due to the elicited stress response (Billard et al., 1981). A summary of these effects is presented in Table 2.

Another of the tertiary effects of stress at the whole animal level, and one that is presently receiving serious attention as a method of biological monitoring for environmental quality, is fish disease incidence (Esch and Hazen, 1980; Walters and Plumb, 1980; Ellis, 1981). Experience has shown that the mere presence of most fish pathogens, unless they are present in overwhelming numbers, does not result in epizootics unless unfavorable environmental conditions also exist that have compromised the host defense system (Walters and Plumb, 1980). Thus infectious fish diseases are not single-caused events, but are one outcome of the continuing interactions between the aquatic environment, the fish, and their pathogens. If the host-environment-pathogen relationship is balanced, good health, growth, and survival will occur. If it is marginal, disease problems will begin to become evident and reduced fish condition, growth, and survival will result. If it is unbalanced, chronic

Table 2. Environmental Factors Affecting Reproductive Physiology[a]

Environmental factor	Aspect of Reproduction Affected
Acidity	Gametogenesis, fecundity, egg fertility
Food availability	Gonadotrophic hormone production, gametogenesis, egg fertility
Dissolved oxygen	Gonadotrophic hormones, gametogenesis
Population density	Gametogenesis, ovulation, spawning
Contaminants	Gametogenesis, ovulation, spawning, egg fertility

[a] Compiled from Billard et al., 1981.

disease problems can be expected. The stress-mediated diseases that seem to have the most promise as indicators in freshwater environments are those due to facultative bacterial fish pathogens such as Aeromonads, Pseudomonads, and Myxobacteria, which are continuously present in most natural waters. A classic example is bacterial gill disease, which frequently is treated successfully in aquaculture by simply reducing the fish population density. Other infectious fish diseases that indicate that tolerance limits to stress are being exceeded include vibriosis (*Vibrio anguillarum*), bacterial hemorrhagic septicemia (*Aeromonas* and *Pseudomonas* sp.), and protozoan parasitic and fungal infestations such as costiasis (*Costia necatrix*) and *Saprolegnia* (Neish, 1977). As judged by its physiological consequences, viral erythrocytic necrosis may be a potentially useful indicator disease in the estuarine environment (MacMillan et al., 1980). Fish disease incidence can, of course, be more easily used as an index of unfavorable environmental conditions in aquacultural facilities, which have relatively controlled conditions, than in wild fish populations in natural waters. However, certain fish health problems have been identified as offering promise in biological monitoring. These include chromosomal and morphological abnormalities of eggs and larvae, skeletal anomalies, neoplasms, fin erosion, and epidermal ulceration (Malins et al., 1980, 1982; Sindermann, 1980). A summary list of stress-mediated diseases, together with a description of common environmental conditions implicated in their occurrence (in intensive fish culture) is given in Table 3.

For anadromous fishes, a whole-animal response that has good potential as an index of the impact of environmental stress is the parr-smolt transformation. An excellent analysis of the physiological changes occurring during the smoltification of Pacific salmon (*Oncorhynchus* sp.), Atlantic salmon (*Salmo salar*), and steelhead trout (*Salmo gairdneri*) was published by Folmar and Dickhoff (1980). Habitat alterations that cause changes in water chemistry, temperature, or photoperiod may also result in the inhibition of one or more of these functions, which thus can be used as indices (Wedemeyer et al., 1980). For example, the normal development pattern of the gill ATPase enzyme system of coho salmon parr (*Oncorhynchus kisutch*) is affected both by water temperature and by otherwise sublethal amounts of dissolved heavy metals. For example, chronic exposure to copper at only 20–30 μg/l partly or completely inactivates gill ATPase function (Lorz and McPherson, 1976). The consequence of this loss of function is reduced seawater tolerance and thus, reduced marine survival. An equally significant behavioral consequence is the inhibition of normal migratory behavior.

In addition to trace heavy metals, the parr-smolt transformation of juvenile anadromous salmonids is also sensitive to contaminants such as the herbicides and nitrates now increasingly common in surface waters as the result of intensive forest and range management, and agricultural practices. For example, both picloram, and the dimethylamine salt of 2, 4-D used for weed control on noncrop lands inhibit the normal migratory behavior of juvenile coho salmon (Lorz et al., 1978). Other formulations, such as Endothal and the 2, 4-D esters used for control of Eurasian water milfoil (*Myriophyllum spicatum*), may also inhibit seawater tolerance and migratory behavior of smolts (Liguori et al., 1983). These effects, which are otherwise sublethal to individual fish, have considerable potential as biological end points because of their significant implications at the population level.

Table 3.   Fish Diseases That Are Stress Mediated and Have Potential as Indicators in Biological Monitoring for Environmental Quality[a]

Disease	Environmental Factors Predisposing to Disease
Furunculosis (*Aeromonas salmonicida*)	Low oxygen (4 mg/liter for salmonids); crowding; handling in the presence of the pathogen
Bacterial gill disease (Myxobacteria sp.)	Crowding; unfavorable environmental conditions; chronic low oxygen (4 mg/liter for salmonids); elevated ammonia (more than 0.02 mg/liter for salmonids); excessive particulate matter in water
Columnaris (*Flexibacter columnaris*)	Crowding or handling during warm water periods if carrier fish are present; temperature increase if the pathogen is present, even if fish are not crowded or handled
Kidney disease (*Renibacterium salmoninarum*)	Water hardness less than about 100 mg/liter (as $CaCO_3$); unfavorable diet composition
Hemorrhagic septicemias, red-sore disease (*Aeromonas, Pseudomonas*)	External parasite infestations; inadequate pond cleaning; handling; crowding; elevated ammonia; low oxygen; chronic exposure to trace contaminants; elevated water temperatures; handling after overwintering at low temperatures
Vibriosis (*Vibrio anguillarum*)	Handling; dissolved oxygen lower than about 6 mg/liter, especially at water temperatures of 10–15°C; brackish water, 10–15‰ salinity
Parasite infestations	Overcrowding of fry and fingerlings; low oxygen; excessive size variation among fish in ponds
Spring viremia of carp	Handling after overwintering at low temperatures
Fin erosion	Crowding; low dissolved oxygen; nutritional imbalances; chronic exposure to trace contaminants; high total suspended solids; secondary bacterial invasion
Epithelial tumors	Chronic, sublethal contaminant exposure
Epithelial ulceration	Chronic, sublethal contaminant exposure
Skeletal anomalies	Chronic adverse environmental quality, PCB, heavy metals, Kepone, Toxaphene exposure, Vitamin C deficiency

[a] Esch and Hazen, 1980; Sindermann, 1980; Wedemeyer and McLeay, 1981; Malins et al., 1982.

Effects on smoltification can also be used to assess the impact of water temperature alterations. The normal development of smoltification in anadromous rainbow trout (steelhead) is particularly sensitive to the effects of otherwise sublethal water temperature, which thus provides a useful habitat-based predictive index. Gill ATPase development is essentially suppressed at temperatures of 13°C, whereas most other physiological functions are unaffected (Zaugg and Wagner, 1973). The sublethal effects of temperature on Atlantic salmon smoltification are probably similar to that of the steelhead. Coho salmon suffer a retarded pattern of ATPase development at 6°C, and a premature development pattern at 20°C (Zaugg and McLain, 1972). However, temperatures up to 15°C do not adversely affect smoltification, and accelerate growth sufficiently so that minimum size requirements can be met in one rearing season. Unfortunately, the process of parr reversion is also accelerated, which shortens the time period during which the smolts can successfully convert to seawater. The accelerated parr reversion of juvenile chinook salmon (*Oncorhynchus tshawytscha*) due to elevated water temperatures can be substantially reduced if this species is reared at salinities of up to 20‰ (Clarke et al., 1981).

A final environmental stress factor to mention that adversely affects smolt performance capability is population density (Specker and Schreck, 1980; Fagerlund et al., 1981; Sandercock and Stone, 1982). However, the population level effect to be measured is reduced ocean survival, which unfortunately renders it time consuming to use as a biological end point.

3.3. Effects on Performance: Challenge Tests

The application of challenge tests for determining the tolerance limits of fish to adverse conditions is based on the presumptions that: (1) a stress load, if sufficient in magnitude or duration, is debilitating and affects performance capability; and (2) the impact of multiple stressors is cumulative (additive or synergistic). To the extent that reduced performance capability impairs growth, survival, or reproduction of sufficient individual fish, the survival probability of the population as a whole is reduced.

At present, performance challenge tests are at an early stage of development, and there is little agreement on the most suitable approaches. Selected tests are discussed here, together with methodologies for conducting them under standardized conditions. In a few cases (e.g., freshwater challenge tests) the methods have been little used, and we are proposing their development. In addition, the procedures given for conducting challenge tests may require modification to meet specific individual requirements. Also, to use reduced ability to tolerate a second stress factor as a measure of the impact of prior stress, it will be necessary to carefully select those challenges that will best reveal the debilitated performance capacity of significance in a particular situation.

3.3.1. Reduced Temperature Tolerance

Effects on ability to tolerate temperature changes are considered to be of fundamental significance at both the individual fish and fish population levels (Elliott, 1981). Exposure to unfavorable or fluctuating temperatures is well-known to effect characteristic stress reactions in fishes (Wedemeyer, 1973; Strange et al., 1977). Thus a thermal challenge to determine whether prior stress has reduced the ability of fish to mount a compensatory response, or simply to determine if tolerance limits have been narrowed, can provide ecologically significant information.

Probably the most widely used method for determining temperature tolerance zones has been to acclimate groups of fish to the desired temperature and then transfer them directly into similar aquaria set at the water temperatures to be tested. The upper and lower tolerance limits are taken as the temperature at which only 50% of the test fish survive for some set period of time (usually 24 hr). An alternative method, which determines the critical thermal maximum or minimum (CTM), is to acclimate groups of fish at the desired temperature and then warm or cool the water at a constant rate (conventionally 1°C/min) until complete loss of equilibrium ensues (Paladino et al., 1980; Bonin, 1981). In either of these procedures, temperature can be used as a challenge stressor to determine whether the zone of thermal tolerance has been reduced by the impact of prior stressors. In interpreting the results of such tests, the species tested and its prior thermal and photoperiod exposure in the laboratory must be considered because these variables can appreciably alter the thermal performance envelope (McLeay and Gordon, 1978b).

As a specific application, thermal challenge can be used to determine the point at which contaminant exposure begins to reduce normal temperature tolerance limits (McLeay and Gordon, 1980). To conduct such tests, groups of fish are exposed to the desired concentrations of each contaminant and then subjected to increasing water temperatures. In acute exposures, the rate of temperature increase should be slower than 1°C/min in order to allow time for the contaminant to exert its effect. Rates on the order of 1°C/hr are usually adequate. For a variety of industrial effluents and pure chemicals, concentrations that reduce the critical thermal maximum (upper lethal temperature) are substantially less than the acutely lethal (LC50) levels (McLeay and Howard, 1977). However, it should be noted that while some toxicants will markedly reduce the upper temperature tolerance limit, others cause only a slight impairment (McLeay and Gordon, 1980). These differences probably reflect differing effects at the physiological level during the process of thermal death. Although the present understanding of the physiology of thermal death is still fragmentary, tissue anoxia at upper lethal temperatures is reported to occur, at least in salmonids (Hughes and Roberts, 1970; Heath and Hughes, 1973). Thus the impacts of stressors that reduce oxygen exchange at the gills, impair tissue respiration, or increase metabolic rates, may be intensified by elevated temperatures. Testing for the adverse effects of prior stressors by determining ability to withstand a subsequent temperature challenge is therefore especially useful when these physiological changes are involved.

3.3.2. Tolerance to Hypoxia

Ability to withstand an oxygen depletion (the sealed-jar test) was originally proposed by Carter (1962) as a rapid, simple biological end point to evaluate the effects of otherwise sublethal contaminant exposure. Sealed-jar tests have since been applied successfully in identifying sublethal contaminant effects on marine as well as freshwater fish (Ballard and Oliff, 1969; Giles and Klaprat, 1979), and in field monitoring surveys to determine the zones of influence of industrial discharges within receiving waters (McLeay and Gordon, 1978a).

To conduct a test, fish are placed in individual 1 or 2 liter containers at a loading density (for salmonids) of 4–5 g/liter (weight of fish per liter of solution). Ten replicates for each treatment level are usually adequate. Oxygen saturated water is used for both the control and test fish and temperature is held constant. The jars are sealed to exclude air and the time to loss of equilibrium or cessation of opercular movement, and the dissolved oxygen remaining at that time are determined for each fish. Experimental conditions should be adjusted so that time to death is in the range of 5–8 hr at a fish loading of 4–5 g/liter. Additional physiological data may be derived from this test by relating the rate of oxygen consumption to survival time (Gordon and McLeay, 1977).

In cases of stress from contaminant exposure, this test normally reveals that residual oxygen levels at death are directly proportional to the exposure concentration. Threshold-effect concentrations are usually equivalent to, or slightly less than, those that are acutely lethal. That is, the sensitivity is approximately equal to the 96 hr LC50 value (Vigers and Maynard, 1977; Giles and Klaprat, 1979).

Tests of ability to withstand an oxygen depletion can be useful in evaluating the effects of prior stress factors if these have caused pathological changes in the gills that have impaired O_2 uptake or CO_2 unloading, or have caused physiological changes that have impaired oxygen transport or inhibited metabolic pathways. For example, ability to withstand hypoxia is reduced by exposure to contaminants that affect respiration (Vigers and Maynard, 1977), but is unaffected by contaminants that exert their toxic action in other ways (McLeay and Gordon, 1980). Thus this challenge test would probably prove useful for evaluating fish disease therapeutants, such as formalin, which cause gill damage as a treatment side effect.

The use of ability to tolerate oxygen depletions as a measure of performance capacity or to compare the physiological capability of genetic strains is also merited. However, widespread use of tests of tolerance to hypoxia awaits a more detailed definition of the intrinsic and extrinsic factors that are responsible for the effect. For the present, the simplicity and rapidity of this procedure supports its use as a screening method.

3.3.3. Impaired Swimming Performance

Tests of swimming ability have long been used to identify adverse effects of environmental factors on fish health and quality, frequently with mixed results (Horak, 1972). The measurement of swimming performance typically involves determining critical swimming speed, fatigue time (endurance), swimming efficiency, or performance rating. The test chamber can be either a tunnel or a rotary flow device, and frequently is designed to be used as a respirometer as well (Smith and Newcomb, 1970).

Impaired swimming performance has been used to define tolerance limits to a variety of environmental factors including dissolved oxygen, ammonia, and salinity (Davis et al., 1963; Jones, 1971; Kutty and Saunders, 1973). For anadromous species, the change in swimming efficiency that normally occurs during the parr-smolt transformation also has potential as an index of the effects of stress (Smith and Margolis, 1970; Glova and McInerney, 1977).

In practice, the use of effects on stamina as a means of evaluating the impacts of prior stress on fish health has a number of limitations. For example, swimming stamina is probably not a major factor influencing the survival of hatchery-reared rainbow trout (*Salmo gairdneri*) released into lakes or streams (Horak, 1972).

Additionally, differing hatchery practices have little effect on the performance ratings obtained for juvenile Atlantic salmon (McNeish and Hatch, 1978). Finally, although acute exposure to certain contaminants will impair swimming performance, chronic exposures may result in normal or even increased stamina (Waiwood and Beamish, 1978). This return to normal performance may occur even when growth has been inhibited and histopathologies and other evidence for a state of chronic stress are apparent (Larmoyeux and Piper, 1973; Webb and Brett, 1973; McLeay and Brown, 1979).

Thus although certain environmental conditions may impair swimming performance, chronic stress (particularly if it involves hypoxia) may result in the compensatory development of a more efficient cardiorespiratory system that yields swimming performance equal to or better than that of unstressed controls. Perhaps the determination of swimming efficiency, which is based on tail-beat frequency versus swimming speed (body lengths/sec), will ultimately prove to be a more useful measure of fish condition (Smith, 1982b).

3.3.4. Scope for Activity

The difference between the oxygen consumption of fish at rest and undisturbed (standard rate), and the oxygen consumption of fish swimming at maximum sustained speed (active rate) was originally defined by Fry (1947) as "scope for activity." Bioenergetic cost, as indicated by scope for activity, is one means for assessing tolerance limits to environmental stress in terms of the energy remaining for routine activity or growth (Brett, 1958, 1964, 1976). Environmental stressors may either raise the standard metabolic rate, or lower the active metabolic rate. The net effect is to reduce the scope for activity (Fry, 1971). Similarly, the scope for activity is sensitive to certain otherwise sublethal effects of water quality perturbations. Contaminants whose mechanism of action involves respiration strongly affect scope (Sprague, 1971). Thus scope for activity was shown to be a sensitive indicator in detecting rainbow trout that had been previously exposed to recycled hatchery water (Mayer and Kramer, 1973).

Scope for activity has also been used to compare the tolerance limits of salmonid species, or genetic strains within species, to changes in temperature, photoperiod, and food supply (Dickson and Kramer, 1971). Effects on scope for activity can also be used to determine tolerance limits of estuarine species to salinity changes due to alterations in freshwater inflow. Wohlschlag and Wakeman (1978) showed that active and standard metabolism, swimming speed, and scope for activity of the spotted seatrout (*Cynoscion nebulosus*) were all a function of the acclimation salinity, and that the measured salinity optima based on these performance criteria agreed with the zoogeographic distribution for this species. Similar information based on multiple environmental factors (e.g., contaminants, salinity, temperature, oxygen) would greatly assist in developing the data base for the species-habitat relationships required to correctly assess the biological costs or benefits of habitat alterations in the estuarine or marine environment.

3.3.5. Resistance to Fish Diseases

It is now well-accepted that a variety of environmental stress factors predispose fish populations to diseases in both the freshwater and marine environments (Wedemeyer et al., 1976; Sindermann, 1980; Ellis, 1981). Thus the use of standardized disease challenge tests to evaluate environmental alterations in terms of potential to predispose to epizootics has considerable merit. Improved information on the point at which multiple stress factors exceed tolerance limits and begin to affect disease resistance would be especially useful in present-day biological monitoring and surveillance programs for environmental quality (Esch and Hazen, 1978, 1980; Sindermann, 1980).

The standardization of disease challenge tests requires the development of controlled methods for experimentally reproducing the infection, and extensive knowledge about the route of administration, dosage, and test duration required for clinical signs to be manifested. Inroads have been made toward the establishment of such tests. The bacterial kidney disease challenge developed by Iwama (1977) for juvenile chinook salmon provides reliable results in evaluating effects of sodium pentachlorophenate on disease resistance. Similarly, a standardized disease challenge with infectious hematopoietic necrosis virus has been developed for rainbow trout to evaluate sublethal exposure to dissolved copper (Hetrick et al., 1979). The standardized challenges of Groberg et al. (1978) with *Aeromonas salmonicida* and *A. hydrophila* can be used to determine the effects of water temperature on infectivity in anadromous salmonids. The development of reliable models for experimental furunculosis (*A. salmonicida*) in rainbow trout, and vibriosis (*V. anguillarum*) in sockeye salmon (*Oncorhynchus nerka*) have also been reported (Gould et al., 1978; Michel, 1980). Cipriano (1982) has developed a contact challenge method for furunculosis in brook trout (*Salvelinus fontinalis*) that gives consistent results, and Knittel (1981) has reported a successful injection challenge for *Yersinia ruckeri*. Similar models for additional fish pathogens would greatly assist in applying infectious disease challenge tests to the determination of defensible tolerance limits to environmental stressors.

For the noninfectious diseases, recent work on the development of standardized procedures for assessing susceptibility to carcinogens indicates that these procedures can also provide a means for characterizing the influence of environmental factors (McLeay, 1979). The pioneering work of Sinnhuber et al. (1977), Grieco et al. (1978), and Wales et al. (1978) has provided the basis for future efforts in this regard.

3.3.6. Reduced Tolerance to Reference Toxicants

Although reduced ability to tolerate standardized toxicants has been little used as a tool to evaluate the impacts of prior stress factors, this approach has a certain amount of potential as a practical test procedure. Adelman and Smith (1976) showed that an otherwise sublethal parasite infestation reduced the tolerance of goldfish (*Carassius auratus*) to sodium chloride exposure. Alexander and Clark (1978) reported that a phenol challenge differentiated unstressed rainbow trout from others that had been fasted, subjected to adverse temperatures, or to sublethal concentrations of chlorine. However, fish subjected to crowding stress tolerated a subsequent phenol challenge as well as unstressed controls. Similarly, resistance to sodium dodecyl sulphate exposure was not reduced by the stress of hauling and handling (Pessah et al., 1976).

A number of chemicals have been evaluated and proposed as suitable reference toxicants. Among the most promising are DDT, sodium dodecyl sulphate, sodium pentachlorophenate, sodium chloride, sodium azide, and phenol. The following criteria, originally proposed by Alderdice (1963), are still valid prerequisites for an effective reference chemical: it must be a general stressor, resulting in a nonspecific integrated biological response; it must reveal impaired fish condition (i.e., the effects of prior stress) by effecting a deviant physiological response in such fish; it must yield reproducible results for healthy fish; it should be soluble and stable in water, easily measured, and effective at low concentrations; and finally, it must be sufficiently uncommon in nature to preclude the likelihood of prior exposure of the test fish. Although no single chemical has yet met all of these criteria, three—sodium pentachlorophenate, sodium chloride, and phenol—can reveal otherwise nonapparent differences in fish condition and thus appear to be the best of the presently available candidates (Adelman and Smith, 1976; Alexandar and Clarke, 1978).

Conventionally, tests with reference toxicants are conducted to determine the 24 or 96 hr LC50, or the time to 50% mortality of the test fish. Water quality characteristics, especially the pH, hardness, temperature, and dissolved oxygen, can markedly affect toxicity by influencing the solubility, ionization, and uptake of the reference chemical, and therefore must be held constant in all such tests.

As the foregoing discussion indicates, it is unlikely that a single response, for example, death, to a chemical challenge will be able to reveal more than general information about the debilitation of fish health resulting from prior stress. Nevertheless, appropriate reference toxicant challenges may prove to be of some value for this purpose. As a practical matter, standardization for each fish species will be required in order to establish a baseline from which comparisons can be made.

3.3.7. *Effects on Ability to Osmoregulate: Salinity Challenge Tests*

The determination of ability to tolerate an osmoregulatory challenge as a measure of the impact of prior stress requires that a number of variables be considered. These include physiological type (stenohaline, anadromous, euryhaline), developmental stage (parr, smolt, juvenile, adult), season, water chemistry and temperature, and test salinity (Eddy, 1981). In the case of freshwater-acclimated fish, direct transfer to seawater of the appropriate salinity can be used to elicit osmoregulatory dysfunction, or a generalized stress response, either of which can be used as an indicator (Hirano, 1969; Singley and Chavin, 1975; Park et al., 1982). Three biological end points are available for use: 24–48 hr survival, ability to regulate blood sodium within 24 hr, or ability to continue to grow and develop normally for extended periods in seawater (Wedemeyer et al., 1980; Clarke, 1982; Park et al., 1982). For anadromous species, the seawater challenge test can be used to reveal the presence of latent infections, effects of prior fish disease treatments, and effects of stress from otherwise sublethal environmental conditions (Clarke, 1982). The salinity challenge tests currently under development by Park et al. (1982), to determine impacts of stress from fish-passage facilities on the performance capacity of smolts, have excellent potential for assessing tolerance limits to a variety of the habitat alterations affecting juvenile anadromous fishes.

Clarke and Blackburn (1977, 1978) have developed the salinity challenge test into a seawater challenge–blood sodium test for monitoring the progress of smoltification in juvenile anadromous salmonids and predicting their potential for seawater growth. In this test, fish are transferred directly from freshwater into seawater (28‰) at the same temperature, and periodic blood sodium analyses are performed. Functional coho salmon smolts with good long-term seawater growth potential are able to regulate their plasma sodium levels to 170 meq/liter or less within 24 hr. Incompletely functional smolts suffer a substantial hypernatremia that may persist for an extended time. Although such fish may be able to survive in seawater and may eventually even succeed in regulating their blood electrolytes back into the normal range, long-term growth will be impaired (Clarke, 1982).

The use of salinity challenge tests to determine impacts of stress on ability of stenohaline or euryhaline species to osmoregulate is merited. However, tests to determine the optimum test conditions, and to calibrate the effects of environmental alterations under controlled conditions on the measured response will be needed.

The development of an analogous freshwater challenge test for assessing the impact of stress factors on the ability of estuarine and marine fishes to osmoregulate is also merited. Such tests would assist in providing the improved biological information required to correctly assess tolerance limits to estuarine and marine habitat alterations. For maximum utility, freshwater challenge tests should be calibrated in terms of effects on long-term survival, reproduction, and growth of the species in question.

3.3.8. Tolerance to Crowding

Crowding, or exceeding species population density tolerance limits, is both a physiological stress factor commonly encountered by fishes and an easily conducted challenge test (Wedemeyer, 1976). Excessive crowding consistently elicits a species-dependent generalized stress response (Burton and Murray, 1979), which may result in adverse effects on growth, reproduction and behavior if it becomes chronic. In the case of anadromous species, reduced early marine survival may result from excessive population densities during freshwater rearing in the absence of any effect on freshwater growth (Fagerlund et al., 1981; Sandercock and Stone, 1982).

For challenge tests, acute crowding will normally elicit a usable generalized stress response. To conduct such tests (with salmonids), groups of fish to be tested are acclimated to a population density (PD) of 0.04 kg/liter for a minimum of 2 weeks and then subjected to the stress factor in question. To initiate a challenge, the PD is increased to the desired value, usually 0.2–0.6 kg/liter, for a fixed period; usually 96 hr. During this time, the severity of the generalized stress response and the time needed for compensation are evaluated and compared with values from control groups. Physiological testing of 10 fish subsamples is usually adequate if hyperglycemia or hypochloremia are monitored because of the low coefficients of variation of these physiological parameters. Test conditions such as water temperature and exchange rates are chosen on the basis of the species and environmental stress factors to be tested. The water exchange rate is normally set to keep the dissolved oxygen near saturation and the un-ionized ammonia below about 0.02 mg/liter. Such experimental conditions would be considered adequate for developing stress mitigation procedures, or to assess effects of contaminant exposure or interspecific interactions.

4. ASSESSING POPULATION TOLERANCE: THE PROBLEMS AND METHODS OF MONITORING

Populations have characteristic properties and responses to stress that are in addition to those of the individual members. These include density-dependent mortality, natality, genetic diversity, and such features as age distribution and population growth rate (r). The numerical status of a population (n) at any given time (t) is governed by the balance between the instantaneous rates of birth (b) and death (d) of its individual members. For example, $b > d$ results in population growth; $b = d$, no change; $b < d$, population decline. In general, the intrinsic growth rate ($r = b - d$) will be 0 for most naturally spawning populations, since they obviously have not grown to infinite size and are also not extinct. If one or more stress factors cause the birth rate to decrease or the death rate to increase, then it is reasonable to expect that, in the absence of genetic adaptation, the population will begin to decline and will continue to do so until the stressor is removed. However, a change in either the birth rate or the death rate will often result in a compensatory change in the other parameter and little or no effect on the population may actually occur (Goodyear, 1980, 1982). Such compensating changes in population parameters are the result of density-dependent feedback mechanisms that are responsible for homeostasis at the population level of biological organization.

Many different density-dependent mechanisms have been documented in fish populations (Goodyear, 1980). Such processes involve both inter- and intraspecific interactions that influence the probabilities for survival and growth of the individuals that comprise the population. These mechanisms make it certain that resource limitations will ultimately limit population growth if some factor causes the intrinsic growth rate to become positive. Unfortunately, there is no certainty that they will act to restore $b = d$ if an environmental change causes r to become negative.

When density-dependent processes do act to offset the reduced survival or reproductive success caused by stress, there are several possible mechanisms (Goodyear, 1980). For example, an increase in mortality at one life stage that reduces the size of the population makes more resources available to the survivors. These increased resources can increase the fecundity of surviving fish, allow faster growth, and thus reduce vulnerability to predators. Many fish populations are able to persist despite substantial mortality due to fishing pressure because of just such density-dependent mechanisms (Jones, 1973). These same mechanisms may be able to compensate for the effects of stress on individual fish and result in no observable effects on the population at large.

A further complicating factor is that the size of natural populations is often subject to large year-to-year variations. Even large shifts in true population size because of stress may thus be impossible to detect without an impractically long time series of data (McCaughran, 1977). When there is a detectable change in abundance, it will often be impossible to assign that change to any single cause because of the complex interactions among the many variables in the biotic and abiotic environment. Additionally, factors such as fishing may simultaneously be acting to depress population size, so that the contribution of individual stress factors to a measured change in the population will be difficult to isolate. These phenomena will tend to conceal the effects of stress at the population level, even when rather substantial effects on individuals in the population are involved.

To resolve these problems, it is first necessary to understand that there are several possible pathways through which a stressing agent can produce an effect on a particular species. The stress can result in direct mortality, or in an indirect mortality due to changes in growth rates or physiological tolerances. Alternatively, the most important impact of a stress factor may result from its effect on the other components of the ecosystem that influence food availability, or the abundance of competitors or predators.

The most frequently used tests of the impacts of environmental stress involve measuring changes in physiological functions, growth, or survival of exposed individuals in a laboratory. The effect of impaired fish health on the ultimate size of a population depends on both the magnitude of the impairment and on the life stage involved. Assuming that the stress factor is selective and affects only the species of interest, an increase in mortality would be little different in effect from a similar increase in mortality due to any other cause (at least with respect to population size and persistence). Any difference that existed would relate to possible community level changes that might occur if the biomass associated with the mortality were channeled into different ecosystem components that themselves influenced the growth or survival of individuals of the affected species. Such changes could, for example, provide a competitive advantage to an otherwise unaffected species.

If such is not the case, the increase in mortality will result, of necessity, in a decrease in the existing compensatory reserve of the population. This, in turn, provides a measure of the ability of the population to withstand any additional stress (Goodyear, 1977). In any case, the immediate effect on the actual size of the population depends on the nature of the mechanisms that control the relation between stock size and recruitment. For example, it is possible for decreased hatching success due to stress to actually increase the equilibrium stock size if the stock-recruitment relation can be adequately described by a population model where the density-dependent mortality is keyed to the initial density of fry—as might occur with density-dependent starvation (Goodyear, 1982). The actual outcome of decreased hatching success would depend on the degree of change in this mortality, and its timing in relation to the operation of the normal density-dependent population control processes. It is theoretically possible for a population to decrease, change very little, or—if the density-dependent processes are sufficiently intense—to actually increase as the result of a directly lethal response, at the individual level, to a stressing agent.

Even in simple cases where a stress factor is directly lethal to individuals and no other components of the community are directly affected, the interpretation of the effect at the population level presents an imposing technological problem. Two types of information are required. First, the timing and extent of the excess mortality must be determined. This information can sometimes be derived from laboratory studies. Second, the nature, timing, and intensity of the density-dependent processes that regulate the size of the population must be understood. This latter information requirement is among the more elusive of the problems of fishery science in general (Cushing, 1968; Hunter, 1976). The problem is further complicated by the fact that under field conditions, the stress effect itself will rarely be selectively lethal—that is, with no sublethal effects and no effects on other species in the community.

It is quite possible for the apparently sublethal effects of stress to have a greater effect at the population level than would be expected from laboratory studies of individuals. Relatively small changes in neuroendocrine mechanisms affecting growth or behavior can substantially modify the long-term survival probability and reproductive capabilities of the individuals in the population (George, 1977). The ultimate effects of such changes may involve subtle, yet powerful, interactions of the affected individuals with other trophic levels or with the physical environment that are all but undetectable with present monitoring or surveillance technology, and defy quantitative measurement. For example, impaired predator avoidance behavior has been reported for several species of fish subjected to otherwise sublethal stress in the laboratory (Goodyear, 1972; Coutant, 1973; Kania and O'Hara, 1974; Sullivan et al., 1978; Coutant et al., 1979). Such altered behavior can, in turn, alter prey selection by predators of the stressed species, and thus indirectly increase mortality rates in the population. Again, however, the actual outcome of this change in mortality will depend on its timing and magnitude with respect to the density-dependent processes that regulate the size of the population.

Stress can also cause reduced feeding (Broderius and Smith, 1979) or induce physiological responses that reduce food conversion efficiency (Gilderhus, 1966). The accompanying reduced rates of growth or development can have profound effects on total mortality, especially if early life stages are involved (Shuter et al., 1980). For example, if the daily mortality for a population of larval fish is 0.50 during some critical period of development and if the duration of this period is lengthened by only one day due to slower growth, then total survival would decrease by 50% due to the increase in the duration of the mortality exposure alone. If the daily mortality rate was also increased by impaired predator avoidance or by the directly lethal effects of a stress factor, then the total increase in mortality would be much greater.

Another indirect result of stress is to reduce the reproductive potential of individuals in a population. For example, reduced growth can also reduce fecundity (LeCren, 1958; Bagenal, 1963; Jensen, 1971; Shuter et al., 1980), or increase the age of first maturity (Lett and Kohler, 1976; Shuter and Koonce, 1977). Both of these responses would result in a decrease in the expected lifetime egg production of females recruited into the population. Thus population stability would require a compensatory change such as an increase in the survival probability of the reduced number of eggs produced.

Stress can also reduce the ability of individuals to withstand fluctuations in normal environmental parameters such as temperature (Cairns and Scherr, 1963; Eisler,

1971; Smith and Heath, 1979), dissolved oxygen (Middaugh et al., 1975), and salinity (Hazel et al., 1971; Brannon and Tate, 1977). Thus the actual level of the excess mortality caused by a particular stress factor in these instances could be controlled by the environmental conditions prevailing at the time. Locally unfavorable conditions could conceivably enhance overall population growth by reducing competition for resources. Conversely, the average effect of stress on individuals might have little relationship to the population effect if survival was dependent upon encounters with randomly distributed lethal conditions. Thus spatial and temporal variations in existing environmental conditions may produce spatial and temporal variations in the influence of the stress factor in question on individual survival probabilities. Such effects have profound implications for the measurement of the effects of stress on age specific survival rates for a population as a whole.

Thus an interpretation at the population level of the indirect effects of stress on a particular species requires quantification of the response in terms of a change in survival probability or a change in reproductive rates, and knowledge of the density-dependent processes that control population size. In contrast to the situation where only directly lethal effects on individuals occur, none of the required information is particularly amenable to measurement. Furthermore, few stress factors are species specific; most produce effects on other community members as well.

From the preceding discussion, it is clear that there are a large number of potential pathways through which environmental stress may act to alter population size or persistence in a community. It is also possible for stressors to alter populations of species that are not themselves sensitive to its direct influence in much the same way that the fluctuation in dominant species may influence subordinate species (Skud, 1982). Similarly, a stress that is sublethal to a particular species may nevertheless cause significant reductions in its population size by reducing the availability of food to some critical life stage (Johnson, 1968). Of course, the effect of the stress factor may be either on the prey population or on its food source. In either case, if the calories available to the species of interest are reduced, impaired growth and an increased probability of predation, starvation, and disease can be expected.

In contrast to the above situation, if the detrimental effect of a stressor on the population of a principal predator exceeds the initial direct effect on the population of the species of interest, mortality in the population of the species of interest may actually decline. The ultimate result could thus be a population increase as a result of the adverse effect on predator densities. Similarly, stressors that cause a significant reduction in a competitor population could result in an increase in the amount of food available to the individuals of another population. This situation could result in increased growth and decreased mortality in the population of interest and thereby increase its size. Alternatively, a decline in predators of the competitor could increase the population of the competitor species to the detriment of the population of interest.

As seen from the preceding discussion, even in the relatively well-studied instances where the link between fish health problems and the presence of environmental contaminants is the question of interest, cause and effect relationships are difficult to establish unless the epizootics are obviously catastrophic.

However, the difficulty of assessing the population-level significance of identified fish health problems in no way diminishes the need to obtain such information. To achieve this, surveillance and monitoring programs to determine biological impacts should be designed to partition the population response into its component parts. In situations where contaminant monitoring *per se* is the question of interest, better scientific information is needed in two general areas. The first is the requirement for improved biological end points to assist in predicting the likely impacts of the many agricultural and industrial chemicals that find their way into the aquatic environment. The second need is for improved quantitative information, and methods of monitoring in cases of existing population declines.

With respect to the need for improved biological end points, a key consideration is that the effects on populations begin with effects on individuals. If there is no effect on individuals (including effects on individuals of other populations that prey upon, compete with, or provide food for the population of interest), there will be no effect on the population. This biological principle supports the concept that sufficiently sensitive laboratory tests can be adequately interpreted when negative results are obtained (Mount, 1980).

In cases where information on the relationship of existing water pollution to an observed fishery problem must be obtained, the design of the project must reflect the complex interactions discussed above. Thus the field work required will be an extensive undertaking. The first step is to determine the extent to which the population is being exposed to contaminants. This requires a statistically adequate sampling and analytical program. If contaminants are found that are known to be toxic to the species of interest, the next step is to determine their relevance as a cause of the identified resource problem. This step requires an analysis of all processes controlling population size, including an identification of other factors that may be currently stressing the population. If the identified contaminants are shown to be nontoxic to the species in question. it is still possible that they are indirectly responsible through actions on other populations. Unfortunately, to resolve this question completely, information on the dynamics of the entire community would be required. Studies to obtain such information are usually not practical and, in addition, are not likely to achieve clear-cut results. However, the possibility of an ambiguous outcome does not make it a certainty, and incremental progress that is made will help advance knowledge so that future problems can be solved more easily. The lack of such basic information is currently hindering the adequate definition of the management alternatives available for solving many of our present day fishery resource problems.

REFERENCES

Adelman, I. R. (1980). Uptake of ^{14}C-glycine by scales as an index of fish growth: Effect of fish acclimation temperature. *Trans. Amer. Fish. Soc.*. **109**, 197–194.

Adelman, I. R. and Smith, L. L., Jr. (1976). Fathead minnows (*Pimephales promelas*) and goldfish (*Carassius auratus*) as standard fish in bioassays and their reaction to potential reference toxicants. *J. Fish. Res. Board Can.* **33**, 209–214.

Alderdice, D. F. (1963). Some effects of simultaneous variation in salinity, temperature and dissolved oxygen on the resistance of juvenile coho salmon (*Oncorhynchus kisutch*) to a toxic substance. Ph.D. thesis, Univ. of Toronto, Ont. 177 pp.

Alexander, D. G. and Clarke, R. (1978). The selection and limitations of phenol as a reference toxicant to detect differences in sensitivity among groups of rainbow trout (*Salmo gairdneri*). *Water Res.* **12**, 1085–1090.

Anderson, R. O. (1980). Proportional stock density (PSD) and relative weight (W_r): Interpretive indices for fish populations and communities. In: S. Gloss and B. Shupp (Eds.), *Practical fisheries management*. Amer. Fish. Soc. NY Chapter, Workshop Proc., processed rep.

Bagenal, T. B. (1963). Variations in plaice fecundity in the Clyde area. *J. Mar. Biol. Assoc. U.K.* **43**, 391–399.

Ballard, J. A. and Oliff, W. D. (1969). A rapid method for measuring the acute toxicity of dissolved materials to marine fishes. *Water Res.* **3**, 313–333.

Barton, B. A., Peter, R. E., and Paulencu, C. R. (1980). Plasma cortisol levels of fingerling rainbow trout (*Salmo gairdneri*) at rest, and subjected to handling, confinement, transport, and stocking. *Can. J. Fish. Aquat. Sci.* **37**, 805–811.

Bayne, B. L. (1980). Physiological measurements of stress. *Rapp. P.-V. Reun. Cons. Int. Explor. Mer.* **179**, 56–61.

Billard, R., Bry, C. and Gillet, C. (1981). Stress, environment, and reproduction in teleost fish. In A. Pickering (Ed.), *Stress and fish*. Academic Press, London, pp. 185–208.

Bonin, J. D. (1981). Measuring thermal limits of fish. *Trans. Amer. Fish. Soc.* **110**, 662–664.

Brannon, D. P. and Tate, L. G. (1977). Effect of salinity on toxicity and tissue accumulation of cadmium in the sailfin molly, *Poecilia latipinna* (Poeciliidae). *Bull. Assoc. Southeastern Biol.* **24**, 3939.

Brett, J. R. (1958). Implications and assessments of environmental stress. In: P. A. Larkin (Ed.), *The investigations of fish-power problems*, H. R. MacMillan Lectures in Fisheries, University of British Columbia, Vancouver, pp. 69–83.

Brett, J. R. (1964). The respiratory metabolism and swimming performance of young sockeye salmon. *J. Fish. Res. Board Can.* **21**, 1183–1226.

Brett, J. R. (1976). Scope for metabolism and growth of sockeye salmon, *Oncorhynchus nerka*, and some related energetics. *J. Fish. Res. Board Can.* **33**, 307–313.

Broderius, S. J. and Smith, L. L. Jr. (1979). Lethal and sublethal effects of binary mixtures of cyanide and hexavalent chromium, zinc or ammonia to the fathead minnow (*Pimephales promelas*) and rainbow trout (*Salmo gairdneri*). *J. Fish. Res. Board Can.* **36**, 164–172.

Bulow, F. J. (1970). RNA-DNA ratios as indicators of recent growth rates of a fish. *J. Fish. Res. Board Can.* **27**, 2343–2349.

Bulow, F. J., Zeman, W. E., Winingham, J. R., and Hudson, W. F. (1981). Seasonal variations in RNA-DNA ratios and in indicators of feeding, reproduction, energy storage, and condition in a population of bluegill, *Lepomis macrochirus* Rafinesque. *J. Fish Biol.* **18**, 237–244.

Burton, C. B. and Murray, S. A. (1979). Effects of density on goldfish blood—I. Hematology. *Comp. Biochem. Physiol.* **62A**, 555–558.

Cairns, J., Jr. and Scherr, A. (1963). Environmental effects upon cyanide toxicity to fish. *Not. Nat.* **361**, 1–11.

Carter, L. (1962). Bioassay of trade wastes. *Nature (Lond.)* **196**, 1304.

Casillas, E. and Smith, L. S. (1977). Effect of stress on blood coagulation and haematology in rainbow trout (*Salmo gairdneri*). *J. Fish Biol.* **10**, 481–491.

Chavin, W. (1973). *Responses of fish to environmental change*. Charles C. Thomas Co., New York, 459 pp.

Cipriano, R. (1982). Furunculosis in brook trout: Infection by contact exposure. *Prog. Fish-Cult.* **44**, 12-14.

Clarke, W. C. (1982). Evaluation of the seawater challenge test as an index of marine survival. *Aquaculture* **28**, 177–183.

Clarke, W. C. and Blackburn, J. (1977). A seawater challenge test to measure smolting of juvenile salmon. Fish. Mar. Serv. Canada. Res. Dev. tech. rep. 705, 11 pp.

Clarke, W. C. and Blackburn, J. (1978). Seawater challenge tests performed on hatchery stocks of chinook and coho salmon in 1977. Fish. Mar. Serv. Canada. Res. Dev. tech. rep. 76. 19 pp.

Clarke, W. C., Shelbourn, J. E., and Brett, J. R. (1981). Effect of artificial photoperiod cycles, temperature and salinity on growth and smolting in underyearling coho (*Oncorhynchus kisutch*), chinook (*O. tshawytscha*), and sockeye (*O. nerka*) salmon. *Aquaculture* **22**, 105–116.

Coutant, C. C. (1973). Effect of thermal shock on vulnerability of juvenile salmonids to predators. *J. Fish. Res. Board Can.* **30**, 965–973.

Coutant, C. C., McLean, R. B., and DeAngelis, D. L. (1979). Influences of physical and chemical alterations on predator-prey interactions. In: H. Clepper (Ed.), *Predator-prey systems in fisheries management*. Sport Fishing Inst., Washington, DC, pp. 57–68.

Cushing, D. H. (1968). Fisheries biology: A study in population dynamics. Univ. of Wisconsin Press, Madison.

Davis, G. E., Foster, J., Warren, C. E., and Doudoroff, P. (1963). The influence of oxygen concentration on the swimming performance of juvenile Pacific salmon at various temperatures. *Trans. Amer. Fish. Soc.* **92**, 111–124.

Dickson, I. W. and Kramer, R. H. (1971). Factors influencing scope for activity and active and standard metabolism of rainbow trout (*Salmo gairdneri*). *J. Fish. Res. Board Can.* **28**, 587–596.

Donaldson, E. M. (1981). The pituitary-interrenal axis as an indicator of stress in fish. In: A. Pickering (Ed.), *Stress and fish*. Academic Press, London, pp. 11–48.

Donaldson, E. M. and McBride, J. R. (1974). Effect of ACTH and salmon gonadotropin on interrenal and thyroid activity of gonadectomized adult sockeye salmon (*Oncorhynchus nerka*). *J. Fish. Res. Board Can.* **31**, 1211–1214.

Eddy, F. B. (1981). Effects of stress on osmotic and ionic regulation in fish. In: A. Pickering (Ed.), *Stress and fish*. Academic Press, London, pp. 77–102.

Eisler, R. (1971). Cadmium poisoning in *Fundulus heteroclitus* (Pisces: Cyprinodontidae) and other organisms. *J. Fish. Res. Board Can.* **28**, 1225–1234.

Ellis, A. E. (1981). Stress and the modulation of defence mechanisms in fish. In: A. Pickering (Ed.), *Stress and fish*. Academic Press, London, pp., 147–170.

Elliott, J. M. (1981). Some aspects of thermal stress on freshwater teleosts. In: A. Pickering (Ed.), *Stress and fish*. Academic Press, London, pp. 209–245.

Esch, G. W. and Hazen, T. C. (1978). Thermal ecology and stress: A case history for red-sore disease in largemouth bass. In: J. Thorp and J. Gibbons (Eds.), *Energy and environmental stress in aquatic sytems*. U.S. Dep. of Energy, Symp. ser. 48, Nat. Tech. Inf. Serv., Publ. CONF-7711111, U.S. Dep. of Commerce, Springfield, VA, pp. 331–363.

Esch, G. W. and Hazen, T. C. (1980). Stress and body condition in a population of largemouth bass: Implications for red-sore disease. *Trans. Amer. Fish. Soc.* **109**, 532–536.

Fagerlund, U. H. M., McBride, J. R., and Stone, E. T. (1981). Stress-related effects of hatchery rearing density on coho salmon. *Trans. Amer. Fish. Soc.* **110**, 644–649.

Folmar, L. C. and Dickhoff, W. C. (1980). The parr-smolt transformation (smoltification) and seawater adaptation in salmonids. *Aquaculture* **21**, 1–37.

Fry, F. E. J. (1947). Effects of the environment on animal activity. University of Toronto Studies, Biology Ser., No. 55 (Publ. Ontario Fish. Res. Lab., No. 68), Univ. of Toronto, pp. 1–62.

Fry, F. E. J. (1971). The effect of environmental factors on the physiology of fish. In: W. S. Hoar and D. J. Randall (Eds.), *Fish physiology*, Vol. VI. Academic Press, New York, pp. 1–98.

George, C. J. (1977). The implication of neuroendocrine mechanisms in the regulation of population character. *Fisheries* **2**(3), 14–19.

Gilderhus, P. A. (1966). Some effects of sublethal concentrations of sodium arsenite on bluegills and the aquatic environment. *Trans. Amer. Fish. Soc.* **95**, 289–296.

Giles, M. A. and D. Klaprat (1979). The residual oxygen test: A rapid method for estimating the acute lethal toxicity of aquatic contaminants. In: E. Sherer (Ed.), *Toxicity tests for freshwater organisms.* Can. Spec. Publ. Fish. Aquat. Sc. No. 44., Dept. Fish. and Oceans, Canada, pp. 37–45.

Glova, G. J. and McInerney, J. E. (1977). Critical swimming speeds of coho salmon (*Oncorhynchus kisutch*) fry to smolt stages in relation to salinity and temperature. *J. Fish. Res. Board Can.* **34**, 151–154.

Goodyear, C. P. (1972). A simple technique for determining effects of toxicants or other stresses on a predator-prey interaction. *Trans. Amer. Fish. Soc.* **101**, 367–370.

Goodyear, C. P. (1977). Assessing the impact of power plant mortality on the compensatory reserve of fish populations. In: W. Van Winkle (Ed.), *Proceedings of the conference on assessing the effects of power plant induced mortality on fish populations.* Pergamon Press, New York, pp. 186–195.

Goodyear, C. P. (1980). Compensation in fish populations. In: C. H. Hocutt and J. R. Stauffer, Jr. (Eds.), *Biological monitoring of fish.* Lexington Books, Lexington, MA, pp. 253–280.

Goodyear, C. P. (1983). Measuring effects of contaminants on fish. In: W. E. Bishop, R. D. Cardwell, and B. B. Heidolph (Eds.), Aquatic Toxicology and Hazard Assessment: 6th Symposium, ASTM, STP-802, Am. Soc. for Test. and Materials, Philadelphia, PA, pp. 414–424.

Gordon, M. R. and McLeay, D. J. (1977). Sealed-jar bioassays for pulpmill effluent toxicity: Effects of fish species and temperature. *J. Fish. Res. Board Can.* **34**, 1389–1396.

Gordon, M. R. and McLeay, D. J. (1978). Effect of photoperiod on seasonal variations in glycogen reserves of juvenile rainbow trout (*Salmo gairdneri*). *Comp. Biochem. Physiol.* **60**, 349–351.

Gould, R. W., Antipa, R. and Amend, D. F. (1978). Immersion vaccination of sockeye salmon (*Oncorhynchus nerka*) with two pathogenic strains of *Vibrio anguillarum. J. Fish. Res. Board Can.* **36**, 222–225.

Grieco, M. P., Hendricks, J. D., Scanian, R. A., Sinnhuber, R. O., and Pierce, D. A. (1978). Carcinogenicity and acute toxicity of dimethyl nitrosamine in rainbow trout (*Salmo gairdneri*). *J. Nat. Cancer Inst.* **60**, 1127–1130.

Groberg, W. J., Jr., McCoy, R. H., Pilcher, K. S., and Fryer, J. L. (1978). Relation of water temperature to infections of coho salmon (*Oncorhynchus kisutch*), chinook salmon (*O. tshawytscha*), and steelhead trout (*Salmo gairdneri*) with *Aeromonas salmonicida* and *A. hydrophila. J. Fish. Res. Board Can.* **35**, 1–7.

Harman, B. J. Johnson, D. L., and Greenwald, L. (1980). Physiological responses of Lake Erie freshwater drum to capture by commerical shore seine. *Trans. Amer. Fish. Soc.* **109**, 544–551.

Hattingh, J. (1976). Blood sugar as an indicator of stress in the freshwater fish, *Labeo capensis* (Smith). *J. Fish Biol.* **10**, 191–195.

Hazel, C. R., Thomsen, W., and Meith, S. J. (1971). Sensitivity of striped bass and stickleback to ammonia in relation to temperature and salinity. *Calif. Fish Game* **57**, 154–161.

Heath, A. G. and Hughes, G. M. (1973). Cardiovascular and respiratory changes during heat stress in rainbow trout (*Salmo gairdneri*). *J. Exp. Biol.* **59**, 323–338.

Hetrick, F. M., Knittel, M. D., and Fryer, J. L. (1979). Increased susceptibility of rainbow trout to infectious hematopoietic necrosis virus after exposure to copper. *Appl. Environ. Microbiol.* **37**, 198–201.

Hirano, T. (1969). Effects of hypophysectomy and salinity change on plasma cortisol concentration in the Japanese eel, *Anguilla japonica. Endocrinol. Jap.* **16**, 557–560.

Hodgins, H. O., McCain, B. B., and Hawkes, J. (1977), Marine fish and invertebrate diseases, host disease resistance, and pathological effects of petroleum. In: D. Malins (Ed.), *Effects of petroleum on arctic and subarctic marine environments and organisms. II. Biological effects.* Academic Press, New York, pp. 95–148.

Horak, D. L. (1972). Survival of hatchery-reared rainbow trout (*Salmo gairdneri*) in relation to stamina tunnel ratings. *J. Fish. Res. Board Can.* **29**, 1005–1009.

Hughes, G. M. and Roberts, J. C. (1970). A study of the effect of temperature changes on the respiratory pumps of the rainbow trout. *J. Exp. Biol.* **52**, 177–192.

Hunn, J. B. (1982). Urine flow rate in freshwater salmonids: A review. *Prog. Fish-Cult.* **44**, 119–125.

Hunter, J. R. (Ed.) (1976). Report of a colloquium on larval fish mortality studies and their relation to fishery research, January 1975. NOAA tech. rep. NMFS CIRC-395. Nat. Mar. Fish. Serv., Dep. of Commerce, Seattle, WA.

Ishioka, H. (1980a). Stress reactions in the marine fish—I. Stress reactions induced by temperature changes. *Bull. Jap. Soc. Sci. Fish.* **46**, 523–531.

Ishioka, H. (1980b). Stress reactions induced by environmental salinity changes in Red Sea bream. *Bull. Jap. Soc. Sci. Fish.* **46**, 1323–1331.

Ivanovici, A. M. (1980). Application of adenylate energy charge to problems of environmental impact assessment in aquatic organisms. *Helgolander Meeresunters* **33**, 556–565.

Iwama, G. (1977). Some aspects of the interrelationship of bacterial kidney disease infection and sodium pentachlorophenate exposure in juvenile chinook salmon (*Oncorhynchus tshawytscha*). M. Sc. thesis. Univ. of British Columbia, Vancouver, 106 pp.

Jensen, A. L. (1971). Response of brook trout (*Salvelinus fontinalis*) populations to a fishery. *J. Fish. Res. Board Can.* **28**, 458–460.

Johansson-Sjöbeck, M. L., Goran, D., Larsson, A., Lewander, K., and Lindman, U. (1978). Haematological effects of cortisol in the European eel, *Anguilla anguilla* L. *Comp. Biochem. Physiol.* **60A**, 165–168.

Johnson, D. W. (1968). Pesticides and fishes—A review of selected literature. *Trans. Amer. Fish. Soc.* **97**, 398–424.

Jones, D. R. (1971). The effect of hypoxia and anemia on the swimming performance of rainbow trout (*Salmo gairdneri*) *J. Exp. Biol.* **55**, 541–551.

Jones, R. (1973). Density dependent regulation of the numbers of cod and haddock. In: B. B. Parish (Ed.), *Fish stocks and recruitment*. Int. Council N. Atlantic Fish., Copenhagen, Denmark, pp. 156–173.

Kania, H. J. and O'Hara, J. (1974). Behavioral alterations in a simple predator-prey system due to sublethal exposure to mercury. *Trans. Amer. Fish. Soc.* **103**, 134–136.

Knittel, M. D. (1981). Susceptibility of steelhead trout *Salmo gairdneri* Richardson to redmouth infection *Yersinia ruckeri* following exposure to copper. *J. Fish Dis.* **4**, 33–40.

Kutty, M. N. and Saunders, R. L. (1973). Swimming performance of young Atlantic salmon (*Salmo salar*) as affected by reduced ambient oxygen concentration. *J. Fish. Res. Board Can.* **30**, 223–227.

Larmoyeux, J. D. and Piper, R. G. (1973). Effects of water reuse on rainbow trout in hatcheries. *Prog. Fish-Cult.* **35**, 2–8.

LeCren, E. D. (1958). Observations of the growth of perch (*Perca fluviatilis* L.) over twenty-two years with special reference to the effects of temperature and changes in population density. *J. Anim. Ecol.* **27**, 187–334.

Lett, P. F. and Kohler, A. C. (1976). Recruitment: A problem of multispecies interaction and environmental perturbations, with special reference to Gulf of St. Lawrence Atlantic herring (*Clupea harengus harengus*). *J. Fish. Res. Board Can.* **33**, 1353–1371.

Liguori, V., Zakour, V., Landolt, M., and Felton, S. (1983). Toxicity of the herbicide Endothal to juvenile chinook salmon (*Oncorhynchus kisutch*). In: W. E. Bishop, R. D. Cardwell and B. B. Heidolph (Eds.), Aquatic toxicology and hazard assessment: 6th symp. ASTM, STP, publication no. 04-802000-16, 530-544.

Lorz, H. and McPherson, B. P. (1976). Effects of copper or zinc in fresh water on the adaptation to sea water and ATPase activity, and the effects of copper on migratory disposition of coho salmon (*Oncorhynchus kisutch*). *J. Fish. Res. Board Can.* **33**, 2023–2030.

Lorz, H., Glenn, S., Williams, R., Kunkel, C., Norris, L., and Loper, B. (1978). Effects of selected herbicides on smolting of coho salmon. U.S. Environmental Protection Agency, Grant rep. R-804283. Oregon Dept. Fish and Wild., Corvallis, OR, 482 pp.

Lugo, A. E. (1978). Stress and ecosystems. In: J. Thorp and J. Gibbons (Eds.) *Energy and environmental stress in aquatic systems*. U.S. Dep. of Energy, Symp. Ser. 48, Nat. Tech. Inf. Serv., Publication conf.-7711111, U.S. Dept. of Commerce, Springfield, VA, pp. 62–101.

MacMillan, J. R., Mulcahy, D., and Landolt, M. (1980). Viral erythrocytic necrosis: Some physiological consequences of infection in chum salmon (*Oncorhynchus keta*). *Can. J. Fish. Aquat. Sci.* **37**, 799–804.

Malins, D. C., McCain, B., Brown, D., Sparks, A., and Hodgins, H. O. (1980). Chemical contaminants and biological abnormalities in central and southern Puget Sound. NOAA tech. memo. OMPA-2, Nat. Oceanic and Atmospheric Admin., Boulder, CO, 292 pp.

Malins, D. C., McCain, B., Brown, D., Sparks, A., Hodgins, H., and Chan, S. (1982). Chemical contaminants and abnormalities in fish and invertebrates from Puget Sound. NOAA tech. memo. OMPA-19, Nat. Oceanic and Atmospheric Admin., Boulder, CO, 165 pp.

Mayer, F. L., Jr. and Kramer, R. H. (1973). Effects of hatchery water reuse on rainbow trout metabolism. *Prog. Fish-Cult.* **35**, 9–10.

Mazeaud, M. M. and Mazeaud, F. (1981). Adrenergic responses to stress in fish. In: A. Pickering (Ed.), *Stress and fish*. Academic Press, London, pp. 49–76.

Mazeaud, M. M., Mazeaud, F., and Donaldson, E. M. (1977). Primary and secondary effects of stress in fish: Some new data with a general review. *Trans. Amer. Fish. Soc.* **106**, 201–212.

McCaughran, D. A. (1977). The quality of inferences concerning the effects of nuclear power plants on the environment. In: W. Van Winkle (Ed.), *Proceedings of the conference on assessing the effects of power plant-induced mortality on fish populations*. Pergamon Press, New York, pp. 229–242.

McFarlane, G. A. and Franzin, W. G. (1978). Elevated heavy metals: A stress on a population of white suckers, *Catostomus commersoni*, in Hammel Lake, Saskatchewan. *J. Fish. Res. Board Can.* **35**, 963–970.

McLeay, D. J. (1973). Effects of a 12-hr and 25-day exposure to kraft pulp mill effluent on the blood and tissues of juvenile coho salmon (*Oncorhynchus kisutch*). *J. Fish. Res. Board Can.* **30**, 395–400.

McLeay, D. J. (1975a). Variations in the pituitary-interrenal axis and the abundance of circulating blood-cell types in juvenile coho salmon (*Oncorhynchus kisutch*), during stream residence. *Can. J. Zool.* **53**, 1882–1891.

McLeay, D. J. (1975b). Sensitivity of blood cell counts in juvenile coho salmon (*Oncorhynchus kisutch*) to stressors including sublethal concentrations of pulpmill effluent and zinc. *J. Fish. Res. Board Can.* **32**, 2357–2364.

McLeay, D. J. (1977). Development of a blood sugar bioassay for rapidly measuring stressful levels of pulpmill effluent to salmonid fish. *J. Fish. Res. Board Can.* **34**, 477–485.

McLeay, D. J. (1979). Potential impact of mutagenicity contained in aquatic discharges on Canadian fisheries. Summary document, B.C. res. project rep. no. 1-05-126. Dep. of Fish. and Oceans, Ottawa, Ont.

McLeay, D. J. and Brown, D. A. (1974). Growth stimulation and biochemical changes in juvenile coho salmon (*Oncorhynchus kisutch*) exposed to bleached kraft pulpmill effluent for 200 days. *J. Fish. Res. Board Can.* **31**, 1043–1049.

McLeay, D. J. and Brown, D. A. (1975). Effects of acute exposure to bleached kraft pulpmill effluent on carbohydrate metabolism of juvenile coho salmon (*Oncorhynchus kisutch*) during rest and exercise. *J. Fish. Res. Board Can.* **32**, 753–760.

McLeay, D. J. and Brown, D. A. (1979). Stress and chronic effects of untreated and treated beached kraft pulpmill effluent on the biochemistry and stamina of juvenile coho salmon (*Oncorhynchus kisutch*). *J. Fish. Res. Board Can.* **36**, 1049–1059.

McLeay, D. J. and Gordon, M. R. (1977). Leucocrit: A simple hemotological technique for measuring acute stress in salmonid fish, including stressful concentrations of pulpmill effluent. *J. Fish. Res. Board Can.* **34**, 2164–2175.

McLeay, D. J. and Gordon, M. R. (1978a). Field study of the effects to fish of treated, partially treated and untreated kraft pulp mill effluent. B.C. res. project rep. no. 1-05-902. Council of Forest Industries of British Columbia, Vancouver, B.C.

McLeay, D. J. and Gordon, M. R. (1978b). Effect of seasonal photoperiod on acute toxic responses of juvenile rainbow trout (Salmo gairdneri) to pulpmill effluent. J. Fish. Res. Board Can. 35, 1388–1392.

McLeay, D. J. and Gordon, M. R. (1979). Concentrations of Krenite causing acute lethal, avoidance and stress responses with salmon. B.C. research project report no. 1-04-156. B.C. Res., Vancouver, B.C.

McLeay, D. J. and Gordon, M. R. (1980). Short-term sublethal toxicity tests to assess safe levels of environmental contaminants. B.C. res. project rep. no. 1-11-299. Dep. Fish. and Oceans, Environ. Canada, Ottawa, Ont.

McLeay, D. J. and Howard, T. E. (1977). Comparison of rapid bioassay procedures for measuring toxic effects of bleached kraft mill effluent to fish. In: W. R. Parker (Ed.), Proc. of the 3rd aquatic toxicology workshop. Environm. Prot. Sev. tech. rep. no. EPS-5-AR-77-1. Halifax, N.S., pp. 141–155.

McNeish, J. D. and Hatch, R. W. (1978). Stamina tunnel tests on hatchery-reared Atlantic salmon. Prog. Fish-Cult. 40, 116–117.

Michel, C. (1980). A standardized model of experimental furunculosis in rainbow trout (Salmo gairdneri). Can. J. Fish. Aquat. Sci. 37, 746–750.

Middaugh, D. P., Davis, W. R., and Yoakum, R. L. (1975). The response of larval fish, Leiostomus xanthurus, to environmental stress following sublethal cadmium exposure. Contrib. Mar. Sci. 19, 13–19.

Mount, D. I. (1980). Needs of toxicity tests to meet specific regulations. In: C. Hocutt and J. Stauffer (Eds.), Biological monitoring of fish. Lexington Books, Lexington, MA, pp. 33–42.

Nakano, T. and Tomlinson, N. (1967). Catecholamine and carbohydrate concentrations in rainbow trout (Salmo gairdneri) in relation to physical disturbance. J. Fish. Res. Board Can. 24, 1701–1715.

Neish, G. A. (1977). Observations on saprolegniasis of adult sockeye salmon, Oncorhynchus nerka (Walbaum). J. Fish Biol. 10, 513–522.

Noakes, D. L. G. and Leatherland, J. (1977). Social dominance and interrenal cell activity in rainbow trout, Salmo gairdneri (Pisces, Salmonidae). Environ. Biol. Fishes 2, 121–136.

Paladino, F. V., Schubauer, J. P., and Kowalski, K. T. (1980). The critical thermal maximum: A technique used to elucidate physiological stress and adaptation in fishes. Rev. Can. Biol. 39, 115–122.

Park, D. L., Smith, J. R., Matthews, G. M., Harmon, J. R., Achord, S., and Monk, B. H. (1982). Transportation operations and research on the Snake and Columbia Rivers, 1981. Nat. Mar. Fish. Serv. Northwest and Alaska Fish. Center, Seattle, WA. Processed rep., 55 pp.

Pearson, C. M., Clements, P. J., and Yu, D. T. Y. (1978). The effects of corticosteroids on lymphocyte functions. Eur. J. Rheumatol. Inflammation 1, 216–225.

Pessah, E., Wells, P. G., and Schneider, J. R. (1976). Dodecyl sodium sulphate (DSS) as an intralaboratory reference toxicant in fish bioassays. In: G. Craig (Ed.), Proc. of the 2nd aquatic toxicity workshop, Ministry of the Environ., Toronto, Ont., pp. 93–121.

Peters, G. (1979). Zur Interpretation des Begriffs "Stress" beim Fisch. Fisch und Tierschutz, Fisch und Umwelt, Heft. 7. Fischer-Verlag, New York.

Peters, G., Delventhal, H., and Klinger, H. (1980). Physiological and morphological effects of social stress in the eel, (Anguilla anguilla L.). Arch. Fischereiwiss. 30, 157–180.

Peters, G., Delventhal, H., and Klinger, H. (1981). Stress diagnosis for fish in intensive culture systems. In: K. Tiews (Ed.) Aquaculture in heated effluents and recirculation systems, Vol. II. H. Heenemann Gmb H and Co., Berlin, Germany, 666 pp.

Pickering, A. D. (1981). Introduction: The concept of biological stress. In: A. Pickering (Ed.), Stress and fish. Academic Press, London, pp. 1–9.

Pickford, G. E., Srivastava, A. K., Slicher, A., and Pang, P. K. T. (1971). The stress response in the abundance of circulating leucocytes in the killifish *Fundulus heteroclitus*. *J. Exp. Zool.* **177**, 89–118.

Ramos, F. and Smith, A. C. (1978a). The C-reactive protein test for detection of early disease in fishes. *Aquaculture* **14**, 261–266.

Ramos, F. and Smith, A. C. (1978b). Ketone bodies in fish skin mucus as an indicator of starvation: A prelimary report. *J. Fish Biol.* **12**, 105–108.

Reinert, R. E. and Hohreiter, D. (1984). Adenylate energy charge as a measure of stress in fish. In this volume, Chapter 11.

Rosenthall, H. and Alderdice, D. F. (1976). Sublethal effects of environmental stressors, natural and pollutional, on marine fish eggs and larvae. *J. Fish. Res. Board Can.* **33**, 2047–2065.

Ryan, P. M. and Harvey, H. H. (1977). Growth of rock bass, *Ambloplites rupestris*, in relation to the morphoedaphic index as an indicator of an environmental stress. *J. Fish. Res. Board Can.* **34**, 2079–2088.

Sandercock, F. K. and Stone, E. T. (1982). A progress report on the effect of rearing density on subsequent survival of Capilano coho. In: R. Neve and B. Melteff (Eds.), *Proc. N. Amer. aquaculture symp.* Sea grant rep. 82-2. Univ. of Alaska, Fairbanks, AK, pp. 82–90.

Schreck, C. B. (1981). Stress and compensation in teleostean fishes: Response to social and physical and factors. In: A. Pickering (Ed.), *Stress and fish*. Academic Press, London, pp. 295–321.

Selye, H. (1950). Stress and the general adaptation syndrome. *Brit. Med. J.* **1**, 1383–1392.

Selye, H. (1973). The evolution of the stress concept. *Amer. Sci.* **61**, 692–699.

Shuter, B. J. and Koonce, J. F. (1977). A dynamic model of the western Lake Erie walleye (*Stizostedion vitreum vitreum*) population. *J. Fish. Res. Board Can.* **34**, 1972–1982.

Shuter, B. J., MacLean, J. A., Fry, F. E. J., and Regier, H. A. (1980). Stochastic simulation of temperature effects on first-year survival of smallmouth bass. *Trans. Amer. Fish. Soc.* **109**, 1–34.

Silbergeld, E. K. (1975). Blood glucose: A sensitive indicator of environmental stress in fish. *Bull. Environ. Contam. Toxicol.* **11**, 20–25.

Sindermann, C. J. (1980). The use of pathological effects of pollutants in marine environmental monitoring programs. In: A. McIntyre and J. Pearce (Eds.), *Biological effects of marine pollution and the problems of monitoring*. Int. Council for the Exploration of the Sea, Copenhagen, Denmark, Rapp. P.-v. reun. cons. int. explor. mer, vol. 179, pp. 129–134.

Singley, J. A. and Chavin, W. (1975). The adrenocortical-hypophyseal response to saline stress in the goldfish, *Carassius auratus* L. *Comp. Biochem. Physiol.* **51A**, 749–756.

Sinnhuber, R. O., Hendrix, J. D., Wales, J. H., and Putnam, G. B. (1977). Neoplasms in rainbow trout, a sensitive animal model for environmental carcinogenesis. *Ann. N.Y. Acad. Sci.* **298**, 389–408.

Skud, B. E. (1982). Dominance in fishes: The relation between environment and abundance. *Science* **216**, 144–149.

Smith, A. C. and Ramos, F. (1976). Occult hemoglobin in fish skin mucus as an indicator of early stress. *J. Fish Biol.* **9**, 537–541.

Smith, A. C. and Ramos, F. (1980). Automated chemical analyses in fish health assessment. *J. Fish Biol.* **17**, 445–450.

Smith, H. D. and Margolis, L. (1970). Some effects of *Eubothrium salvelini* on sockeye salmon, *Oncorhynchus nerka*, in Babine Lake, British Columbia. *J. Parasitol.* **56**, 321–322.

Smith, L. S. (1982a). *Introduction to fish physiology*. TFH Publications, Neptune, NJ, 352 pp.

Smith, L. S. (1982b). Decreased swimming performance as a necessary component of the smolt migration in salmonids of the Columbia River. *Aquaculture* **28**, 153–161.

Smith, L. S. and Newcomb, T. W. (1970). A modified version of the Blazka respirometer and exercise chamber for large fish. *J. Fish. Res. Board Can.* **27**, 1321–1324.

Smith, M. J. and Heath, A. G. (1979). Acute toxicity of copper, chromate, zinc, and cyanide to freshwater fish: Effect of different temperatures. *Bull. Environ. Contam. Toxicol.* **22**, 113.

Snieszko, S. F. (1974). The effects of environmental stress on outbreaks of infectious diseases of fishes. *J. Fish Biol.* **6**, 197–208.

Soivio, A., and Nikinmaa, M. (1981). The swelling of erythrocytes in relation to the oxygen affinity of the blood of the rainbow trout, *Salmo gairdneri* Richardson. In: A. Pickering (Ed.), *Stress and fish*. Academic Press, London, pp. 103–120.

Specker, J. L. and Schreck, C. B. (1980). Stress responses to transportation and fitness for marine survival in coho salmon (*Oncorhynchus kisutch*) smolts. *Can. J. Fish. Aquat. Sci.* **37**, 765–769.

Sprague, J. B. (1971). Measurement of pollutant toxicity to fish—III Sublethal effects and "safe" concentrations. *Water Res.* **5**, 245–266.

Srivastava, A. K. and Agrawal, U. (1977). Involvement of pituitary-interrenal axis and cholinergic mechanisms during the cold-shock leucocyte sequence in a fresh water tropical teleost, *Colisa fasciatus*. *Arch. Anat. Microsc.* **66**, 97–108.

Strange, R. J., Schreck, C. B., and Golden, J. T. (1977). Corticoid stress response to handling and temperature in salmonoids. *Trans. Amer. Fish. Soc.* **106**, 213–218.

Strange, R. J., Schreck, C. B., and Ewing, R. D. (1978), Cortisol concentrations in confined juvenile chinook salmon (*Oncorhynchus tshawytscha*). *Trans. Amer. Fish. Soc.* **107**, 812–819.

Sullivan, J. F., Atchison, G. J., Kolar, D. J., and McIntosh, A. W. (1978). Changes in the predator-prey behavior of fathead minnows (*Pimephales promelas*) and largemouth bass (*Micropterus salmoides*) caused by cadmium. *J. Fish. Res. Board Can.* **35**, 446–451.

Vigers, G. A. and Maynard, A. W. (1977). The residual oxygen bioassay: A rapid procedure to predict effluent toxicity to rainbow trout. *Water Res.* **11**, 343–346.

Waiwood, K. G. and Beamish, F. W. H. (1978). Effects of copper, pH and hardness on the critical swimming performance of rainbow trout (*Salmo gairdneri* Richardson). *Water Res.* **12**, 611–619.

Wales, J. H., Sinnhuber, R. O., Hendricks, J. D., Nixon, J. E., and Eisele, T. A. (1978). Aflatoxin B$_1$ induction of hepatocellular carcinoma in the embryos of rainbow trout (*Salmo gairdneri*). *J. Nat. Cancer Inst.* **60**, 1133–1139.

Walters, G. R. and Plumb, J. A. (1980). Environmental stress and bacterial infection in channel catfish, *Ictalurus punctatus* Rafinesque. *J. Fish Biol.* **17**, 177–185.

Webb. P. W. and Brett, J. R. (1973). Effects of sublethal concentrations of sodium pentachlorophenate on growth rate, food conversion efficiency, and swimming performance in underyearling sockeye salmon (*Oncorhynchus nerka*). *J. Fish. Res. Board Can.* **30**, 499–507.

Wedemeyer, G. (1973). Some physiological aspects of sublethal heat stress in the juvenile steelhead trout (*Salmo gairdneri*) and coho salmon (*Oncorhynchus kisutch*). *J. Fish. Res. Board Can.* **30**, 831–834.

Wedemeyer, G. (1976). Physiological response of coho salmon (*Oncorhynchus kisutch*) and rainbow trout (*Salmo gairdneri*) to handling and crowding. *J. Fish. Res. Board Can.* **33**, 2699–2702.

Wedemeyer, G. and Yasutake, W. T. (1977). Clinical methods for the assessment of the effects of environmental stress on fish health. U.S. Fish and Wildl. Serv. Tech. Pap. 89. Washington, D.C., 18 pp.

Wedemeyer, G. and McLeay, D. (1981). Methods for determining the tolerance of fishes to environmental stressors. In: A. Pickering (Ed.), *Stress and fish*. Academic Press, London, pp. 247–268.

Wedemeyer, G. A., Meyer, F. P., and Smith, L. (1976). *Environmental stress and fish diseases*. TFH Publications, Neptune, NJ, 200 pp.

Wedemeyer, G. A., Saunders, R. L., and Clarke, C. C. (1980). Environmental factors affecting smoltification and early marine survival of anadromous salmonids. *U.S. Nat. Mar. Fish. Serv. Mar. Fish. Rev.* **42**, 1–14.

Wohlschlag, D. E. and Wakeman, J. M. (1978). Salinity stresses, metabolic responses, and distribution of the coastal spotted seatrout, *Cynoscion nebulosus*. Contrib. Mar. Sci. 21. Univ. of Texas Mar. Sci. Inst., Galveston, TX, pp. 172–185.

Zaugg, W. S. and McLain, L. R. (1972). Changes in gill adenosine-triphosphatase activity associated with parr-smolt transformation in steelhead trout, coho, and spring chinook salmon. *J. Fish. Res. Board Can.* **29**, 161–171.

Zaugg, W. S. and Wagner, H. H. (1973). Gill ATPase activity related to parr-smolt transformation and migration in steelhead trout (*Salmo gairdneri*): Influence of photoperiod and temperature. *Comp. Biochem. Physiol.* **45**, 955–965.

13

THE EFFECTS OF POLLUTANTS AND CONTAMINANTS ON STEROIDOGENESIS IN FISH AND MARINE MAMMALS

H. C. Freeman, G. B. Sangalang, and J. F. Uthe

Fisheries Research Branch
Department of Fisheries and Oceans
Halifax, Nova Scotia

1. INTRODUCTION

It is difficult to determine the chronic effects of sublethal quantities of chemical contaminants on an organism, yet these trace quantities are capable of eliminating a species as effectively as a single lethal dose. It is only necessary that a contaminant cause the reproductive capacity of a species to be lowered below maintenance levels or affect a species adversely so its members cannot successfully compete with other organisms in its environment.

Early reports of the adverse effects of contaminants on steroidogenesis in vivo and in vitro in fish and marine mammals (Hill and Fromm, 1968; Lockhart et al., 1972; Sangalang and O'Halloran, 1972; Grant and Mehrle, 1973; Sangalang and Freeman, 1974; Donaldson and Dye, 1975; Freeman and Idler, 1975; Freeman and Sangalang, 1977) have stimulated research on steroid hormone metabolism in fish in search of sensitive indicators of sublethal effects of pollutants.

The steroid hormones present in fish and their metabolic pathways are well-known (Idler and Truscott, 1972; Ozon, 1972). Changes in steroid hormone metabolism can give some insight into physiological changes in fish; in the case of mineralo-corticosteriods, osmoregulation; in the case of androgens and estrogens, reproduction; and in the case of adrenocorticosteroids, stress. These changes in hormone metabolism in fish may be useful as an early warning of a problem caused by contaminants.

The mammalian stress response has been described by Selye (1937) and is characterized by three stages: (1) the alarm reaction; (2) the stage of resistance, where adjustments and accommodations could be made; and (3) the stage of exhaustion, where death eventually occurs. It is desirable to detect pollution problems while they are still reversible, that is, before the stage of exhaustion. One technique by which this can be done is the determination of the effect of pollutants on steroidogenesis in a test animal.

In the present paper we would like to present some examples of studies in our laboratory on the effects of trace quantities of chemical contaminants administered in vivo and in vitro on steroid hormone metabolism in fish and in a marine mammal, the seal.

2. EFFECTS OF CONTAMINANTS ON STEROIDOGENESIS IN FISH

2.1. Cadmium

An altered steroid hormone metabolism was demonstrated (Fig. 1) when brook trout (*Salvelinus fontinalis*) were exposed for 24 hr to soft freshwater containing 25 ppb cadmium (Cd) and then killed 6 days after Cd treatment (Sangalang and O'Halloran, 1973). This figure is an autoradiogram of metabolites produced when testicular incubations of treated and untreated control trout were carried out in the presence of a radioactive steroid precursor, $[^{14}C]$-pregnenolone. It is clear from the radioactive profile that there was a considerable alteration in steroidogenesis in the testes of treated fish. For example, there was 24% less conversion of precursor to

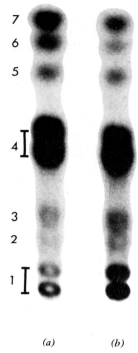

(a)   *(b)*

Figure 1.   Autoradiogram of the thin layer chromatogram (TLC) of one-fifth of dichloromethane extracts of Cd-damaged (*a*) and control (*b*) testes of brook trout incubated with $[4\text{-}^{14}C]$-pregnenolone in vitro. 1–7 indicate positions of radioactive metabolites. Areas 1, 3, and 6 indicate positions of unknown products; areas 2, 4, 5, and 7 were isopolar with 11β-OHT, 11-KT, T, and Preg (pregnenolone), respectively (Sangalang and O'Halloran, 1973).

steroid metabolites by the testes from the treated fish when compared with testes of untreated fish. There was also considerable hemorrhaging in the testes from the Cd-treated fish (Fig. 2). A cross section of the same testes showed that the testicular damage was extensive.

The effects of 1 ppb of Cd in water on 11-ketotestosterone (11-KT) and testosterone (T) in brook trout were determined during the various stages of sexual maturation (Fig. 3) (Sangalang and Freeman, 1974). Blood T and 11-KT levels correlated with the approaching period of functional maturity in control fish and reached peaks of 0.25 and 2.68 μg/100 ml of plasma, respectively, and gradually declined with onset of testicular regression. 11-KT levels of 2.34–3.96 μg/100 ml of plasma in the Cd-treated fish were higher than in the controls and remained elevated during testicular regression. In the treated fish T levels increased 2 weeks later than the controls and also remained elevated throughout. The testes of Cd-treated fish regressed at least 2 weeks earlier than the controls. The results suggested an impairment in the clearance and/or utilization of T and 11-KT by the Cd-treated fish.

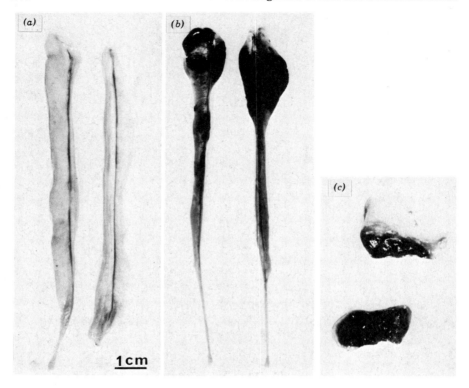

Figure 2. (*a*) Gross appearance of normal mature testes of brook trout. (*b*) Damaged testes of fish exposed to 25 ppb Cd for 24 hr and killed on day 7. (*c*) Top—external appearance of the anterior section of the left testis in (*b*). Bottom—cut surface showing extensive discoloration suggestive of massive hemorrhage (Sangalang and O'Halloran, 1973).

An autoradiogram of the thin layer chromatogram (TLC) of dichloromethane extracts from normal testes of brook trout incubated with $[4\text{-}^{14}C]$-pregnenolone in the presence of 0, 10, 100, and 1000 μg Cd/g of testes (Fig. 4) (Sangalang and O'Halloran, 1973) demonstrated that the altered steroid hormone metabolism increased as the concentration of Cd increased. The concentrations of the compounds chromatographing with T, 11β-hydroxytestosterone (11β-OHT), and unreacted $[4\text{-}^{14}C]$-pregnenolone increased with the concentration of Cd while the concentration of the compound isopolar with 11-KT decreased.

2.2. PCBs

High levels (10 μg/g wet weight) of polychlorinated biphenyls (PCB) occur in livers of Atlantic cod (*Gadus morhua*) in certain areas of the North Atlantic ocean. PCB and other chlorinated organic compounds are known to interfere with steroidogenesis and reproduction in a number of species, (Porter and Wiemeyer, 1969; Freeman and Idler, 1975). Therefore, it was important to determine the effects of PCB on steroidogenesis in the Atlantic cod.

Atlantic cod in groups of 12–15 fish were fed three times weekly, at 3–4% body weight, a diet of 10 g pieces of herring containing either 0, 1, 5, 10, 25, or 50 μg/g of gelatin-encapsulated PCB (as Aroclor 1254) (Freeman et al., 1982). Blood samples were taken regularly for steroid hormone analysis at the start (0 time) through to the termination of the feeding experiment (91 days). Tissues were also taken at the end of the experiment for PCB analysis and histological examination. (The results of the histology are not included in this report.) The capacities of the head kidneys and gonads of each sex to synsthesize steroids hormones in vitro were determined at each treatment level using [4-^{14}C]-progesterone as a precursor.

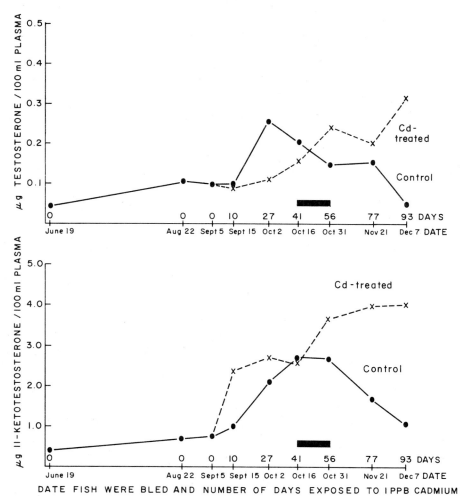

Figure 3.   Effect of 1 ppb Cd on peripheral blood plasma levels of T and 11-KT in brook trout. Solid bar indicates peak period of functional maturity (Sangalang and Freeman, 1974).

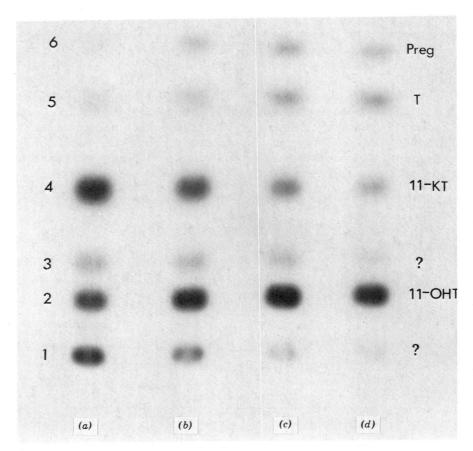

Figure 4. Autoradiogram of the thin layer chromatogram of one-tenth of dichloromethane extracts of normal testes of brook trout incubated with [4-^{14}C]-pregnenolone in vitro. (a)–(d) represent 1 g testes samples incubated with 0, 10, 100, and 1000 μg Cd, respectively.

2.2.1. Kidney Steroid Hormone Metabolism

The autoradiogram (Fig. 5) of the TLC of the female cod head kidney (adrenal homologue) incubation extracts showed altered steroid hormone metabolism at all PCB feeding levels. In the control group there were only trace quantities of bio-synthesized compounds isopolar with cortisol (F) and corticosterone (B), while in the treated groups, especially at the 5 and 10 μg/g PCB feeding levels, there were high yields of these compounds as well as additional metabolites, the greatest number and yields of which occurred at the 10 μg/g feeding level. This biochemical change at the 10 μg/g feeding level may be comparable to the "stage of resistance" described by Selye (1937). At the 50 μg/g feeding level (Fig. 5), there was only one major metabolite and a considerable amount of the unconverted precursor,

[^{14}C]-progesterone, remained. A marked degree of impairment of steroid hormone metabolism at this level suggested that these fish were in the "stage of exhaustion" and that they would have eventually died.

In the male head kidney incubations there was also a great difference in the numbers and quantities of products biosynthesized at all PCB feeding levels (Fig. 6) compared to the controls (0 μg PCB/g). Similar to that observed in the female head kidney incubation, there was also an increased number of products formed at the 25 and 10 μg PCB/g feeding levels in the male.

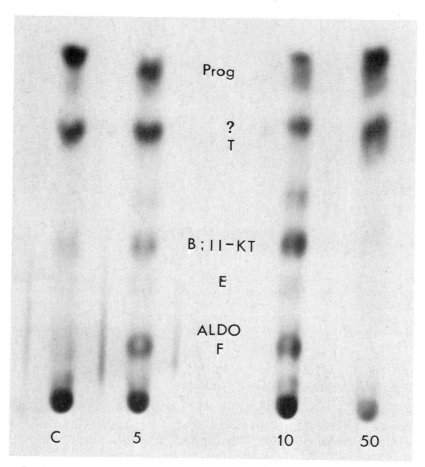

Figure 5. Autoradiogram of the TLC of one-tenth of dichloromethane extracts of female cod head kidneys incubated with [4-^{14}C]-progesterone in vitro. Head kidney tissues were from cod fed with 0 (C, control) and 5, 10, and 50 μg Aroclor 1254/g. The locations of nonradioactive carrier steroids are indicated: F(cortisol), E(cortisone), ALDO(aldosterone), 11-KT(11-ketotestosterone), B(corticosterone), T(testosterone), and Prog(progesterone) (Freeman and Sangalang, 1977).

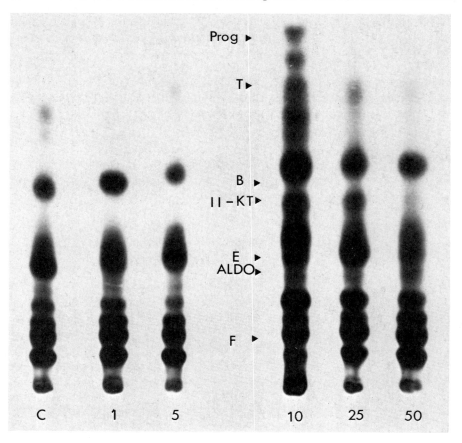

Figure 6. Autoradiogram of the TLC of one-tenth of dichloromethane extracts of male cod head kidneys incubated with [4-^{14}C]-protesterone in vitro. Tissues were from cod fed with 0 (C, control), 1, 5, 10, 25, and 50 μg Aroclor 1254/g herring for 91 days. The positions of nonradioactive carrier steroids are indicated and compared with the positions of the radioactive metabolites: F(cortisol), E(cortisone), ALDO(aldosterone), B(corticosterone), 11-KT, T, and Prog(progesterone) (Freeman and Sangalang, 1977).

2.2.2. Testes Steroid Hormone Metabolism

The autoradiogram of the extracts (Fig. 7) of the incubations of the testes showed a number of PCB-induced changes compared with control fish. In the control group (C), there were two radioactive areas isopolar with two male sex steroids, testosterone (T) and 11-ketotestosterone (11-KT) (Fig. 7). The other area (P) is isopolar with the substrate, (^{14}C)-progesterone. At the 1 μg/g feeding level there was little change from the control in products formed isopolar with T and 11-KT. The concentrations of PCB in the testes of the control cod was about the same as that in the testes from the cod in the 1 μg/g feeding group. At the 5 μg/g feeding level there was great stimulation in biosynthesis accompanied by increased production of metabolites isopolar with T and 11-KT and the appearance of a high concentration of an unknown metabolite (?) more polar than T. This is similar to what occurred in other species where organochlorine compounds induced activities of enzymes capable of hydroxylating steroid hormones with the formation of polar compounds (Peakall, 1967; Litterst and Van Loon, 1972; Platonow et al., 1972).

At the 10 and 50 µg/g feeding levels, where the mean concentrations of PCB in the testes were 0.45 and 2.9 µg/g, respectively, there was an inhibition of synthesis of products isopolar with 11-KT. The unknown metabolite (?) was also biosynthesized at a lower level than at the 5 µg/g level but its concentration was greater than in the control and the 1 µg/g feeding levels. Thus PCB treatment in vivo caused an altered steroid hormone metabolism in vitro in cod testes.

2.2.3. Plasma Androgens

Plasma T and 11-KT levels in the control fish showed a positive correlation with the approach of functional sexual maturity (Figs. 8 and 9). The androgen levels were low in the early stages of the experiment, and as the fish approached sexual maturity, the levels peaked.

T and 11-KT production were both stimulated in vivo at the 1 and 5 µg/g feeding levels. Above the 5 µg/g feeding level, plasma T and 11-KT levels were lower than the controls with an apparently lowered or decreased rate of production and a late arrival or absence of a peak period. Some fish in the higher PCB feeding groups had plasma 11-KT and T levels that remained low throughout the PCB

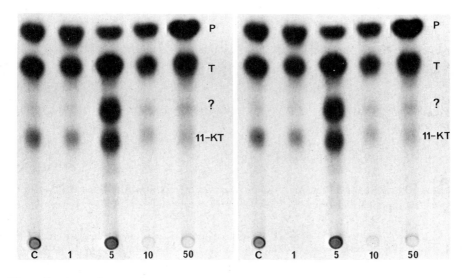

Figure 7.   Autoradiogram of the TLC of dichloromethane extracts of cod testes incubated with [4-¹⁴C]-progesterone in vitro. Tissues were from cod fed with 0 (C, control), 1, 5, 10, and 50 µg Aroclor 1254/g herring for 91 days. The positions of nonradioactive carrier steroids 11-KT, T, and P(progesterone) are compared with those of the radioactive metabolites (Freeman and Sangalang, 1977).

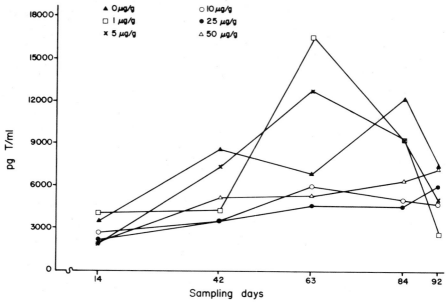

Figure 8. Levels of T in the blood (plasma) of male cod fed without and with Aroclor 1254 (Freeman and Sangalang, 1977).

feeding experiments, indicating interference of PCB with androgen production and/ or utilization. It is significant that these fish had high levels of PCB in their testes with the highest being 3.72 µg/g (wet weight).

3.   EFFECTS OF MERCURY ON HORMONE METABOLISM IN HARP SEALS

3.1.   Kidney Steroid Hormone Metabolism

Figure 10 is an autoradiogram of the radioactive metabolites biosynthesized in vitro from [³H]-pregnenolone plus [¹⁴C]-progesterone by adrenals of harp seals (*Pagophilus groenlandicus*) fed a diet of 3–5% body weight of herring containing 0 (control C) and 0.25 mg/kg (treated) of body weight of methyl mercury (MeHg) twice daily (Freeman et al., 1975). It showed that the biosynthesis of cortisol (F) and a compound isopolar with aldosterone (ALDO) at 4 hr was greater in the control than in the MeHg-treated seal, indicating an altered steroid hormone metabolism in the latter. The MeHg-treated seal was apparently unable to regulate its saltwater metabolism and had suffered from edema and died. This suggested that MeHg intefered in the biosynthesis of mineralocorticosteroids, such as aldosterone, by the adrenals.

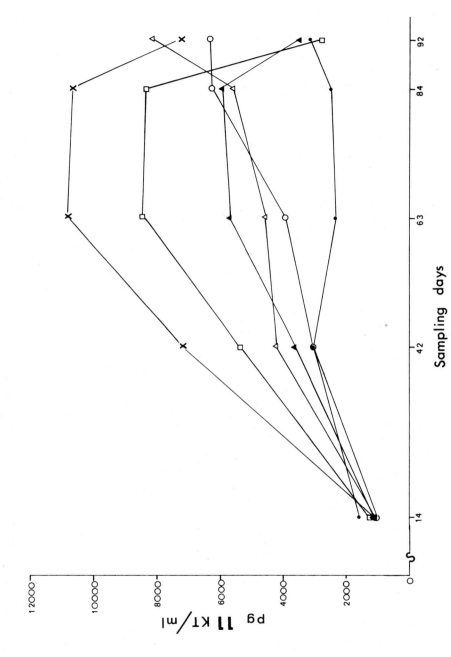

Figure 9. Levels of 11-KT in the blood of male cod fed without and with Aroclor 1254 (Freeman and Sangalang, 1977).

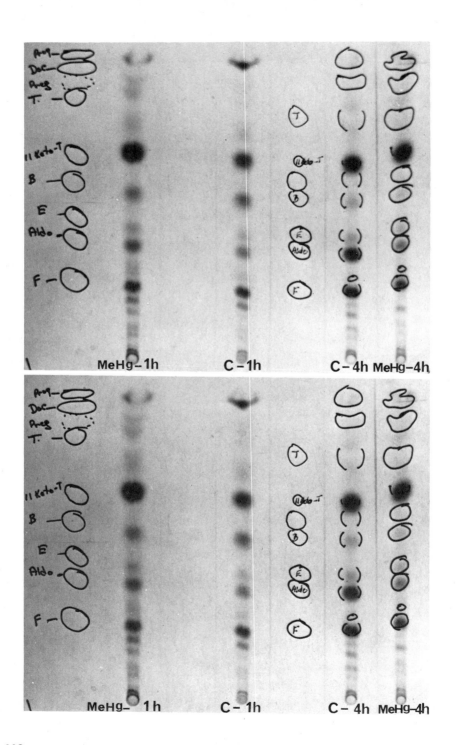

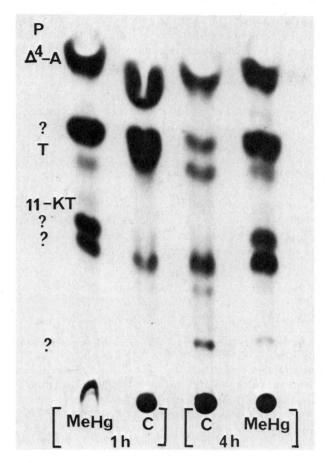

Figure 11. Autoradiogram of the TLC of one-tenth of the neutral fractions of steroid extracts of ovaries from untreated (C) and MeHg-treated harp seal. The tissues were each incubated with [³H]-pregnenolone plus [4-¹⁴C]-progesterone for 1 hr and 4 hr in vitro. Steroid symbols indicate the relative positions of nonradioactive carrier steroids T, 11-KT, and P (progesterone) (Freeman et al., 1975).

3.2. Ovary Steroid Hormone Metabolism

Ovaries from the control and treated seals were also incubated with [³H]-pregnenolone plus [¹⁴C]-progesterone in vitro. The yields of estrogens were less than 1% in both. The autoradiographic profiles of the chromatograms of the estrogen fractions indicated very low radioactivities and it was noted that metabolites isopolar with estradiol—17β and estrone—were visible in the extract from the treated but not from the

Figure 10. Autoradiogram of the chromatogram of one-tenth of the steroid extracts of adrenal tissues from untreated (C) and MeHg-treated harp seal. The tissues were each incubated with [³H]-pregnenolone plus [4¹⁴C]-progesterone for 1 hr and 4 hr in vitro. Symbols indicate the relative positions of ultra violet fluorescent carrier steroids (Freeman et al., 1975).

control. An autoradiogram of the chromatogram of the neutral steroid fractions of the ovaries (Fig. 11) showed that the principal metabolites were in these fractions. The autoradiogram indicated marked differences in biosynthetic patterns between the treated and control seals in both the 1 hr and 4 hr incubations. There was stimulation of metabolites (unknown) slightly more polar than 11-KT and less polar than T in the ovaries from the treated seal when compared with the control. These alterations were, presumably, of relevance to estrogen biosynthesis since neutral C-19 steroids (e.g., T, Δ4-androstenedione) are immediate precursors of estrogens.

4. CONCLUSION

In conclusion, we have shown with the above examples how changes in steroid hormone metabolism may be used as sensitive indicators of chemically induced pollution problems. These changes may be determined by studying the capacities of certain organs (e.g., adrenals and gonads) to biosynthesize critical steroid hormones using radioactive precursors.

REFERENCES

Donaldson, E. M. and Dye, H. M (1975). Corticosteroid concentrations in sockeye salmon (*Oncorhynchus nerka*) exposed to low concentrations of copper. *J. Fish Res. Board Can.* **32**, 533–539.

Freeman, H. C. and Idler, D. R. (1975). The effect of polychlorinated biphenyl on steroidogenesis and reproduction in the brook trout (*Salvelinus fontinalis*). *Can. J. Biochem.* **53**, 666–670.

Freeman, H. C. and Sangalang, G. B. (1977). Changes in steroid hormone metabolism as a sensitive method of monitoring pollutants and contaminants. E.P.S. tech. rep. no. EPS-5-AR-77-1, Halifax, N.S., pp. 123–132.

Freeman, H. C., Sangalang, G. B., Uthe, J. F. and Ronald, K. (1975). Steroidogenesis *in vitro* in the harp seal (*Pagophilus groenlandicus*) without and with methyl mercury treatment *in vivo*. *Environ. Physiol. Biochem.* **5**, 428–439.

Freeman, H. C., Sangalang, G. B., and Flemming, B. (1982). The sublethal effects of polychlorinated biphenyl (Aroclor 1254) diet on the Atlantic cod (*Gadus morhua*). *Sci. Total Environ.* **24**, 1–11.

Grant, B. F. and Mehrle, P. M. (1973). Endrin toxicosis in rainbow trout (*Salmo gairdnerii*). *J. Fish. Res. Board Can.* **30**, 31–40.

Hill, C. W. and Fromm, P. O. (1968). Response of the interrenal gland of rainbow trout (*Salmo gairdnerii*) to stress. *Gen. Comp. Endocrinol.* **11**, 69–77.

Idler, D. R. and Truscott, B. (1972). Corticosteroids in fish. In: D. R. Idler (Ed.), *Steroids in nonmammalian vertebrates*. Academic Press, New York, pp. 127–252.

Litterst, C. L. and Van Loon, E. M. (1972). Enzyme induction by polychlorinated biphenyls relative to known inducing agents (36867). *Proc. Soc. Exp. Biol. Med.* **141**, 765–768.

Lockhart, W. L., Uthe, J. F., Kenny, A. R., and Mehrle, P. M. (1972). Methyl mercury in northern pike (*Esox lucius*), distribution elimination and some biochemical characteristics of contaminated fish. *J. Fish. Res. Board Can.* **29**, 1519–1523.

Ozon, R. (1972). Androgens in fish, amphibians, reptiles and birds. In: D. R. Idler (Ed.), *Steroids in nonmammalian vertebrates*. Academic Press, New York, pp. 329–389.

Peakall, D. B. (1967). Pesticide-induced enzyme breakdown of steroids in birds. *Nature* **216**, 505–506.

Platonow, N. S., Lystrap, R. M., and Geissinger, H. D. (1972). The distribution and excretion of polychlorinated biphenyls (Aroclor 1254) and their effect on urinary gonadal steroid levels of the boar. *Bull. Environ. Contam. Toxicol.* **7**, 358–365.

Porter, R. D. and Wiemeyer, S. N. (1969). Dieldrin and DDT: Effects on sparrow hawk eggshell reproduction. *Science* **165**, 199–200.

Sangalang, G. B. and Freeman, H. C. (1974). Effects of sublethal cadmium on maturation and testosterone and 11-ketotestosterone production *in vivo* in brook trout. *Biol. Reprod.* **11**, 429–435.

Sangalang, G. B. and O'Halloran, M. J. (1972). Cadmium-induced testicular injury and alterations of androgen synthesis in brook trout. *Nature* **240**, 470–471.

Sangalang, G. B. and O'Halloran, M. J. (1973). Adverse effects of cadmium on brook trout testes and on *in vitro* testicular androgen synthesis. *Biol. Reprod.* **9**, 394–403.

Selye, H. (1937). Studies on adaptation. *Endocrinology* **21**, 169–188.

14

ASPECTS OF THE ENDOCRINE STRESS RESPONSE TO POLLUTANTS IN SALMONIDS

Edward M. Donaldson, Ulf H. M. Fagerlund, and J. R. McBride

West Vancouver Laboratory
Fisheries Research Branch
Department of Fisheries and Oceans
West Vancouver, British Columbia

1. INTRODUCTION

This report delineates the means by which the endocrine stress response can be used to detect and quantify the acute and chronic effects of water pollutants on fish. The classical stress response described in mammals by Selye (1950) is essentially an endocrine response to external or internal stressors. The response to stress in the teleost is patterned along the same lines. It is centered around two endocrine systems, the chromaffin tissue, the homologue of the mammalian adrenal medulla, which produces catecholamines (Mazeaud, 1981), and the interrenal tissue, the

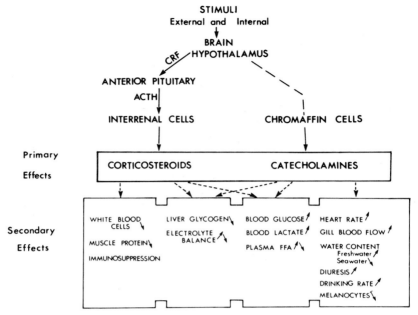

Figure 1. Scheme incorporating current information on the interrelationship between primary and secondary effects of stress in fish (reproduced with permission from Mazeaud et al., 1977).

homologue of the mammalian adrenal cortex, which produces corticosteroid hormones (Donaldson, 1981). The increase in the concentration of these two classes of hormones has been referred to as the primary stress response (Mazeaud et al., 1977). This endocrine activity in turn triggers a number of changes in biochemical and physiological parameters termed secondary responses (Fig. 1) (Mazeaud et al., 1977). The determination of plasma catecholamine concentration provides a potential technique for assessment of the primary stress response to pollutants; however, the scope of this paper is restricted to a discussion of the use of the other primary stress parameter, namely the interrenal response, for evaluating the effect of pollutants on fish.

2.   HYPOTHALAMUS-PITUITARY-INTERRENAL AXIS

The hypothalamus-pituitary-interrenal (HPI) axis is the control mechanism for interrenal activity. In response to a stressor and through the mediation of the central nervous system, an as yet unidentified corticotropin releasing factor (CRF) (Vale et al., 1981) passes from the hypothalamus into the pituitary gland, where it stimulates the release of corticotropin (ACTH) from the lead-hematoxylin-stainable (corticotrope) cells of the pars distalis. ACTH carried by the blood stimulates the synthesis and release of cortisol from the interrenal tissue. This tissue is associated with the cardinal veins in the head kidney. The corticosteroid concentration in the blood in turn may influence ACTH release through negative feedback at the hypothalamic and/or pituitary level as has been demonstrated by administration of dexamethasone (Donaldson and McBride, 1967). This acute response of the HPI axis does not produce changes in the associated endocrine glands that can be immediately detected by observation with the light microscope. Only after the relatively prolonged influence of a stressor are structural alterations noted in these glands, usually proceeding from enlargement of the cells (hypertrophy) to an increase in the number of cells (hyperplasia). The presence of a pituitary-interrenal axis in the salmonid has been demonstrated by hypophysectomy, feedback inhibition using the synthetic corticosteroid dexamethasone, injection of mammalian and piscine ACTH preparations, and treatment with an 11β-hydroxylation inhibitor (for review see Donaldson, 1981).

The activity of the HPI axis can potentially be evaluated at each of its three levels (Fig. 2). The assessment of hypothalamic and pituitary activity involves histological and biochemical procedures that are relatively difficult to perform (Donaldson, 1981). Consequently such assessment is mainly carried out for the purpose of studying endocrine mechanisms. Histological and biochemical techniques for assessing activity at the third level, the interrenal, are sensitive, reproducible, and while requiring some expertise, are not difficult to master. The histological assessment is most conveniently carried out by measuring the diameter of a sufficient

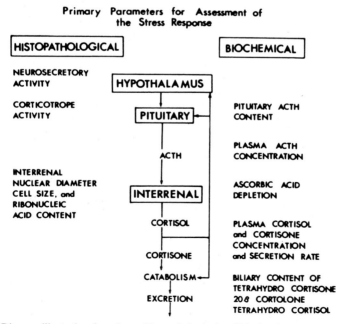

Figure 2.   Diagram illustrating the primary histopathological and biochemical parameters that may be used for the assessment of the activity of the HPI axis (reproduced with permission from Donaldson, 1981).

number of cell nuclei. In our experience a total of 25 determinations obtained from 3–5 groups of cells per fish is adequate to reflect the state of the tissue. In cases where the nuclei are not circular in shape an average value of the measurements for each axis is used. The quantitative assessment of numbers of cells (hyperplasia) is more difficult (McBride and van Overbeeke, 1969). Biochemical assessment at the interrenal level may be approached by measuring the following: (1) the concentration of cortisol (the main biologically active corticosteroid hormone synthesized by the interrenal in most teleosts) in peripheral blood by competitive protein binding (Fagerlund, 1970; Donaldson and Dye, 1975) or by radioimmunoassay (McBride et al., 1979); (2) the rate of cortisol synthesis by in vitro incubation of interrenal tissue with a radioactive precursor such as progesterone (Leloup-Hatey et al., 1981); and (3) cortisol secretion rate by determining the clearance rate of radioactive cortisol from plasma in vivo (Donaldson and Fagerlund, 1968, 1969, 1972). The first of these biochemical methods is the most appropriate to the study of stress responses to pollutants because of the sensitivity and relative simplicity of current radioimmunoassay techniques. However, determination of the effect of pollutants on plasma cortisol concentrations is complicated by the sensitivity of the interrenal to the effects of handling. This necessitates that blood samples be obtained either from animals in situ with a minimum delay (5 min), for example, by spear fishing (Fagerlund, 1967), or from rested animals exposed to pollutants under controlled laboratory conditions. In our studies acclimation of individual or pairs of fish is accomplished by holding the fish for 72 hr in black painted glass aquaria with a black plexiglass cover. This results in low basal cortisol concentrations in the fish and has facilitated the detection of stress responses to sublethal levels of a variety of pollutants. Other techniques such as determination of cortisol secretion rate by single injection or infusion to constant specific activity of labeled cortisol would be of value for the in-depth evaluation of stress responses.

3. THE INTERRENAL RESPONSE TO POLLUTANTS

The HPI axis has been used to measure the stress induced in salmonids by culture procedures,disease, and water pollutants (Donaldson, 1981). Only the latter type of stressor is discussed here. Many of humanity's domestic and primary and secondary industrial activities have the potential to affect water quality for salmonids and other teleosts. Investigations have shown that several categories of pollutants including heavy metals, pesticides, resin acids, ammonia, landfill leachate, and municipal sewage, when present at appropriate concentrations, induce a stress response in salmonids that can be detected by evaluation of the activity of the HPI axis. It should be noted that the use of the HPI axis as an index of the stress response to xenobiotics is a relatively new concept and as a consequence, the number of pollutants tested to date is small. In cases where a literature search indicates that a particular pollutant has not been assessed for an HPI axis response, a laboratory pre-screening is necessary in order to determine level of susceptibility.

Measurement of interrenal nuclear diameters provides an effective means of evaluating the response of the HPI axis to the chronic presence of stressors (Fig. 3), while measurement of plasma cortisol effectively evaluates the response to acute stressors. For these reasons the two methods are particularly effective when used together. Histological assessment can also be applied in very small fish where the lack of adequate blood volume prevents a plasma cortisol evaluation.

To date, we have observed increases in interrenal nuclear diameter and/or plasma cortisol concentration in salmonids exposed to dehydroabietic acid, a component of kraft pulp mill effluent (Dye and Donaldson, 1974), copper (Donaldson and Dye, 1975), sanitary landfill leachate (McBride et al., 1979), ammonia (Donaldson, Dye, and Haywood, 1980, unpublished, cited in Donaldson, 1981), the butoxyethanol ester of 2,4-dichlorophenoxyacetic acid (BEE 2,4-D) (McBride et al., 1981), and municipal sewage (Fagerlund and McBride, 1982, unpublished).

The time course of cortisol response to a pollutant stressor can follow several distinct patterns. In salmon exposed to 635 ppb copper, cortisol concentrations increased rapidly over an 8 hr period and all fish died before the 24 hr sample. On the other hand, at 63.5 ppb the cortisol concentration peaked at 2 hr and declined to 8 hr, suggesting an adaptation response. It then rose again at 24 hr. At 6.35 ppb

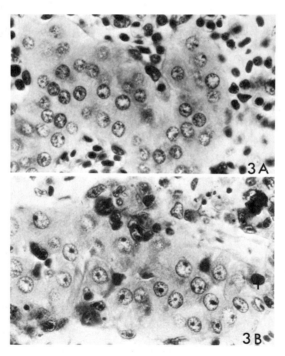

Figure 3. (A) Interrenal cells of yearling coho salmon (control). Cells are closely packed and the nucleoli are inconspicuous. Mean nuclear diameter 5.74 ± 0.21 u. H. & E., 250×. (B) Hypertrophied interrenal cells in yearling coho exposed to a marine waste sludge for 2 days. Note prominent nucleoli. Mean nuclear diameter 6.87 ± 0.32 u. H. & E., 250×.

there was a transitory increase in cortisol, which declined after 4 hr and returned to control levels at 24 hr (Donaldson and Dye, 1975).

Interrenal hypertrophy occurs more gradually than the corticosteroidogenic response. In rainbow trout (Fig. 4) exposed to 5% sanitary landfill leachate there was a significant increase in interrenal nuclear diameter at 2 days (McBride et al., 1979). At 7 days the response was greatest in those fish exposed to 5% leachate and was also significant in fish exposed to 0.5% leachate. The 96 hr LC_{50} for this leachate ranged between 5.8 and 7.5%.

Interestingly, in each of the above histological assessments where a stress response was indicated, the alteration involved a hypertrophy of the interrenal nuclei. Atrophy, or a reduction in nuclear size has, however, been noted in response to certain stressors (acid exposure, Mudge et al., 1977; anoxia, osmotic and temperature stress, Weatherly, 1963).

To our knowledge, studies to delineate the minimum time gap between application of the stressor and the appearance of the first detectable alterations in interrenal tissue structure have not been conducted. Some information regarding the time course of interrenal response was provided by Mudge et al. (1977), who reported structural changes in this tissue of acid-exposed trout after 3–24 hr exposure. In this study a transitory elevation in plasma cortisol levels was noted during the first few hours of exposure, followed by a reduction to control levels after 24 hr of exposure. Allan (1971) reported an elevation in plasma cortisol levels in juvenile coho salmon during the initial 4 days of exposure to cold temperatures, but interrenal nuclear hypertrophy was not noted until after 14 days of exposure. In our experience

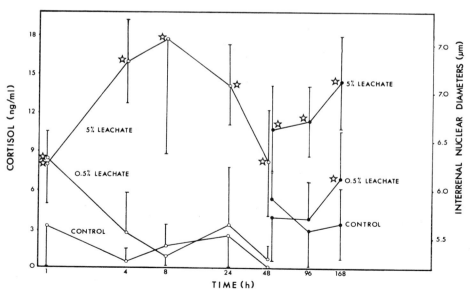

Figure 4. Plasma cortisol concentrations and interrenal nuclear diameters (mean ± SD) of rainbow trout exposed to landfill leachate. Values significantly different ($P < .05$) from controls are denoted by asterisks.

the minimum time lapse between application of a stressor and the appearance of interrenal structural alterations identifiable with the light microscope is, generally, in the order of 24–48 hr.

While a number of toxic xenobiotics from several distinct classes of chemicals have been shown to induce a stress response in the HPI axis, there are some individual chemicals, for example, cadmium (Schreck and Lorz, 1978), that do not appear to have this effect. However, this cadmium study was characterized by high and variable cortisol concentrations (30–60 ng/ml) in the control fish, which may have masked a moderate corticosteroid response.Despite the existence of this and possibly other exceptions, the monitoring of the HPI axis to detect acute or chronic sublethal stress responses in fish shows considerable promise. This is especially true where fish are exposed to a complex mixture of pollutants, for example, landfill leachate or municipal sewage, or are exposed simultaneously to two or more sublethal stressors, for example, temperature and chemical, because the HPI stress response integrates the effect of the individual components.

Exposure to several acute or chronic low-level stressors at the same time or in relatively quick succession may result in a cumulative response that is pathological. The response of fish to such conditions involving multiple stressors has received only scant attention. A special case of response to multiple stressors involves the question of increased predisposition to disease of previously stressed animals. Knittel (1980) noted that exposure of steelhead trout to concentrations of zinc as high as 98 ppb did not increase susceptibility to *Yersinia ruckeri* infection, but exposure to cadmium or copper at concentrations of 0.9 ppb and 2.0 ppb, respectively, did increase the susceptibility of the trout to the pathogen. The relationship between stress and increased susceptibility to disease in fish, however, remains largely unknown. Obviously, with the threat of increasing pollution of the aquatic environment the need to address the latter topic is urgent.

4.   CONCLUSIONS

Results obtained to date using the HPI axis to assess the response of fish to pollutants suggest two avenues of approach for the utilization of this technique. First of all, both plasma cortisol concentration and interrenal nuclear diameter can be used under controlled laboratory conditions to assess whether or not individual xenobiotics or complex mixtures of pollutants induce an endocrine stress response over a range of concentrations. Secondly, the measurement of interrenal nuclear diameter has potential as a field technique for assessment of stress in teleosts. This cytological response is particularly suited to field studies because it is not immediately influenced by the process of fish capture and sampling. The use of plasma cortisol concentration in the field may be possible in those situations where wild fish can be rapidly sampled or where test fish can be carefully exposed in live holding cages. The essential element here is the need for adequate controls held under similar conditions in nonpolluted water.

REFERENCES

Allan, G. D. (1971). Measurement of plasma cortisol and histometry of the interrenal gland of juvenile, pre-smolt coho salmon (*Oncorhynchus kisutch* Walbaum) during cold temperature acclimation. M.Sc. thesis, Univ. of British Columbia, Vancouver, 63pp.

Donaldson, E. M. (1981). The pituitary-interrenal axis as an indicator of stress in fish. In: A. D. Pickering (Ed.), *Stress and fish*. Academic Press, London, Chap 2, pp. 11–47.

Donaldson, E. M. and Dye, H. M. (1975). Corticosteroid concentrations in sockeye salmon (*Oncorhynchus nerka*) exposed to low concentrations of copper. *J. Fish. Res. Board Can.* **33**, 533–539.

Donaldson, E. M. and Fagerlund, U. H. M. (1968). Changes in the cortisol dynamics of sockeye salmon (*Oncorhynchus nerka*) resulting from sexual maturation. *Gen. Comp. Endocrinol.* **11**, 552–561.

Donaldson, E. M. and Fagerlund, U. H. M. (1969). Cortisol secretion rate in gonadectomized female sockeye salmon (*Oncorhynchus nerka*): Effects of estrogen and cortisol treatment. *J. Fish. Res. Board Can.* **26**, 1789–1799.

Donaldson, E. M. and Fagerlund, U. H. M. (1972). Corticosteroid dynamics in Pacific salmon. Sixth int. symp. on Comp. Endocrinol., Banff, June 13–19, 1971. *Gen. Comp. Endocrinol. Suppl.* **3**, 254–265.

Donaldson, E. M. and McBride, J. R. (1967). The effects of hypophysectomy in the rainbow trout *Salmo gairdnerii* (Rich.) with special reference to the pituitary-interrenal axis. *Gen. Comp. Endocrinol.* **9**, 93–101.

Dye, H. M. and Donaldson, E. M. (1974). A preliminary study of corticosteroid stress response in sockeye salmon to a component of kraft pulp mill effluent dehydroabietic acid. Fish. Mar. Ser. Res. Dev. tech. rep. 461, 18 pp.

Fagerlund, U. H. M. (1967). Plasma cortisol concentration in relation to stress in adult sockeye salmon during the freshwater stage of their life cycle. *Gen. Comp. Endocrinol.* **8**, 197–207.

Fagerlund, U. H. M. (1970). Determining cortisol and cortisone simultaneously in salmonid plasma by competitive protein binding. *J. Fish. Res. Board Can.* **27**, 596–601.

Knittel, M. D. (1980). Heavy metal stress and increased susceptibility of steelhead trout (*Salmo gairdneri*) to *Yersinia ruckeri* infection. In: J. G. Eaton, P. R. Parrish and A. C. Hendricks (Eds.), *Aquatic toxicology*. ASTM spec. tech. publ. 707. Amer. Soc. for Test. and Mater., 1980, pp. 321–327.

Leloup-Hatey, J., Hardy, A., Martelly, E., and Fontaine, Y.-A. (1981). Anguilles contaminees par les hydrocarbures apres l'echange de l'Amoco Cadiz. Modifications du fonctionnement de l'interrenal. *C. R. Acad. Sci. Paris* **292**, Ser. 3, 461–464.

Mazeaud, M. M. (1981). Adrenergic responses to stress in fish. In: A. D. Pickering (Ed.), *Stress and fish*. Academic Press, London, Chap. 3, pp. 49–75.

Mazeaud, M. M., Mazeaud, F., and Donaldson, E. M. (1977). Primary and secondary effects of stress in fish: Some new data with a general review. *Trans. Amer. Fish. Soc.* **106**, 201–212.

Mudge, J. E., Dively, J. L., Neff, W. H., and Anthony, A. (1977). Interrenal histochemistry of acid-exposed brook trout, *Salvelinus fontinalis* (Mitchill). *Gen. Comp. Endocrinol.* **31**, 208–215.

McBride, J. R. and van Overbeeke, A. P. (1969). Hypertrophy of the interrenal tissue in sexually maturing sockeye salmon (*Oncorhynchus nerka*) and the effect of gonadectomy. *J. Fish. Res. Board Can.* **26**, 2975–2985.

McBride, J. R., Donaldson, E. M., and Derksen, G. (1979). Toxicity of landfill leachates to underyearling rainbow trout (*Salmo gairdneri*). *Bull. Environ. Contam. Toxicol.* **23**, 806–813.

McBride, J. R., Dye, H. M., and Donaldson, E. M. (1981). Stress response of juvenile sockeye salmon (*Oncorhynchus nerka*) to the butoxyethanol ester of 2,4-D. *Bull. Environ. Contam. Toxicol.* **27**, 877–884.

Schreck, C. B. and Lorz, H. W. (1978). Stress response of coho salmon (*Oncorhynchus kisutch*) elicited by cadmium and copper and potential use of cortisol as an indicator in stress. *J. Fish. Res. Board Can.* **35**, 1124–1129.

Selye, H. (1950). *Stress*. Acta Inc., Montreal, 822 pp.

Vale, W., Spiess, J., Rivier, C., and Rivier, J., (1981). Characterization of a 41 residue ovine hypothalamic peptide that stimulates the secretion of corticotropin and β-endorphin. *Science* **213**, 1394.

Weatherley, A. H. (1963). Thermal stress and interrenal tissue in the perch *Perca fluviatilis* (Linnaeus). *Proc. Zool. Soc. London* **141**, 527–555.

15

COMPARATIVE EPIDEMIOLOGY: THE USE OF FISHES IN ASSESSING CARCINOGENIC CONTAMINANTS

R. A. Sonstegard

Departments of Biology and Pathology
McMaster University
Hamilton, Ontario

J. F. Leatherland

Department of Zoology
University of Guelph
Guelph, Ontario

1. INTRODUCTION

Our biosphere contains a spectrum of natural and synthetic environmental toxicants. Their nature is often unknown and their effects on living organisms undetermined. The effects may be subtle, indirect, and have prolonged latent periods. Furthermore, the health effects of these toxicants may depend on the unique characteristics of the affected organism (e.g., age, sex, and genetic constitution), which alter the nature or extent of effects.

Concern for the effects on human health of environmental toxicants, particularly carcinogens, is increasing rapidly. Their hazards to humans may be assessed in two general ways: epidemiological studies and laboratory testing in vivo and in vitro. Epidemiology has been one of the most productive areas of human cancer research. Epidemiology has conclusively linked environmental factors to cancer in humans (Epstein, 1974; Hammond, 1975). The standard approach is to compare the incidence of each type of neoplasm in humans among various populations, both nationally and internationally. These studies are confounded, however, by a 20–30 year latent period before cancer becomes evident in a given human population. Therefore, the effects of environmental problems we encounter today may not be detectable in human populations as a clinical disease for two or three decades, at which time they may be widespread.

In recent years there have been rapid developments in the knowledge of cellular and subcellular biology and molecular structure and function. The results have been several rapid, cheap, and sensitive in vitro tests for carcinogens (Ames et al., 1975; Hart et al., 1977). These tests provide an invaluable tool for screening compounds for carcinogenesis, but they are designed for testing individual "purified" chemicals. It is increasingly clear that environmental carcinogenesis in humans is not necessarily caused by an individual chemical, but by a complex of chemicals through dietary and atmospheric exposure.

Clearly, a major problem in testing individual compounds is to show how they behave in the presence of other materials. The few cases in which the compounds have been shown to interact involve cigarette smoking combined with exposure to some other chemical (e.g., cigarette smoking and asbestos exposure; cigarette smoking and the inhalation of radon; cigarette smoking and the use of alcohol) (Selikoff et al., 1968; Rothman and Keller, 1972). All of these combinations produce risks that are more than additive. For example, heavy smokers have about two times the risk of nonsmokers of dying of cancer of the oral cavity. The heavy drinker-heavy smoker has about fifteen times the risk as the nonsmoker-nondrinker.

With the possible exception of arsenic and benzene, the known human carcinogens are also carcinogenic in most test animals (Tomatis, 1979). Therefore, in vivo studies of carcinogenesis in laboratory animals are useful for predicting effects in humans, and are essential for assessing the hazard of new and existing chemicals. Similarly, epidemiology is necessary to detect errors arising from in vitro and/or in vivo studies. Thus epidemiology is the first and last line of defense and provides a means of verifying and adjusting the conclusions of animal studies.

Humans share the biosphere with a variety of animals. Often these animals are exposed to levels of pollutants much higher than those to which people are exposed.

Consequently, the pathobiology of these animals is likely to reflect chemical contamination of the ecosystem and may serve as a "sentinel" of environmental health. The premise that underlies this hypothesis is that disease processes in humans and other animals are similar. There is good evidence that this is generally true, since virtually every form of human cancer has an experimental counterpart; it would appear that every multicellular animal (invertebrates and vertebrates alike) is subject to cancer. Although there are differences in susceptibility between species and between individuals of the same species, carcinogenic compounds do affect most test species. Conversely, there is increasing evidence to indicate that toxicants producing cancer in animals will similarly affect humans (Tomatis, 1979).

Therefore, a program of comparative epidemiology (epizootiology) may provide valuable "sentinel organisms" to identify sites and species at risk to environmental carcinogens. The occurrence of each pathological type in selected indicator animals would be compared to various geographical areas. Although biological monitoring of environmental effects is attractive, much work needs to be done to establish its relevance. For example, there is a need for parallel field and laboratory studies to show the biological significance of observed pathologies and to establish cause-effect relationships. However, even if cause-effect studies are not available, well defined models of feral animal populations will be an invaluable asset. Some species may be particularly sensitive to environmental toxicants and may identify environmental problems. They could be an early warning system, possibly enabling toxic chemicals to be identified and eliminated before they have deleterious effects on human populations.

2. TUMOR FREQUENCIES IN FERAL AQUATIC ORGANISMS

The aquatic environment is one of the ultimate recipients of pollution. Through bioconcentration and bioaccumulation, aquatic biota are often subject to the same or higher levels of chemical pollutants as humans. Therefore, tumor frequencies in aquatic organisms may be sensitive indicators of what the future holds for humanity.

We have been exploring the utility of monitoring tumor frequencies in feral fishes as "sentinels" to environmental carcinogens. We have made extensive field epizootiological studies of tumor frequencies in fishes inhabiting polluted and nonpolluted waters of the Great Lakes (Superior, Michigan, Huron, Erie, and Ontario). Epizootics of neoplasia that appear to have an environmental etiology were detected in a variety of species.

Gonadal and pituitary tumors were found in cyprinid fishes (carp, *Cyprinus carpio*; goldfish, *Carassius auratus*; and carp × goldfish hybrids) collected from areas throughout the Great Lakes. Tumor frequencies approached 100% in specific age-sex groups. Field and laboratory studies, together with reviews of museum collections made prior to 1952, suggest that the gonadal tumor is induced by an environmental factor discharged into lake systems (Sonstegard, 1974, 1976, 1977, 1978; Sonstegard et al., 1976; Leatherland and Sonstegard, 1977; Sonstegard and Leatherland, 1979c).

Epizootics of benign skin tumors (papillomas) were found in white sucker (*Catostomus commersoni*) populations throughout the Great Lakes basin. However, elevated tumor frequencies (50.8%) were found in fish collected from areas around an industrial complex on Lake Ontario. In addition to the "clustering" of recorded cases, the tumors were predominantly associated with the lips, which have almost constant contact with bottom sediments. This shift in tumor location compared to other sites on the Great Lakes, together with the clustering recorded, strongly suggest that a factor in bottom sediments enhances the expression of a C-type virus associated with the tumor (Sonstegard, 1973, 1976, 1977, 1978).

Epizootics of thyroid hyperplasia (goiter) were found in Great Lakes coho salmon (*Oncorhynchus kisutch*) with a prevalence as high as 82%. Interlake studies of the occurrence of the goiter in coho salmon between 1972 and 1978 indicate that the prevalence of the goiters is increasing and that environmental goitrogens (possibly pollutants) may be involved (Sonstegard and Leatherland, 1976, 1979c; Moccia et al., 1977, 1978; Leatherland et al., 1978; Sonstegard, 1978; Leatherland and Sonstegard, 1980a, 1981).

Similarly, a number of researchers (Brown et al., 1973; McCain et al., 1977; Pierce et al., 1978; Smith et al., 1979; Black et al., 1980, 1981; Varanasi et al., 1980; Baumann et al., 1982) have reported epizootics of neoplasia in feral fishes, which may have an environmental etiology. A comprehensive symposium dealing with this subject has been published by the New York Academy of Sciences (Kraybill et al., 1977).

The problems inherent in extrapolating these data to human health are obvious. Risk assessment and extrapolation are compromised by the limited information that is available on the pathobiology of feral animals, their different metabolic pathways and rates, and the almost total lack of data on the synergistic and antagonistic interactions of suspect agents. It is apparent, however, the field epizootiological studies may focus subsequent laboratory studies of environmental carcinogenesis.

3. BIOACCUMULATION

The ability of aquatic organisms to bioaccumulate environmental xenobiotics is well established. For example, the concentration gradient of polychlorinated biphenyls (PCBs) from water to Great Lakes salmon is in excess of a million (Reinert and Bergman, 1974; Veith et al., 1977; Norstrom et al., 1978; Reinert, 1979; Sonstegard and Leatherland, 1979a, b). Bioconcentration gradients have aided biochemical measurement and identification of a variety of contaminants and have led to the discovery of compounds unsuspected in the environment. Bioconcentration has led to surveillance programs using indicator species (Norstrom et al., 1978) to accurately monitor chemical changes in the environment and to identify point-sources of pollutants. This surveillance has also identified dangerously contaminated fish species that has led to legislation controlling their consumption.

Classical laboratory cause-effect studies have been a valuable tool in studies of carcinogenesis. This approach has linked the pathobiology of feral hatchery fishes to specific chemical insults, for example, the carcinogenic potential of aflatoxins in induction of trout hepatomas (Halver et al., 1968) and the effects of organochlorines on thyroid physiology of salmonids (Leatherland and Sonstegard, 1978, 1979).

To establish specific cause-effect relationships between environmental pollutants and carcinogenesis in feral aquatic animals, it is necessary to measure contaminant levels in the species of interest. This is particularly important since carcinogens may act synergistically.

We have fed feral fishes that exhibit pathobiological effects to laboratory rats to evaluate the effects of contaminant mixtures carried by the fish. These studies show parallel pathobiologic effect responses in both the fish and the rats fed the fish (Sonstegard and Leatherland, 1979a, b; Leatherland and Sonstegard, 1980b, 1982); thus the factors inducing the responses are tissue-associated. This approach is particularly relevant to human health, as it addresses the health hazards of consuming fish contaminated with an array of chemical pollutants. It is in contrast to the classical toxicological approach, which evaluates the effects of single contaminants. The carcinogenic and mutagenic potential of extracts of aquatic organisms may also be assessed by means of bioassays, for example, by Ames testing (Parry et al., 1976; Sonstegard, 1978).

4. AQUATIC ORGANISMS AS LABORATORY TEST ANIMALS

The potential of aquatic organisms as laboratory test animals for detecting carcinogens is best exemplified by aflatoxins. In the early 1960s, trout hatcheries in the Pacific Northwest were plagued by epizootics of hepatomas. Field studies established that mold-contaminated feed was associated with these epizootics, and feeding trials showed that the etiological agent was aflatoxin, a metabolite of the common mold *Aspergilus flavus*. Rainbow trout are the most sensitive animal tested. Hepatomas were induced with 0.05 ppb aflatoxin in diets (Halver et al., 1968). Recently, Sinnhuber and co-workers have reported that immersion of eyed rainbow trout eggs in an aqueous solution of aflatoxin for 60 min induced a high incidence of liver cancer within one year (Sinnhuber and Wales, 1974; Sinnhuber et al., 1977). This technique has particular relevance as a sensitive bioassay tool for evaluating small quantities of potential carcinogens.

In the past five years a number of cytogenetic studies have detected an increased frequency of chromosomal aberrations in fishes held in polluted waters or exposed to waterborne carcinogens (Kligerman et al., 1975; Badr and El-Dib, 1976; Longwell, 1976a,b; Prein et al., 1978; Longwell and Hughes, 1979). Similarly, sister chromatid studies (in vivo and in vitro) suggest that fish may be particularly sensitive bioassay organisms for mutagens and carcinogens (Kligerman and Bloom, 1978a,b; Barker and Rackham, 1979; Kligerman, 1979).

In Japan, marine algae (*Porphyra tenera*) have been found with tumorlike growths that have been linked to carcinogens in industrial discharges (Ishio et al., 1972, 1977). Similarly, cytogenetic studies showed that ferns (*Osmunda regalis*) growing in waters polluted with industrial wastes had higher frequencies of chromosomal aberrations than did those in unpolluted waters (Klekowski and Berger, 1976).

Until recently, the use of aquatic organisms as in vivo test systems for the detection of carcinogens and/or mutagens has been undeveloped. One advantage of these organisms in carcinogenesis testing is that they may be directly exposed to the test chemical in aquaria or placed in cages directly in the environment.

5. NEW ANIMAL MODELS

There is a need for innovative approaches to determine the etiology of human cancer. The fundamental molecular defects in human cancers are not yet known. Discussions of etiology are generally limited to descriptions of environmental and host factors and to speculation concerning their effects at the molecular level. *Homo sapiens* has existed for only a short period of time, and is the product of millions of years of evolution. Neoplastic-related processes are not new phenomena, as shown by the wide phylogenetic distribution of neoplasia. Studies of the pathobiology that leads to neoplasia in phylogenetically older animals may give valuable insights to how these processes arise and function in humans.

6. NEED FOR ENVIRONMENTAL MONITORING

Pragmatically, the human race is an experimental subject in a large-scale program of environmental toxicology and carcinogenesis that no scientist could ethically conduct. An increase in adverse health effects (e.g., cancer) in human populations in 20–30 years will alert the scientific community to new problems and catalyze research and educational programs to limit and/or eliminate human exposure. Field epizootiology of health effects in well defined "sentinel" organisms may provide valuable pointers to humanity's future problems. Such an approach might lead to the earlier detection and elimination of environmental health hazards to humans and may eventually have an extraordinary impact in reducing the incidence of environmentally related diseases.

REFERENCES

Ames, B. N., McCann, J., and Yamasaki, E. (1975). Methods for detecting carcinogens and mutagens with the *Salmonella* mammalian-microsome mutagenicity test. *Mut. Res.* **31**, 347–364.

Badr, E. A. and El-Dib, S. I. (1976). Effects of water pollution on the cell division cycle and chromosome behavior in *Tilapia* sp. *J. Cell. Biol.* **70**, 189.

Barker, C. J. and Rackham, B. C. (1979). The induction of sister chromatid exchanges in cultured fish cells (*Ameca splendens*) by carcinogenic mutagens. ICES workshop on the problems of monitoring biological effects of pollution in the sea, Beaufort, NC.

Baumann, P. C., Smith, W. D., and Ribick, M. (1982). Hepatic tumor rates and polynuclear aromatic hydrocarbon levels in two populations of brown bullheads (*Ictalurus nebulosus*). In: M. Cooke and A. J. Dennis (Eds.), Sixth Int. Symp. on Physical and Biological Chemistry. Battelle Press, Columbus, OH, pp. 93–102.

Black, J. J., Holmes, M., Dymerski, P. P., and Zapisek, W. F. (1980). Fish tumor pathology and aromatic hydrocarbon pollution in a Great Lakes estuary. In: B. K. Afghan and D. Makay (Eds.), *Hydrocarbons and halogenated hydrocarbons in the aquatic environment*, Environ. Sci. Res., Vol. 16. Plenum Press, New York and London, pp. 559–565.

Black, J. J., Hart, T. F., Jr., and Evans, E. (1981). HPLC studies of PAH pollution in a Michigan trout stream. In: M. Cooke and A. J. Dennis (Eds.), *Polynuclear aromatic hydrocarbons: Chemical analysis and biological fate*. Battelle Press, Columbus, OH, pp. 343–356.

Brown, E. R., Hazdra, J. J., Keith, L., Greenspan, J. B. G., and Beamer, P. (1973). Frequency of fish tumors in a polluted watershed as compared to nonpolluted Canadian waters. *Cancer Res.* **33**, 189–198.

Epstein, S. S. (1974). Environmental determinants of human cancer. *Cancer Res.* **34**, 2425–2435.

Halver, J. E., Ashley, L. M., Smith, R. R. and Wogan, G. N. (1968). Age and sensitivity of trout to aflatoxin B_1. *Fed. Proc. Amer. Soc. Exp. Biol.* **27**, 552.

Hammond, E. C. (1975). The epidemiological approach to the etiology of cancer. *Cancer* **35**, 652–654.

Hart, R. W., Hays, S., Brash, D., Daniel, F. B., Davis, M. T., and Lewis, N. J. (1977). *In vitro* assessment and mechanism of action of environmental pollutants. *Ann. N.Y. Acad. Sci.* **298**, 141–158.

Ishio, S., Kawabe, K., and Tomiyama, T. (1972). Algal cancer and its causes. I. Carcinogenic potencies of water and suspended solids discharged to the River Ohmuta. *Bull. Jap. Soc. Sci. Fish.* **38**, 17–24.

Ishio, S., Chen, J. C., Kawasaki, Y., and Ohba, N. (1977). Cell division of *Gyrodinium* sp. and mitotic delay induced by causal substances of algal tumor and carcinogens. *Bull. Jap. Soc. Sci. Fish.* **43**, 507–516.

Klekowski, E. J., Jr. and Berger, B. B. (1976). Chromosome mutations in a fern population growing in a polluted environment: A bioassay for mutagens in aquatic environments. *Amer. J. Bot.* **63**, 239–246.

Kligerman, A. D. (1979). Induction of sister chromatid exchanges in the central mudminnow following *in vivo* exposure to mutagenic agents. *Mut. Res.* **64**, 205–217.

Kligerman, A. D. and Bloom, S. E. (1978a). Increased rates of sister chromatid exchange in the central mudminnow following *in vivo* exposure to alkylating agents. *Mut. Res.* **53**, 75–76.

Kligerman, A. D. and Bloom, S. E. (1978b). An *in vivo* aquatic system for detecting water-borne mutagens. *Can. J. Gen.* **20**, 447.

Kligerman, A. D., Bloom, S. E., and Howell, W. M. (1975). *Umbra limi*: A model for the study of chromosome aberrations in fishes. *Mut. Res.* **31**, 225–233.

Kraybill, H. F., Dawe, C. J., Harshbarger, J. C., and Tardiff, R. G. (1977). Aquatic pollutants and biologic effects with emphasis on neoplasia. *Ann. N.Y. Acad. Sci.* **298**.

Leatherland, J. F. and Sonstegard, R. A. (1977). Structure and function of the pituitary gland in gonadal tumour-bearing and normal cyprinid fish. *Cancer Res.* **37**, 3151–3168.

Leatherland, J. F. and Sonstegard, R. A. (1978). Lowering of serum thyroxine and triiodothyronine levels in yearling coho salmon, *Oncorhynchus kisutch*, by dietary Mirex or PCB's. *J. Fish. Res. Board Can.* **35**, 1285–1289.

Leatherland, J. F. and Sonstegard, R. A. (1979). Effect of dietary Mirex and PCB (Aroclor 1254) on thyroid activity and lipid reserves in rainbow trout, *Salmo gairdneri*. *J. Fish. Dis.* **2**, 43–48.

Leatherland, J. F. and Sonstegard, R. A. (1980a). Seasonal changes in thyroid hyperplasia, serum thyroid hormones and lipid concentrations and pituitary gland structure in Lake Ontario coho salmon (*Oncorhynchus kisutch*): A comparison with coho salmon from Lakes Michigan and Erie. *J. Fish. Biol.* **16**, 539–562.

Leatherland, J. F. and Sonstegard, R. A. (1980b). Structure of the thyroid and adrenal glands in rats fed diets of Great Lakes coho salmon. *Environ. Res.* **23**, 77–86.

Leatherland, J. F. and Sonstegard, R. A. (1981). Thyroid dysfunction in Great Lakes coho salmon, *Oncorhynchus kisutch* (Walbaum): Seasonal and interlake differences in serum T3 uptake and serum totals and free T4 and T3 levels. *J. Fish Dis.* **4**, 413–423.

Leatherland, J. F. and Sonstegard, R. A. (1982). Thyroid responses in rats fed diets formulated with Great Lakes coho salmon. *Bull. Environ. Contam. Toxicol.*, in press.

Leatherland, J. F., Moccia, R. D., and Sonstegard, R. A. (1978). Ultrastructure of the thyroid gland in goitered coho salmon (*Oncorhynchus kisutch*). *Cancer Res.* **38**, 149–158.

Longwell, A. C. (1976a). Chromosome mutagenesis in developing mackerel eggs sampled from the New York Bight. *Amer. Soc. Limnol. Oceanogr., Spec. Symp.* **2**, 337–339.

Longwell, A. C. (1976b). Chromosome mutagenesis in developing mackerel eggs sampled from the New York Bight. U.S. Dep. of Commerce, NOAA tech. memo. ERL-MESA-7, 61 pp.

Longwell, A. C. and Hughes, J. B. (1979). Cytologic, cytogenetic, and embryologic conditions of early-stage Atlantic mackerel eggs in sea surface waters, natural and unnatural stress factors, researching and monitoring biological effects of ocean pollution. ICES workshop on the problems of monitoring biological effects of pollution in the sea. Beaufort, NC.

McCain, B. B., Pierce, K. V., Wellings, S. R. and Miller, B. S. (1977). Hepatomas in marine fish from an urban estuary. *Bull. Environ. Contam. Toxicol.* **18** (11), 1–2.

Moccia, R. D., Leatherland, J. F. and Sonstegard, R. A. (1977). Increasing frequency of thyroid goiters in coho salmon (*Oncorhynchus kisutch*) in the Great Lakes. *Science* **198**, 425–426.

Moccia, R. D., Leatherland, J. F., Sonstegard, R. A., and Holdrinet. M. V. H. (1978). Are goiter frequencies in Great Lakes salmon correlated with organochlorine residues? *Chemosphere* **7**, 649–652.

Norstrom, R. J., Hallett, D. J., and Sonstegard, R. A. (1978). Coho salmon (*Oncorhynchus kisutch*) and herring gulls (*Larus argentatus*) as indicators of organochlorine contamination in Lake Ontario. *J. Fish. Res. Board Can.* **35**, 1401–1409.

Parry, J. M., Tweats, D. J., and Al-Mossawi, M. A. J. (1976). Monitoring the marine environment for mutagens. *Nature* **264**, 538–540.

Pierce, K. B., McCain, B. B., and Wellings, S. R. (1978). Pathology of hepatomas and other liver abnormalities in English sole (*Parophyrs vetulus*) from the Duwamish River estuary, Seattle, Washington. *J. Nat. Cancer Inst.* **50** (6), 1445–1449.

Prein, A. E., Thie, G. M., Alink, G. M., and Koeman, J. H. (1978). Cytogenetic changes in fish exposed to water of the River Rhine. *Sci. Tot. Environ.* **9**, 287–291.

Reinert, R. E. (1979). Pesticide concentrations in Great Lakes fish. *Pestic. Monit. J.* **3**, 233–240.

Reinert, R. E. and Bergman, H. L. (1974). Residues of DDT in lake trout (*Salvelinus namaycush*) and coho salmon (*Oncorhynchus kisutch*) from the Great Lakes. *J. Fish. Res. Board Can.* **31**, 191–199.

Rothman, K. J. and Keller, A. Z. (1972). The effect of joint exposure to alcohol and tobacco on risk of cancer of the mouth and pharynx. *J. Chron. Dis.* **25**, 711–716.

Selikoff, I. J., Hammond, E. C., and Churg, J. (1968). Asbestos exposure, smoking and neoplasia. *J. Amer. Med. Assoc.* **204**, 106–112.

Sinnhuber, R. O. and Wales, J. H. (1974). Aflatoxin B_1 hepatocarcinogenicity in rainbow trout embryos. *Fed. Proc.* **33**, 247.

Sinnhuber, R. O., Hendricks, J. D., Wales, J. H., and Putnnam, G. B. (1977). Neoplasms in rainbow trout, a sensitive animal model for environmental carcinogenesis. *Ann. N.Y. Acad. Sci.* **298**, 389–408.

Smith, C. E., Peck, T. H., Klauda, R. J., and McLaren, J. B. (1979). Hepatomas in Atlantic tomcod (*Microgadus tomcod*) collected in the Hudson River estuary, New York. *J. Fish Dis.* **2**, 313–319.

Sonstegard, R. A. (1973). Relationship between environmental factors and viruses in the induction of fish tumors. In: M. S. Madhy and B. J. Dutka (Eds.), *Viruses in the environment and their potential hazards.* Symposium Proceedings Canada Centre for Inland Waters, pp. 119–129.

Sonstegard, R. A. (1974). Neoplasia incidence studies in fishes inhabiting polluted and non-polluted waters in the Great Lakes of North America. In: *XI Int. Cancer Congr.*, Panel 17, pp. 172–173.

Sonstegard, R. A. (1976). The potential utility of fishes as indicator organisms for environmental carcinogens. In: F. M. D'Itri (Ed.), *Wastewater renovation and reuse*. Marcel Dekker Inc., New York, pp. 561–577.

Sonstegard, R. A. (1977). Environmental carcinogenesis studies in fishes of the Great Lakes of North America. In: H. F. Kraykill, C. J. Dowe, J. C. Harshbarger, and R. G. Tardiff (Eds.), *Aquatic pollutants and biologic effects with emphasis on neoplasia. Ann. N.Y. Acad. Sci.* **298**, 261–269.

Sonstegard, R. A. (1978). Feral aquatic organisms as indicators of waterborne carcinogens. In: O. Hutzinger, L. H. van Lelyveld, and B. C. J. Zoeteman (Eds.), *Aquatic pollutants: Transformation and biological effects*. Pergamon Press, New York, pp. 349–358.

Sonstegard, R. A. and Leatherland, J. F. (1976). Studies of the epizootiology and pathogenesis of thyroid hyperplasia in coho salmon (*Oncorhynchus kisutch*) in Lake Ontario. *Cancer Res.* **35**, 4467–4475.

Sonstegard, R. A. and Leatherland, J. F. (1979a). Growth retardation in rats fed coho salmon collected from the Great Lakes of North America. *Chemosphere* **7**, 903–910.

Sonstegard, R. A. and Leatherland, J. F. (1979b). Hypothyroidism in rats fed Great Lakes coho salmon. *Bull. Environ. Contam. Toxicol.* **22**, 779–784.

Sonstegard, R. A. and Leatherland, J. F. (1979c). Aquatic organism pathobiology as a sentinel system to monitor environmental mutagens. In: B. K. Afghan and D. Mackay (Eds.), *Hydrocarbons and halogenated hydrocarbons in the aquatic environment*. Plenum Press, New York, pp. 513–520.

Sonstegard, R. A., Leatherland, J. F., and Dawe, C. J. (1976). Effects of gonadal tumors on the pituitary-gonadal axis in cyprinids from the Great Lakes. *Gen. Comp. Endocrinol.* **29**, 269.

Tomatis, L. (1979). The predictive value of rodent carcinogenicity tests in the evaluation of human risks. *Ann. Rev. Pharmacol. Toxicol.* **19**, 511–530.

Varanasi, U., Gmur, D. J. and Krahn, M. M. (1980). Metabolism and subsequent binding of benzo(a)pyrene to DNA in pleuronectid and salmonid fish. In: A. Bjorseth and S. J. Dennis (Eds.), *Polynuclear aromatic hydrocarbons: Chemistry and biological effects*. Battelle Press, Columbus, OH, pp. 455–470.

Veith, G. D., Keuhl, K. W., Puglisi, F. A., Glass, G. E., and Eaton, J. G. (1977). Residues of PCB and DDT in Western Lake Superior ecosystem. *Arch. Environ. Contam. Toxicol.* **5**, 478–484.

16

APPRAISING THE STATUS OF FISHERIES: REHABILITATION TECHNIQUES

Peter J. Colby

Ontario Ministry of Natural Resources
Fisheries Research Section
Thunder Bay, Ontario

1. INTRODUCTION

The need to identify the relative effect of multiple disturbances on a fishery is pressing. In the Great Lakes we have seen evidence of changing community structure associated with exploitation, habitat alteration, and species introduction—see SCOL (Loftus and Regier, 1972), PERCIS (Colby and Wigmore, 1977), SLIS (Smith, 1980), and STOCS (Berst and Simon, 1981). Our problem is to quantify the relative impact of each individual and combined stress on fish populations and communities. Observations of Great Lakes populations provide only a limited number of the replications necessary to draw relationships. However, if we observe the responses of the same species of fishes in smaller lakes with different stressors, some distinct population responses emerge (Colby and Nepszy, 1981).

In an effort to improve our diagnostic capabilities I have classified the response of fish populations to a variety of stimuli. This grouping of responses should help define the current status of a fishery as a reference. Once defined, any unusual departure may suggest the origin of the stressor. The classification of population responses is the primary objective of this paper. An additional objective is to discuss approaches to restructuring fish populations and communities to meet societal needs.

2. THEORETICAL CONCEPTS

Historically, many biologists believed that fish populations produced more than enough eggs to replace themselves, regardless of population density. Subsequent changes in population size were caused by environmental effects on egg and fry survival (see Nikolsky, 1965, for a review of this subject).

More recent concepts describe fish populations as self-regulating or cybernetic systems fluctuating within limits. Control is achieved through negative feedback (i.e., changes in growth rate, age at maturity, fecundity, etc.) in response to changes in density (Ricker 1954, 1958; Nikolsky, 1965; McFadden, 1967, 1977; Menshutkin, 1968; Goodyear, 1980). The limits to which feedback can maintain populations in equilibrium with their predators and prey are largely dependent on energy, nutrient (Colby and Nepszy, 1981), and habitat availability.

Generalized relationships between parental-stock density and the production of progeny have been described by Ricker (1954, 1958), McFadden (1967, 1977), Peterman (1977), and Goodyear (1980). All suggest that within certain limits of abundance, the parent stock produces excess progeny. The number of survivors produced per parent depends on additional mortality factors, resulting in a new equilibrium density. Thus surplus production occurs within a range of stock densities and varies with the magnitude of the stress (see McFadden, 1977 and Goodyear, 1980 for recent reviews).

Critics of the compensatory concept might argue that self-regulating populations must exhibit a stock-recruitment relationship, which is not always apparent. Strong year classes have arisen from relatively low stock densities for lake herring (*Coregonus artedii*) in Lake Erie (Van Oosten, 1946, 1949) and walleye (*Stizostedion vitreum vitreum*) in Green Bay (Hile, 1950). These observations do not necessarily refute stock-recruitment relationships. They show that occasionally environmental conditions prevail that permit excellent but unusual reproduction and year class survival for these species. Conversely, a large stock size does not necessarily ensure a large year class.

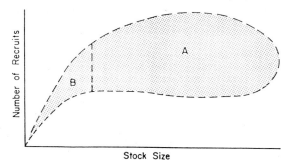

Figure 1. Schematic representation to illustrate: *A*, the great variability typical of the yearly number of recruits for different stock sizes; and *B*, the decrease in the numbers of recruits observed at low stock levels when populations may be in danger (Anonymous, 1980).

Recently, an ad hoc Group of the Ocean Science Board (Anonymous, 1980) proposed that the cause was the great variability in normal recruitment (Fig. 1). However continued commercial fishing on very low density pelagic stocks causes reduced recruitment and an irreversible trend toward smaller populations (Fig. 1, area *B*). Therefore Fig. 1 may be a single curve with considerable variance or a superposition of a set of curves, determined by the environment. Thus a single curve for recruitment is invalid and implies a regularity in nature that contradicts the evidence (Anonymous, 1980). Figure 1 supports our observations of walleye in Lake Erie. Stock-recruitment relationships became obvious at a time when stocks were being greatly reduced by exploitation (area *B*). Increased exploitation of Western Lake Erie walleyes during the 1950s and 1960s reduced population abundance to very low levels over a relatively short time (Shuter and Koonce, 1977; Shuter et al., 1979). Data collected during this period was analyzed by these authors. Their analyzes using two indices of breeding stock size produced similar results: a strong dependence of recruitment on stock size and the spring rate of warming.

The stock-recruitment relationship for walleye is more easily observed in Lake Erie where few age groups exist and the abiotic environment favors consistently good reproduction. In contrast, northern populations exhibit many age groups (Colby and Nepszy, 1981) and recruitment occurs over several years, probably due to the effect of temperature fluctuation on walleye spawning and feeding of recently hatched fish. Serns (1982) found significant links in Escanaba Lake, Wisconsin, between the fall density of age-0 walleyes and both the standard deviation and coefficient of variation of May water temperatures.

Another factor contributing to the cloud of scattered points in Fig. 1 is the determination of recruitment by sampling fish that are too old (W. Lysack, Manitoba Department of Natural Resources, personal communication). Lysack assumes that depensatory mortality occurs at a very early age. Therefore a more exact stock-recruitment relationship can be derived by trawling for young-of-the-year or yearling fish than by sampling 2 and 3 year old fish with gillnets. Lysack speculates that different species or stocks of the same species may react differently to continuous commercial fishing Both critical and noncritical depensatory (Peterman, 1977) mortality can occur in area *B* of Fig. 1 (Fig. 2). Noncritical depensatory mortality

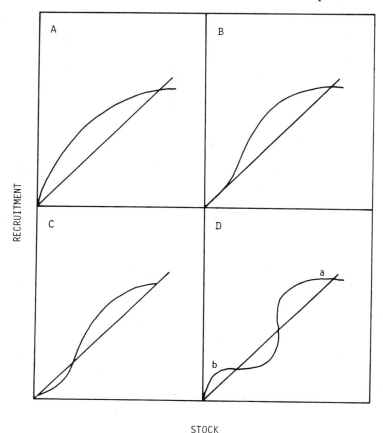

STOCK

Figure 2.   Hypothetical stock-recruitment (S-R) relationships that may operate in heavily fished lakes. (A) Normal stock-recruitment relationship. (B) Noncritical depensation causing an irreversible trend to very small stocks. (C) Critical depensation causing a collapse of the stock. (D) Reduction of the stock to a lower domain of stability. Points *a* and *b* are the respective upper and lower domain of stability described by Peterman (1977).

(curve *B*) is not sufficient to prevent replacement of adults. Critical mortality (curve *C*), however, is sufficient to reduce the biomass and subsequent population fecundity below that required for the surviving adults to replace themselves. The stock either collapses or is reduced to a lower domain of stability (Peterman, 1977).

Peterman (1977) suggests that depensation mortality caused by predation on young fish may cause more than one stability domain in the Ricker stock-recruitment model (Fig. 2, curve *C*). In this case, high exploitation rates could drastically reduce the size of a population to a lower level of abundance. The reduced population now fluctuates about a new equilibrium within a lower domain of stability (Fig. 2, curve *D*, point *b*). (W. Lysack, Manitoba Department of Natural Resources, personal communication) believes lake whitefish, *Coregonus clupeaformis*, in Manitoba are represented by either *A*- or *B*-type curves, while walleyes are represented by *B*- and *C*-type curves.

An underlying stock-recruitment relationship could also be obscured by the dependence of subordinate species abundance on that of the dominant species (Skud, 1982). Skud's evaluation of changes in abundance of dominant and subordinate marine fishes in response to changes in the physical environment were as follows: (1) dominant species responded positively to factors that improved their survival, and subordinate species responded negatively to the same factors; (2) the responses changed when dominance changed, indicating the abundance of the subordinate species is controlled by the abundance of the dominant species. Skud (1982) believes the dependence of subordinate abundance on that of the dominant species explains the lack of a stock-recruitment relationship for certain species.

Clearly, recruitment is a density-dependent function responding to stock size, exploitation, and a number of environmental factors. In addition, accurate recruitment estimates may be influenced by sampling gear and inappropriate selection of recruitment indices.

Unfortunately, recruitment failure may also be influenced by chemically induced reproductive dysfunction. Fisheries toxicologists have developed long lists of substances, singly or in combination, that can interfere with hormone biosynthesis, vitellogenesis, or spermatogenesis, or that can induce skeletal deformities, impaired yolk utilization, and delayed hatching in developing embryos. The effects of chemical and density-dependent stressors on recruitment are indistinguishable. Evidence to distinguish between stressors will require baseline data on density-dependent or environmental factors and may only be circumstantial at best.

3. DIAGNOSTIC VARIABLES

A variety of measurements have been used to determine the responses of fisheries to various stresses, primarily exploitation. Some commonly used diagnostic variables are: growth rate, mortality rate, size at age, age at maturity, mean age, longevity, fecundity, gonadal somatic index, condition factor, sex ratio, catch per unit effort (CUE), yield, particle size, and species composition of the aquatic community.

It is important that a given parameter realistically characterizes the population and is not an artifact of sampling bias. For example, Abrosov's (1969) suggestion that mean age of the catch should exceed mean age to maturity by 1 ½ years to maintain stable populations assumes an unbiased sample of mature fish. This may not be the case if fish are caught by angling. The relative abundance of younger fish in the catch is often greater than their relative abundance in the population.

In Escanaba Lake, the mean age of the angling catch for 1980 and 1981, respectively, was 3.7 and 3.2 years, whereas the mean age to maturity was 5.1 and 5.6 (S. Serns, Wisconsin Department of Natural Resources, personal communication). The Escanaba walleye population did not collapse as predicted by the biased sample. Errors of this nature may cause erroneous conclusions about angling exploitation effects on a lake trout fishery (C. Olver, Ministry of Natural Resources, personal

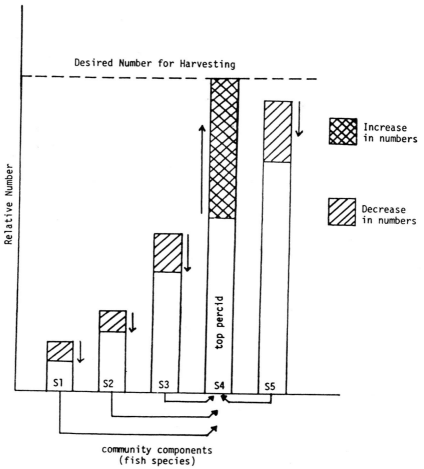

Figure 3. Hypothetical model illustrating possible changes in fish community structure due to a moderately intensive angling fishery for a target species (S4) such as walleye (Colby and Olver, 1978). Arrows indicate the transfer of energy within a fish community as a result of angling.

communication). If the relative abundance of brood stock was measured by trapnetting during spawning, a very different age and maturity structure would be observed, indicating a lower vulnerability to angling. The important consideration is the interaction between methods of harvest and the different ages of fish that cause unique and predictable changes in the population dynamics.

Monitoring key population variables may identify the major stresses impacting a fish population or community. Exploitation may affect populations differently, depending on the method of harvest and fishing pressure. For example, commercial gillnet fisheries for older fish can adversely effect the reproductive potential of a population by culling the more fecund females. This results in reduced recruitment and low yields (Adams and Olver, 1977). Conversely, an angling fishery may stimulate compensatory mechanisms by selectively removing intermediate-size fish. This permits adequate escapement of brood stock and increased recruitment and subsequent yield (Serns and Kempinger, 1981). In the latter case, energy entering the ecosystem may be channeled through the target species at the expense of other species (Fig. 3). If several species are similarly exploited, the total yield for recreational fisheries may exceed that for commercial fisheries (e.g., Elsey and Thomson, 1977; Kempinger and Carline, 1977; Lewis, 1979). The total yields would exceed the theoretical potential yield determined from the morphoedaphic index (MEI) (Ryder, 1965) by 100–200%. Though variables reflecting energy and nutrient availability are of importance in determining total fish yield, the "method of harvest" parameter should not be underestimated, especially when determining individual species yields (Ontario Ministry of Natural Resources, 1982).

Fisheries biologists often equate yield to the health of a population, but since yield is very much a function of exploitation, its interpretation could be misleading. Conceivably a fishery impacted by a stressor (a toxicant) could provide a yield indicative of a healthy stock in response to a given harvest strategy. The impact of the stressor could be masked by the strategy. The real test would be a comparison between very similar stocks (stressed vs. unstressed) in similar environments subjected to the same method of harvest.

4. GENERAL RESPONSE PATTERNS

Fish populations react to environmental stressors in a varied, but often predictable and categorical manner, depending on the nature and magnitude of the stimulus. I have grouped population responses into five generalized response patterns (GRPs). Although these categories may be reduced or expanded with additional information, they organize our knowledge, and help assess additional disturbances.

4.1. Group 1 Responses

This generalized response has six characteristics:

1. Mean age decreases as younger fish are recruited into the fishery. Unselective mortality of the adult population does not affect the age composition or the mean age. If all age groups of adults are equally vulnerable to the fishing gear, the mean age will remain relatively stable as the stock size is reduced (Nikolsky, 1965).
2. CUE or yield initially increases due to increasing fishing effort. Both subsequently decline because a large fraction of the original standing stock is removed.
3. Maturity occurs at an earlier age.
4. Growth rates increase in response both to reduced stock size and to increased food availability.

5. Survival and possible longevity is reduced, assuming that longevity is inversely related to growth rate.

6. Individual fecundity at age increases due to reduced density and an increased food supply.

The group 1 response is the characteristic compensatory reaction of a fish population to initial exploitation stress (Fig. 4). Although changes in growth rate, age at maturity, and fecundity are normal feedback responses to changes in food supply, they are also responses to exploitation. Examples of this response for Great Lakes species are: lake trout (Power and Gregoire, 1978), lake whitefish (Healey, 1975; Lysack, 1980; Jensen, 1981a), yellow perch, *Perca flavescens* (Schneider and Leach,

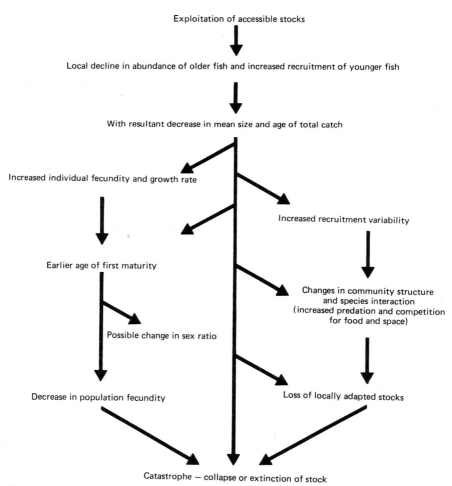

Figure 4. Diagram showing chronological effects of overexploitation on a fish population (modified from Colby and Olver, 1978).

1977), walleye (Lysack, 1980; see review by Colby and Nepszy, 1981), and sauger, *Stizostedion canadense* (Lysack, 1980). Schneider and Leach (1977) reported a shift to the right in the size versus maturity ogive for female walleyes in Lake Erie. Lysack (1980, 1981) found the age versus maturity ogive to shift to the left for walleye and sauger in Lake Winnipeg following heavy exploitation. These shifts are accompanied by a decreased mean age. When occurring early in the "fishing-up" process, these dynamic changes are characterized by a faster rate of growth and earlier maturity.

Power and Gregoire (1978) found that lake trout suffering predation by freshwater harbor seals (*Phoca vitulina*) had different population characteristics than those in neighboring lakes. Lake trout in populations exploited by seals were smaller, their growth rate was faster, longevity was reduced, age at sexual maturity was halved, and individual fecundity per unit weight had increased at the expense of egg diameter. The less vulnerable brook trout (*Salvelinus fontinalis*) were dominant in the lake with harbor seals, suggesting a change in fish community structure in response to predation. Healey (1978) found increased individual fecundity among experimentally harvested populations of both lake trout and lake whitefish and a similar pattern has been reported for walleye (D. Baccante, Ontario Ministry of Natural Resources, unpublished data).

In the "fishing-up" process the exploitation rate is greatly accelerated. Consequently, the diagnostic variables will change as indicated in Table 1 until the capacity to compensate is exceeded. This compensation capacity is determined primarily by energy and nutrient and habitat availability.

A few authors have described methods for measuring group 1 responses and some have developed criteria for estimating limits to compensation. For example, Thompson (1950) applied CUE as an index of normal yield, while Abrosov (1969) used mean age of the catch and at first maturity. Goodyear (1980) proposed the compensation ratio, an index of the degree of compensation required for an exploited population to reach an equilibrium, and Colby and Nepszy (1981) developed the "crisis curve." All are examples of techniques useful for regulating yield within compensatory limits.

4.2. Group 2 Responses

The second generalized response pattern has three characteristics:

1. Size at age increases due to reduced stock size.
2. Mean age increases due to reduced recruitment
3. CUE decreases due to decreasing stock size

This pattern is typical of a large loss of recruitment resulting in collapse, replacement, or extinction of the stock in a relatively short time (the longevity of the species). Indirect causes such as eutrophication and habitat alteration may provide a competitive advantage to other species. Direct causes include interspecific predation or competition due to species introductions. Although exploitation can cause collapse or replacement

Table 1. Generalized Response Pattern (GRP) of Population Variables to Various Disturbances

Suspected Stressor	GRP	Mean Age	Age at Maturity	Growth Rate (size at age)	Density (CUE, yield)	Biomass	Fecundity Individual	Fecundity Population	Sex Ratio	Condition (K)
Fishing Pressure										
Commercial										
Low	1	−	−	+	+	−	+	+?	−♀?	+
Intermediate (genetic selection)	3	+	+	−[b]	−	−	+	−	?	−
Heavy	1	−	−	+	−	−	+	−	−♀	+
Angling										
Low	1	−	−	+,−	−,+	−,+	+,−	+,?	?	+
Heavy	1	−	−	+	−	−	+	−	−♂	+

Stresses causing recruitment failure; species interactions (predation, competition); pollution −(toxicants), etc.

having a deleterious effect on the early stages of fish	2	+	0	+	−	−	−	+	?	+?
Increase in population size (i.e., introductions or reduced predation) leading to a relative rapid decline in food supply	4	?	+	−	+	+	−	−	?	−
Community interactions; restructuring due to habitat changes brought about by eutrophication; introductions; exploitation causing gradual reduction in niche dimension	5	?	+	−	−	−	−	−?	−?	?	0

[a] Positive and negative designate increasing and decreasing or higher and lower values. Zeros designate no suspected change. Commas separate early and late responses.
[b] Although selection may cause a decrease in size at age, the slower growing fish may increase in growth rate. Thus the range may change over time; the minimum size may become larger.

of a population, it is loss of habitat or species introduction, often combined with exploitation, that brings about extinction. The group 2 response typifies a population in its waning years. Recruitment failure among walleyes has occurred in the Moon River, Georgian Bay, and may be caused by unfavorable flow regimes for early life history stages due to upstream reservoirs (Winterton, 1975).

Recruitment failure may also be associated with interspecific competition. Schiavone (1981, 1983) and Mills and Schiavone (1982) report a concomitant decrease in walleye recruitment and increasing populations of black crappie (*Promoxis nigromaculatis*). Sedimentation and eutrophication promote the buildup of aquatic macrophytes, which favors the survival of black crappie at the expense of rock bass (*Ambloplites rupestris*) and walleye. In Black Lake the demise of the walleye was quite drastic (Table 2) and is believed to result from black crappie. They compete with walleye for food, thus increasing the walleye's vulnerability to angling, and may prey on young-of-the-year walleye (Schiavone, 1983). Schiavone (New York State Department of Environmental Conservation, personal communication) also believes some walleye still exist in the more mesotrophic waters but at a lower level of abundance or domain of stability (Fig. 2D, area b). This is characteristic of a group 5 response.

Recruitment failure may also be caused by pollutants that could harm early life history stages of fishes. McKim (1977) reported that in 56 life-cycle toxicity tests with 34 organic and inorganic chemicals and 4 species of fish, the embryo-larval and early juvenile life stages were generally the most sensitive. However, care must be taken to separate contaminant effects from physical environmental changes, such as temperature or eutrophication, or interspecific competition.

Gunn and Keller (1981) demonstrated that acidic conditions within gravel beds were lethal to the early developmental stages of lake trout and brook trout. Hatching appeared to be particularly critical for lake trout and the authors report similar problems for Atlantic salmon (*Salmo salar*), brook trout, brown trout (*Salmo trutta*), and Eurasian perch (*Perca fluviatilis*).

Evans (1978) found recruitment failure among lake whitefish in Lake Simcoe, Ontario, associated with the process of eutrophication. The size structure of the

Table 2. Changes in Size and Mean Age of Male Walleyes in Black Lake, New York[a]

Year	Mean Length (total in mm)	N	Mean Age (years)	N
1953	414	115	—	—
1961	391	298	5.13	175
1962	406	361	5.57	190
1963	371	116	5.05	175
1964	386	241	5.60	171
1970	419	591	6.13	76
1971	432	237	—	—
1975	481	61	9.70	60

[a] After Schiavone, 1981.

spawning lake whitefish populations has increased in recent years. The modal length categories were 300–400 mm (1963), 390–400 mm (1966–1967), 510–520 mm (1976), and 520–530 (1977). The small increase in mean size of individual fish from 1976–1977 indicates that no recruitment has taken place and that a static population has simply grown one year older (Evans, 1978). Also the angling catch declined by 98% from 1968 to 1977. The decline in angler harvest is strongly correlated with population size (mark-recapture estimates) and indicates a 92% reduction in the lake whitefish population since 1968.

In addition, the mean age of lake whitefish increased from 1963 to 1977, indicating continued recruitment failure. If the population stabilizes at a lower level, supported by reduced recruitment, the mean age of the population should drop and then stabilize. This response is characteristic of group 5. However, continued recruitment failure reflected by increases in mean age will lead to population extinction and are characteristic of a group 2 response. Two possible causes are predation or competition by a fish species more adapted to the changing environment (as described by Schiavone, 1983) or deterioration of spawning or nursery habitat. The cause of reduced recruitment or failure in Lake Simcoe is not known. The decline of the lake whitefish is associated with lower water quality, which is considered inadequate for their natural reproduction and subsequent recruitment.

4.3.   Group 3 Responses

Exploitation may produce unique responses:

1.   Size at age, phenotypic variance of condition K ($K = W/L^3$), individual fecundity, and CUE decrease.
2.   Mean age and age to maturity increase.

This response has been described for lake whitefish in Lesser Slave Lake (Handford et al., 1977) and is completely contrary to the expected response of whitefish to exploitation (Healey, 1975). Handford et al. (1977) found that lake whitefish in Lesser Slave Lake, Alberta, responded to exploitation by an initial increase in growth followed by a gradual decrease, reduced condition, and an increase in mean age. They hypothesized that gillnet fisheries are highly selective of large, heavy, fast growing individuals and culling of this sort will provide an evolutionary response in the stock. They argue that size and weight are genetically determined and that the selective mortality represented by gillnet fishing is likely to permanently change the genetic composition of the stock.

Handford et al. (1977) were also able to demonstrate the decline in the phenotypic variance of condition that would be expected if selection is occurring. Thus selection through exploitation will eventually produce slower growing individuals that live longer before reaching a harvestable size. Consequently, the mean age at maturity will increase, and the age-specific fecundity will be diminished, insofar as these are determined more by size than age, and this in turn will decrease their rate of replacement to the disadvantage of the population. The authors suggest that this phenomenon of reversing the expected trend toward a younger age-distribution is only to be expected in an intermediate range of fishing intensities. They warn it is especially serious because the size-specific selection associated with gillnet fisheries is capable of reducing the capacity of the population to maintain itself in the face of continued exploitation. Also in a very lightly fished population, the selection differential will be too slight to provide an effective force, while in heavily exploited populations, the selective effect will be overpowered by the usual demographic shift toward younger age-groups caused by density-dependent effects on individual growth.

Group 3 responses could also be affected by a reduction in food supply. Handford et al. (1977) described the effect of siltation associated with oil exploration on the food supply of lake whitefish. The authors questioned whether a reduction in food supply that depressed growth-rate would also produce a decline in the variance of condition factor. The authors disregarded strong year classes that would not likely bring about long periods of directional change in mean age and could easily be followed through the fishery.

It is quite probable that certain subpopulations or stocks in larger lakes may be responding differentially to a stressor. Lysack (Manitoba Department of Natural Resources, personal communication) believes group 1 and 3 responses now exist simultaneously in the walleye and sauger stocks of southern Lake Winnipeg. Approximately 10–15% of these fish caught in experimental gillnets consist of relatively old, immature, or extremely thin fish.

4.4. Group 4 Responses

This pattern is characterized by:

1. Decreasing size at age.
2. Increase in CUE or yield.
3. Later age at maturity of the exploited stock.

All of these responses are due to an increase in stock size and occur when a population increases to the point where food is limiting.

Walleyes introduced into Colorado reservoirs respond in this manner (Puttman and Weber, 1980). The decline in size at age in Boyd Reservoir (Fig. 5), a turbid water body with sparse vegetation, was associated with a decline in black crappie soon after introduction.

If predation pressure is removed, centrarchid populations (primarily *Lepomis* sp.) will often increase until stunting occurs, resulting in high CUEs of undesirable size fish. Stunted prey species characterize a disturbed or unbalanced fish community (at least from the anglers' point of view), and usually results from high fishing pressures being exerted on the predators (Anderson and Weithman, 1978). Conversely, the exploitation of predatory walleye in Boyd Reservoir was insufficient to maintain good growth. Group 4 responses reflect an expanding population responding to limitations of carrying capacity rather than a direct perturbation.

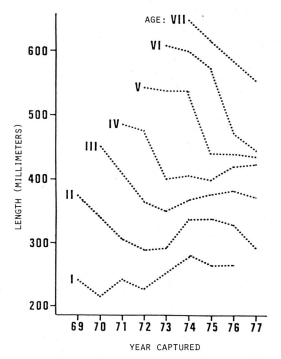

Figure 5.   Average length, by age-group, of angler-caught walleye, Boyd Reservoir, 1969–1967 (Puttman and Weber, 1980).

4.5.   Group 5 Responses

These responses are characterized by the following:

1.   Slow growth.
2.   Later maturity.
3.   Initially decreasing CUE, which stabilizes as the population reaches a lower domain of stability.
4.   The mean age remaining relatively unchanged (mortality in equilibrium with restricted recruitment).
5.   The fish are in good condition.

Group 5 includes populations that suffered a loss in numbers through competition or predation and are unable to regain their former abundance. At the new and lower equilibrium, the population has demographic characteristics of a dominant population at carrying capacity (group 4 responses) but now fluctuates within a lower domain of stability (Fig. 2D, area b). The walleye populations described by Schiavone (1981) living in marginal mesotrophic waters of the Indian River system show typical group 5 responses. The habitat had not changed sufficiently to permit complete replacement by black crappie. A similar example of group 5 responses brought about by habitat change and black crappie replacement of walleye may be found in the Port Severn area of Georgian Bay, Lake Huron (Hogg, 1975, 1976; La France, 1980).

Group 5 responses may result from habitat change and species introductions or both, and can be accelerated by exploitation. These stressors encourage niche shifts that restrict the most vulnerable species to a subordinate role within the lower domain of stability. Other possible examples of group 5 responses are provided by McComish (1981) and Bregazzi and Kennedy (1982). McComish (1981) reports that yellow perch in the Indiana waters of Lake Michigan are decreasing in growth and maturing later without increasing greatly in abundance. This response may be a result of their interaction with alewives (*Alosa psuedoharengus*) and bloaters (*Coregonus hoyi*).

Bregazzi and Kennedy (1982) found unusual responses of Eurasian perch to increasing eutrophication. Instead of the expected increase in growth rate and the associated increase in fecundity and earlier maturation of fish (Le Cren, 1951; Alm, 1959), they reported no changes in these parameters despite a declining population. No dominant year classes were apparent and feeding strategy had not altered since an earlier study. Instead, there was a marked preponderance of young fish, a slight reduction in the growth rate up to the age of III, and a decline in the percentage of early maturing perch. The reduced growth of young fish and their delayed maturity indicates that some compensation in the perch population is occurring within a lower domain of stability. Simultaneously the cyprinid species dominance changed from rudd (*Scardinius erythrophthalmus*) to roach (*Rutilis rutilis*) (Bregazzi and Kennedy, 1982). Presumably the later stages of accelerated eutrophication altered the environment sufficiently to favor roach and reduce the niche dimension of Eurasian perch.

The subtle group 5 responses to accelerated eutrophication and species introductions or change in dominance contrast with the more abrupt group 1 responses to accelerated exploitation. Accelerated exploitation permits sufficient lag time for the normal compensatory feedback mechanism to become operable (especially if the abiotic environment has not been altered). Conversely, the subtle group 5 responses occur gradually. They are at first unnoticed, but reach a state where differences become rapidly apparent and difficult to reverse. Gradual changes in spawning habitat or food availability could produce dynamic changes within the fish community, ending in abrupt displacement and changes in species dominance.

5.   COMMUNITY INTERACTIONS: REHABILITATION TECHNIQUES

Few fish populations exist in complete isolation; some may be geographically confined in small lakes. Others, in very large bodies of water, approach some degree of isolation; but for the most part, fish populations are always interacting. Each community of interacting species is constrained by the abiotic environment and by mechanisms for maintaining a community structure within some limits. Consequently, what happens to one population influences the others. This in turn will influence the measurement and interpretation of responses 1–5. It is unreasonable to monitor long-term trends of a single population and expect it to behave as if in isolation. Alterations in community structure and niche shifts (Werner and Hall, 1976) may bring about changes in growth and age to maturity in a population without necessarily affecting its size. Therefore, when using population variables as diagnostic tools, the "population physician" should be aware of the effects of subtle community changes.

There are several new approaches to describing disturbed aquatic communities. Ryder and Kerr (1978) have defined characteristics of more resilient harmonic and less resilient astatic fish communities. Sprules (1980) and Ryder et al. (1981) review the rationale for using particle sizes within aquatic communities to evaluate community changes. More recently, Mills and Schiavone (1982) used zooplankton size structure and estimates of lake productivity to aid interpretation of fishery data. They found that the size composition of the zooplankton community is closely linked with the growth and size structure of the fish community; the most desirable size structures of fish communities were associated with larger size zooplankton. This may be a simple and economical approach for assessing the status of a fish population and for determining management options.

Anderson and Weithman (1978) proposed an approach to evaluate the status of coolwater fish communities. Their index, proportional stock density (PSD), uses "particle (fish) size" as the primary criterion for determining whether a fish population or community (predator and prey) is balanced, for humanity's benefit. In Ontario, the Walleye Research Unit is manipulating a walleye population by harvesting intermediate-size fish, leaving many larger fish for maintaining good reproduction and keeping the fish community in balance through predation. Multiple slot size harvests (harvesting or protecting certain size groupings) may provide good yields and a quality fishing experience. For example, a hypothetical approach for managing many of Northern Ontario's walleye angling fisheries might be to harvest fish between 30 and 50 cm in length. Those between 50 and 65 cm would be protected, but not trophy-size fish over 65 cm. This assumes that removal of tropy fish will have minimal impact on the dynamics of the populations being harvested. The size of harvestable slots may vary between lakes or regions.

In lakes where walleye angling pressure has caused a compensation resulting in crowding and slower growth, size and "quality" of fishing might be increased as follows. Assume group 4 responses have occurred where the fishing regulations permit the angler to keep 6 walleye over 36 cm; a change to a limit of 8–12 walleye (6–10 between 26 and 34 cm and 2 over 44 cm) may reduce biomass and increase the size frequency distribution of the population. These slot limits can be adjusted upwards as the size of fish within the population increase to some maximum imposed by the system's capacity to support increased growth. Removing fish of intermediate size to reduce density and maximize yield mimics the predator-prey relationship. Removing the smaller members of a prey species when they are abundant maximizes energy flow between trophic levels and maintains the prey biomass without endangering fecund adults and reducing the reproductive capacity.

The quality of the walleye fisheries in Ontario are under review by the Walleye Research Unit using a combination of PSD (Anderson and Weithman, 1978) and CUE as measures of angling satisfaction. For example, a comparison of two Northern walleye lakes illustrates the types of effect of fishing pressure and management strategies on angling satisfaction (Fig. 6).

Escanaba Lake supports an intensive high yield angling fishery primarily for walleye. The lower walleye CUE and PSD relative to Savanne Lake probably results from distributing fish among many fishermen. However, some of the difference is no doubt due to effort being directed to more species. The comparison illustrates the effect on angling quality of maintaining high annual yields providing the maximum numbers of recreational hours. In such a fishery the compensatory response of the target species (Fig. 3) is near maximum and the fish community has been restructured to accommodate a high yield of walleyes. The two parameters of size (PSD) and frequency of catch (CUE) can be combined as follows to provide a quality fishing index (QFI) for the purpose of comparing lakes:

$$QFI = \frac{\text{number of quality size } (\geq 38 \text{ cm})}{\text{number of standard size } (\geq 25 \text{ cm})} \times CUE \times 100$$

As an example, the quality fishing index for the two research lakes (Fig. 5), for the years indicated follows:

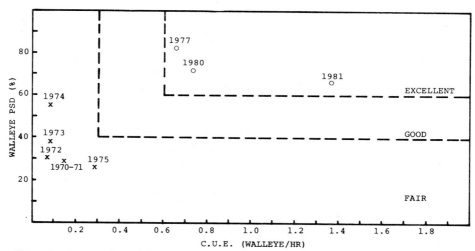

Figure 6. A comparison of the quality of angling experience between two lakes exposed to different angling pressures. Escanaba Lake ($\times$), Wisconsin, receives heavy angling pressure (100–200 hr/ha) and provides one of the highest sustainable (30 yr) yields reported (9 kg/ha·yr) in the literature (Kempinger and Carline, 1977). Controlled walleye removal from Savanne Lake (O), Ontario, began in 1980 where intermediate-size fish were selected for removal (1.23–2.18 kg/ha·yr). Most of the walleye in Savanne Lake were captured with trapnets and only 0.4 kg/ha·yr were removed by angling (0.2–1.24 hr/ha). The figure illustrates a method to compare angling quality and individual satisfaction when high yields are shared by many anglers. The criteria for good and excellent walleye fishing are based on the author's judgement. Walleye creel data for Escanaba Lake, was provided by S. Serns, Wisconsin Department of Natural Resources.

Savanne Lake (1981)
$$QFI = 0.66 \times 1.36 \times 100 = 89.8$$

Escanaba Lake (1975)
$$QFI = 0.27 \times 0.29 \times 100 = 7.8$$

Additional indices for identifying the quality of a commercial fishery were proposed by W. Lysack (Manitoba Department of Natural Resources, personal communication). He suggests that PSD can be replaced by other measurement ratios depending on the type of fishery. For example:

$$\frac{\text{mean biomass per gang (from a commercial net)}}{\text{mean biomass per gang (from an experimental net)}} \times CUE \times 100$$

or

$$\frac{\text{mean individual weight (large mesh fishery)}}{\text{mean individual weight (small mesh fishery)}} \times CUE \times 100$$

These indices, in addition to measuring the quality of fishing, measure the quality of a Northern coolwater fish community dominated by preferred predators. In Northern Ontario, the desired community structure is a high abundance of quality size predators, with prey serving mainly as agents of energy transfer.

One problem with this ratio is that quality fishing is defined by the client: some fishermen prefer lower CUE and smaller sized fish in return for no restrictions on their fishing experience, and in certain regions, some anglers prefer the prey over the predators. However, the index should still be useful for comparing fisheries. Quality fishing on the basis of size may not be sufficient to justify changes in natural reproduction and subsequent yield. A possible reason why walleye natural reproduction and yield is excellent in Escanaba Lake is that the population is being turned over rapidly. This is a good management strategy to provide many recreational hours of acceptable angling.

The purpose of quantifying a quality fishing experience is to identify and set management targets, and to understand the causal factors that contribute to a given index. With additional insights, populations and communities may be manipulated to meet the "demand of the market".

6.   POTENTIAL USE OF DIAGNOSTIC VARIABLES

Population variables should be used to diagnose stressors only with extreme caution. Variables or criteria used in isolation, without additional indicators, could lead to erroneous conclusions.

Van Oosten (1946) warned that low capacity to produce fish, natural fluctuations in abundance, unsatisfactory fishing, uneconomic utilization, reduction in numbers, reduced total yield, scarcity of exploitable fish, declining CUE, reduction in stocks through factors other than fishing, and extermination are not necessarily synonymous with depletion due to overfishing. He also warned that depletion* varies in severity and can exist despite a perfect balance between recruitment and yield. Gradual changes within the community, such as changes in species composition, niche shifts, changes in particle sizes and their variance, and behavioral changes at the population level, may alter the demographic characteristics of populations.

Groups 1, 3, and 4 responses are reversible with current or proposed fishery management techniques. Groups 2 and 5 responses may not be reversible without a major effort on international issues, such as eutrophication, acid rain, or changing species dominance among Great Lakes fishes.

Monitoring population variables may indicate whether stocks are responding to exploitation or habitat change, also, for example, reduced yield caused by exploitation-induced genetic drift. The relative impact of different harvest strategies may possibly be measured by monitoring sex ratios on the spawning ground.

Lake whitefish and walleye males may disappear because they are more heavily exploited than are females (Hile, 1954); however the opposite occurred where walleye were commercially fished in Red Lake, Minnesota (Smith and Pycha, 1961). The proportion of females in gillnets during July and August decreased with increasing age after age VI; at age X only 23.5% of the catch were females. Since females grow faster than males, they may have been more vulnerable to the commercial fishery at an earlier age. Therefore the effects of harvest methods on sex ratios makes the ratios a useful diagnostic tool.

Little Cutfoot Sioux, Minnesota, is a lake receiving high angling pressure, 140 hr/ha (D. Schupp, Minnesota Department of Natural Resources, personal communication). The ratio of male to female walleyes on the spawning grounds (Table 3) decreased from an average of 6.1:1 (1942–1951) to 2.9:1 (1972–1979), and in some years the ratio approached 1:1 (T. Osborne, Minnesota Department of Natural Resources, personal communication). Were the number of males sufficient to fertilize all the eggs? The changing sex ratio may have arisen from a differential angling harvest of males. This hypothesis, however, was not supported by observation in Escanaba Lake, Wisconsin; no differences in sex ratio were found between the angling catch of male and female walleyes (Serns and Kempinger, 1981).

In contrast, the predominance of male walleyes during the spawning run has increased in recent times in Red Lake (B. Strand, Minnesota Department of Natural Resources, personal communication). The average ratio of males to females has increased from 24:1 during the time of Smith and Pycha (1961) to an average of 52:1 for the period 1970–1979 (Table 3). The cause may be commercial gillnetting (Smith and Pycha, 1961).

Lysack (1981) further supports Smith and Pycha's (1961) hypothesis. He found that a commercial gillnet fishery for saugers in Lake Winnipeg selected for larger females. In those areas receiving the least fishing pressure, the sex ratio of males to females was 0.59:1; in two heavily fished areas the sex ratios were 1.61:1 and 1.37:1.

---

* Depletion is defined as the reduction, through overfishing, in the level of abundance of the exploitable segment of a stock that prevents the realization of the maximum productive capacity.

Table 3.   Mean Sex Ratios (M:F) for Two Heavily Exploited Minnesota Walleye Populations (Ranges in Parentheses); the Primary Fishery in Red Lakes Is Commercial While That in Little Cut Foot Sioux Is Angling

Period	Red Lakes[a]	Period	Little Cut Foot Sioux[b]
1950–1959	24.0:1 (10:1–36:1)	1942–1951	6.4:1 (3:1–14:1)
1960–1969	38.9:1 (8:1–189:1)	1952–1961	2.5:1 (0.8:1–6:1)
1970–1979[c]	52.1:1 (16:1–148:1)	1962–1971	3.6:1 (1.2:1–7:1)
		1972–1982	2.9:1 (1.0:1–6:1)

[a] Data supplied by Bob Strand, Minnesota Department of Natural Resources.

[b] Data supplied by Tom Osborn, Minnesota Department of Natural Resources.

[c] No data for 1973, 1974, and 1975.

There are no studies of the effect of social structure on the reproductive success of "mass spawners" like walleye. Such a study is needed since adult sex ratios may affect the well-being of a population.

Comparison of yield and population variables may reflect habitat change. If an equilibrium yield varies while demographic variables (i.e., growth, age to maturity, and survival) are constant, the carrying capacity of the environment may have changed—especially if no new species have become established. For example, Lake Erie's habitat has undergone considerable changes that may be major factors in the decline of walleyes. However, since the ban on fishing due to mercury contamination, the walleye populations have increased. Comparison of contemporary with historic walleye yields, when the population demography is similar, might indicate the degree of subsequent habitat alteration.

Genetic changes in stocks are difficult to verify; one might suspect genetic changes if an expected response scenario does not occur (Handford et al., 1977).

Yield may indicate the effect of stressors on a fish population or community. Several good indices of fish yield have been developed. Fish yield in Northern temperate lakes has been linked to morphoedaphic factors (Ryder, 1965), benthic standing crop (Matusek, 1978), and temperature (Schlesinger and Regier, 1982). Although these relationships are significant, I believe some yields are underestimated, because the effect of harvesting methods has been underestimated. A portion of the sexually mature fish must escape to sustain higher yields of greater quality. New and innovative harvest strategies to protect part of the adult stock have been suggested (Larkin, 1978; Jensen, 1981b; Berkes and Gönenc, 1982); adaptive probing experiments are needed (Walters, 1981).

The use of different harvest strategies, such as "slot size" management, should improve fish yields. Trends in yield coupled with changes in the demography of fish populations should improve estimates of maximum equilibrium yields.

The diagnosis of population status through indicator variables will improve when mechanisms controlling community homeostasis are better understood, for example neuroendocrine (George, 1977). In the interim, categorizing responses will enhance diagnostic capabilities. The diagnostic approach, while subjective and artificial, indicates the effects of perturbations on a system. However, it is essential to understand population responses to environmental variables or management strategies before investigating additional impacts such as contaminants.

The separation of all stress effects on population characteristics is beyond the scope of this paper, but categorizing general responses should refine our diagnostic capability. Expansion of the diagnostic capability at all levels (biochemical, physiological, population, and community) is needed to identify specific stress effects on a fishery.

ACKNOWLEDGMENTS

I wish to thank Vic Cairns, Walter Lysack, Walter Momot, Charlie Olver, Phil Ryan, Steven L. Serns, Dennis H. Schupp, and Peter V. Hodson for their contributions, helpful suggestions, and constructive criticism, and Letizia Tamasi for typing the manuscript and preparing the figures.

REFERENCES

Abrosov, V. N. (1969). Determination of commercial turnover in natural bodies of water. *Prob. Ichthyol.* **9**, 482–489.

Adams, G. F. and Olver, C. H. (1977). Yield properties and structure of boreal percid communities in Ontario. *J. Fish. Res. Board Can.* **34**, 1613–1625.

Alm, G. (1959). Connection between maturity, size, and ages in fishes. *Rep. Inst. Freshwater Res. Drottningholm* **40**, 5–145.

Anderson, R. O. and Weithman, A. S. (1978). The concept of balance for coolwater fish populations. *Amer. Fish. Soc. Spec. Publ.* **11**, 371–381.

Anonymous (1980). *Fisheries ecology: Some constraints that impede advances in our understanding.* Ocean Sci. Board, Nat. Acad. of Sci., Washington, DC. Available from Ocean Sci. Board, 2101 Constitution Avenue, Washington, DC 20418.

Berkes, F. and Gönenc, T. (1982). A mathematical model on the exploitation of northern lake whitefish with gillnets. *North Amer. J. Fish. Manag.* **2**, 176–183.

Berst, A. H. and Simon, R. C. (Eds.) (1981). Proc. of the stock concept int. symp. (STOCS). *Can. J. Fish. Aquat. Sci.* **38**(12).

Bregazzi, P. R. and Kennedy, C. R. (1982). Responses of a perch, *Perca fluviatilis* L. population to eutrophication and associated changes in fish fauna in a small lake. *J. Fish. Biol.* **20**, 21–31.

Colby, P. J. and Nepszy, S. J. (1981). Variation among stocks of walleye (*Stizostedion vitreum vitreum*): Management implications. *Can. J. Fish. Aquat. Sci.* **38**, 1814–1831.

Colby, P. J. and Olver, C. H. (Eds.) (1978). Management implications derived from the PERCIS training session. Ontario Minist. of Nat. Resour. file rep., 51 pp.

Colby, P. J. and Wigmore, R. H. (Eds.) (1977). Proc. of the 1976 Percid int. symp. (PERCIS). *J. Fish. Res. Board Can.* **34**, 1447–1999.

Elsey, C. A. and Thomson, R. T. (1977). Exploitation of Walleye (*Stizostedion vitreum vitreum*) in Lac Des Mille Lacs, Northern Ontario, by Commercial and Sport Fisheries, 1958–75. *J. Fish. Res. Board Can.* **34**, 1769–1773.

Evans, D. O. (1978). An overview of the ecology of the lake whitefish, *Coregonus clupeaformis* (Mitchill) in Lake Simcoe, Ontario with special reference to water quality and introduction of the rainbow smelt, *Osmerus mordax* (Mitchill). Ontario Minist. of Nat. Resour. file rep., 132 pp.

George, C. J. (1977). The implications of neuroendocrine mechanisms in the regulation of population character or on a more christian view of the black box. *AFS Fish.* **2**, 14–19, 30.

Goodyear, C. P. (1980). Compensation in fish populations. In: C. H. Hocutt and J. R. Stauffer, Jr. (Eds.), *Biological monitoring of fish.* Lexington Books, P. C. Heath and Co., Lexington, MA, pp. 253–280.

Gunn, J. M. and Keller, W. (1981). Emergence and survival of lake trout (*Salvelinus namaycush*) and brook trout (*S. fontinalis*) from artificial substrates in an acid lake. Ont. Fish. tech. rep. ser. no. 1: pp. iii and 9.

Handford, P., Bell, G. and Reimchen T. (1977). A gillnet fishery considered as an experiment in artificial selection. *J. Fish. Res. Board Can.* **34**, 954–961.

Healey, M. C. (1975). Dynamics of exploited whitefish populations and their management with special reference to the Northwest Territories. *J. Fish. Res. Board Can.* **32**, 427–448.

Healey, M. C. (1978). Fecundity changes in exploited populations of lake whitefish (*Coregonus clupeaformis*) and lake trout (*Salvelinus namaycush*). *J. Fish. Res. Board Can.* **35**, 945–950.

Hile, R. (1950). Green Bay walleyes, a report on the scientific investigation of the marked increase in abundance of walleye in Green Bay. *The Fisherman* (Grand Haven, Mich.) **18**(3), 5–6.

Hile, R. (1954). Fluctuations in the growth and year-class strength of walleye in Saginaw Bay. *U.S. Dept. Int. Fish. Wildl. J. Fish. Bull.* **91**(56), 5–49.

Hogg, D. M. (1975). Port Severn yellow pickerel project progress report. Ontario Minist. of Nat. Resour. file rep., 21 pp.

Hogg, D. M. (1976). Port Severn yellow pickerel project progress report. Ontario Minist. of Nat. Resour. file rep., 25 pp.

Jensen, A. L. (1981a). Population regulation in lake whitefish *Coregonus clupeaformis* (Mitchill). *J. Fish. Biol.* **19**, 557–573.

Jensen, A. L. (1981b). Optimum size limits for trout fisheries. *Can. J. Fish. Aquat. Sci.* **38**, 657–661.

Kempinger, J. J. and Carline, R. F. (1977). Dynamics of the walleye (*Stizostedion vitreum vitreum*) population in Escanaba Lake, Wisconsin, 1955–72. *J. Fish. Res. Board Can.* **34**, 1800–1811.

La France, W. (1980). Port Severn walleye population study. Ontario Minist. of Nat. Resour. file rep., 16 pp.

Larkin, P. A. (1978). Fisheries management—An essay for ecologists. *Ann. Rev. Ecol. Syst.* **9**, 57–73.

Le Cren, E. D. (1951). The length-weight relationship and seasonal cycle in gonad weight and condition in the perch (*Perca fluviatilis*). *J. Anim. Ecol.* **30**, 201–219.

Lewis, C. A. (1979). 1979 Creel Census Rice Lake and the Otonabee River. Ontario Minist. of Nat. Resour. file rep., 49 pp.

Loftus, K. H. and Regier, H. A. (Eds.) (1972). Proceedings of the symposium on salmonid communities in oligotrophic lakes (SCOL). *J. Fish. Res. Board Can.* **29**, 613–986.

Lysack, W. (1980). 1979 Lake Winnipeg fish stock assessment program. Man. Dep. of Nat. Resour. MS rep. no. 80-30, 118 pp.

Lysack, W. (1981). 1980 Lake Winnipeg fish stock assessment program. Man. Dep. Nat. Resour. MS rep. no. 81-20, 67 pp.

Matusek, J. E. (1978). Empirical predictions of fish yields of large North American lakes. *Trans. Amer. Fish. Soc.* **107**, 385–396.

McComish, T. S. (1981). Yellow perch population characteristics in Indiana waters of Lake Michigan, 1976–79. Ball State Univ., final rep., Federal aid project 3-283-R, Segment 3, U.S. Dep. Commerce, NOAA, NMFS, 97 pp.

McFadden, J. T. (1967). Environmental impact assessment for fish populations. In: R. K. Sharma, J.D. Buffington, and J. T. McFadden, (Eds.), *Proc. of the conf. on the biological significance of environmental impacts.* Sponsored by the U.S. Nucl. Regul. Comm., Washington, DC, reproduced by US Dep. of Commerce, Nat. Tech. Infor. Serv. PB-258-563; Springfield, VA, pp. 89–137.

McFadden, J. T. (1977). An argument supporting the reality of compensation in fish populations and a plea to let them exercise it. In: W. VanWinkle (Ed.), *Assessing the effects of power-plant induced mortality on fish populations.* Pergamon Press, New York, pp. 153–182.

McKim, J. M. (1977). Evaluation of tests with early life stages of fish for predicting long-term toxicity. *J. Fish. Res. Board Can.* **34**, 1148–1154.

Menshutkin, V. V. (1968). A fish population as a self-regulating system. *Prob. Ichthyol.* **8**, 68–74.

Mills, E. L. and Schiavone, A., Jr. (1982). Evaluation of fish communities through assessment of zooplankton populations and measures of lake productivity. *North Amer. J. Fish. Manage.* **2**, 14–27.

Nikolsky, G. V. (1965). *Theory of fish population dynamics as the biological background for rational exploitation and management of fishery resources.* Nanka Press, Moscow, 328 pp.; English edition, Oliver and Boyd, Edinburgh, 1969, 323 pp.

Ontario Ministry of Natural Resources (1982). Partitioning yields estimated from the morphoedaphic index into individual species yields. Final rep. of SPOF working group no. 12, 71 pp.

Peterman, R. M. (1977). A simple mechanism that causes collapsing stability regions in exploited salmonid populations. *J. Fish. Res. Board Can.* **34**, 1130–1142.

Power, G. and Gregoire, J. (1978). Predation by freshwater seals on the fish community of Lower Seal Lake, Quebec. *J. Fish. Res. Board Can.* **34**, 844–850.

Puttman, S. J. and Weber, D. T. (1980). Variable walleye fry stocking rates in Boyd Reservoir, Colorado. Colorado Div. Wildl. tech. publ. no. 33, 47 pp.

Ricker, W. E. (1954). Stock and recruitment. *J. Fish. Res. Board Can.* **11**, 558–623.

Ricker, W. E. (1958). Handbook of compilations for biological statistics of fish populations. Fish. Res. Board Can. Bull. 119, 300 pp.

Ryder, R. A. (1965). A method for estimating the potential fish production of north-temperate lakes. *Trans. Amer. Fish. Soc.* **94**, 214–218.

Ryder, R. A. and Kerr, S. R. (1978). The adult walleye in the percid community—A niche definition based on feeding behavior and food specificity. *Amer. Fish. Soc. Spec. Publ.* **11**, pp. 39–51.

Ryder, R. A., Kerr, S. R., Taylor, W. W., and Larkin, P. A. (1981). Community consequences of fish stock diversity. *Can. J. Fish. Aquat. Sci.* **38**, 1856–1866.

Schiavone, A., Jr. (1981). Decline of the walleye population in Black Lake. *N.Y. Fish Game J.* **28**(1), 68–72.

Schiavone, A., Jr. (1983). The Black Lake, New York, fish community: 1931 to 1979. *N.Y. Fish Game J.* **30**(1), 78–90.

Schlesinger, D. A. and Regier, H. A. (1982). Climatic and morphoedaphic indices of fish yields from natural lakes. *Trans. Amer. Fish. Soc.* **111**, 141–150.

Schneider, J. C. and Leach, J. H. (1977). Walleye (*Stizostedion vitreum vitreum*) fluctuations in the Great Lakes and possible causes, 1800–1975. *J. Fish. Res. Board Can.* **34**, 1878–1889.

Serns, S. L. (1982). Influence of various factors on density and growth of age-0 walleyes in Escanaba Lake, Wisconsin, 1958–1980. *Trans. Amer. Fish. Soc.* **111**, 299–306.

Serns, S. L. and Kempinger, J. J. (1981). Relationship of angler exploitation to the size, age, and sex of walleyes in Escanaba Lake, Wisconsin. *Trans. Amer. Fish. Soc.* **110**, 216–220.

Shuter, B. J. and Koonce, J. F. (1977). A dynamic model of the western Lake Erie walleye (*Stizostedion vitreum vitreum*) population. *J. Fish. Res. Board Can.* **34**, 1922–1928.

Shuter, B. J., Koonce, J. F., and Regier, H. A. (1979). Modeling the western Lake Erie walleye population: A feasibility study. Great Lakes Fish. Comm. tech. rep. no. 34, pp. 39.

Skud, B. E. (1982). Dominance in fishes: The relation between environment and abundance. *Science* **216**, 144–149.

Smith, B. R. (Ed.) (1980). Proc. of the sea lamprey int. symp. (SLIS). *Can. J. Fish. Aquat. Sci.* **37**, 1585–2215.

Smith, L. L., Jr. and Pycha, R. L. (1961). Factors related to commercial production of the walleye in Red Lakes, Minnesota. *Trans. Amer. Fish. Soc.* **90**, 190–217.

Sprules, G. W. (1980). Zoogeographic patterns in the size structure of zooplankton communities, with possible applications to lake ecosystem modeling and management. In: W. C. Kerfoot (Ed.), *Evolution and ecology of zooplankton communities.* The Univ. Press of New England, Hanover, NH.

Thompson, W. F. (1950). The effects of fishing on stocks of halibut in the Pacific. Fish. Res. Inst., Univ. of Washington, Seattle, WA, 60 pp.

Van Oosten, J. (1946). A definition of depletion of stocks. *Trans. Amer. Fish. Soc.* **76**, 283–289 (reprint published in TAFS in 1949).

Van Oosten, J. (1949). The present status of the United States commercial fisheries of the Great Lakes. In: *Trans. 14th North Amer. Wildl. Conf.*, pp. 319–330.

Walters, C. J. (1981). Optimum escapements in the fact of alternative recruitment hypotheses. *Can. J. Fish. Aquat. Sci.* **38**, 678–689.

Werner, E. E. and Hall, D. J. (1976). Niche shifts in sunfishes: Experimental evidence and significance. *Science* **191**, 404–406.

Winterton, G. K. (1975). Structure and movement of a spawning stock of walleye, *Stizostedion vitreum vitreum* (Mitchill), in Georgian Bay. M.Sc. thesis, Univ. of Guelph, Guelph, Ontario, 95 pp.

17

MODELING APPROACHES FOR ASSESSING THE EFFECTS OF STRESS ON FISH POPULATIONS

D. S. Vaughan, R. M. Yoshiyama, J. E. Breck, and D. L. DeAngelis

Environmental Sciences Division
Oak Ridge National Laboratory
Oak Ridge, Tennessee

1. INTRODUCTION

Fish populations are subjected to a variety of natural and human-induced stresses including fishing, thermal and chemical pollution, and entrainment/impingement at power plants. The effect of each new stress on the health of a population is determined, in part, by the level of preexisting stresses. The effect of preexisting stresses may, to some extent, be ameliorated by compensatory processes that prevent a population from becoming extinct whenever environmental conditions change. Goodyear (1980) provides a detailed discussion of compensation in fish populations. However, the compensatory reserve of a population may be exhausted by these preexisting stresses, so that the addition of a new stress would lead to the decline, and possible extinction, of the population. This paper provides a review and comparison of five modeling approaches that may be useful for quantifying the effects of stress (multiple or otherwise) on fish populations. Traditional fisheries management approaches to modeling and other approaches are addressed, including: (1) surplus production models, (2) yield or yield-per-recruit models, (3) stock-recruitment models, (4) Leslie matrix models, and (5) bioenergetics models.

For each modeling approach, the kinds of data required to implement the basic model and ways that the effects of stress might be incorporated into the model structure are described. The assumptions, advantages, and usefulness of the various modeling approaches are compared. The ability of a modeling approach to represent and to be influenced by three basic population processes (mortality, growth, and reproduction) is emphasized in the comparisons.

2. MODELING APPROACHES

2.1. Surplus Production Models

Surplus production models can describe the dynamics of exploited populations without requiring knowledge of recruitment, individual growth, and mortality characteristics of the populations. General descriptions and applications of surplus production models are given by Schaefer (1968), Pella and Tomlinson (1969), and Fox (1970). These models require a time series of data consisting of total weight of catch from the population by year, and total fishing effort in standardized units by year. Catch per unit of standardized effort is assumed proportional to the population size, while fishing effort is assumed proportional to fishing mortality. Furthermore, the growth in population biomass in the absence of fishing mortality is assumed to be a function of population biomass. This function is such that no growth occurs when the population biomass is at zero or at some maximum value, while maximum growth in population biomass occurs at some intermediate level of population

Dr. Vaughan's present address: National Marine Fisheries Service, Southeast Fisheries Center, Beaufort Laboratory, Beaufort, North Carolina 28516.

biomass. The general form of the surplus production model (including fishing mortality, qf) is

$$\frac{dP(t)}{dt} = P(t) \cdot R[P(t)] - qfP(t) \tag{1}$$

where: $dP(t)/dt$ is the rate of change in population biomass, $P(t)$, over time; $R[P(t)]$ is the regulatory function; q is the catchability coefficient; and f is the fishing effort. The expression $P(t)R[P(t)]$ regulates the growth in population biomass of the unexploited stock, while $qfP(t)$ represents the yield to the fishery. The effect of stress on a population would probably be through a change in the regulatory function, although fishing mortality might also be altered (e.g., increased susceptibility to fishing). Changes in mortality, individual growth, and/or reproduction would be reflected in changes in the parameters, or possibly in the form, of the regulatory function. When $dP(t)/dt$ equals zero, the equilibrium yield is given by

$$Y(t) = qfP(t) = P(t) \cdot R[P(t)] \tag{2}$$

The maximum, or optimal, equilibrium yield, Y^*, occurs when the growth rate of a population is at its maximum, P^*. Maximum growth rate is obtained by differentiating the expression $P(t)R[P(t)]$ with respect to $P(t)$, setting the result equal to zero, and solving for $P (= P^*)$. Maximum equilibrium yield is then given by

$$Y^* = q \cdot f^* \cdot P^* \tag{3}$$

where f^* is obtained from Eq. (2); for example,

$$f^* = \frac{R(P^*)}{q} \tag{4}$$

The regulatory function for the Schaefer surplus production model (Schaefer, 1954, 1957) is

$$R[P(t)] = r \frac{1 - P(t)}{K} \tag{5}$$

where r is the intrinsic growth rate and K is the carrying capacity of the environment. Substituting Eq. (5) into Eq. (1) results in a parabolic equation in terms of the population biomass (Fig. 1a). For the unexploited stock ($f = 0$), the maximum growth rate of the population biomass occurs midway between $P = 0$ and $P = K$. The intersection of this curve (the solid parabola in Fig. 1a) with the yield line (dashed line in Fig. 1a) gives the equilibrium yield, and the intersection of the yield line at the maximum value of the parabola gives the maximum equilibrium yield. Because $P^* = K/2$ for the Schaefer model and $f^* = r/2q$, then $Y^* = q \cdot f^* \cdot P^* = rK/4$. The population trajectory (100 yr) is given in Fig. 1b for a population

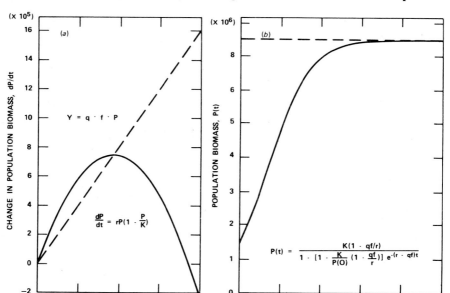

Figure 1. (a) Hypothetical Schaefer surplus production model where solid curve represents the dynamics of the unexploited stock and the dashed line represents the yield to the fishery. Intersection of the solid and dashed lines represents the sustainable yield, with maximum sustainable yield occurring at the maximum value of the solid curve. Parameter values are $r = 0.2$, $K = 15$ million units of biomass, and $qf = 0.2$ per yr. (b) Population trajectory for the Schaefer (solid line) surplus production model in (a) undergoing maximum sustainable yield (MSY). Dashed line represents the equilibrium population size (7.5 million) for population under MSY.

producing at maximum equilibrium yield (with $r = 0.2$, $K = 15$ million, and $qf = 0.1$).

Some important assumptions for this approach include the following:

1. Equilibrium population and stable age structure are attained instantaneously for each level of fishing effort.
2. No appreciable lags occur in population processes that mediate density-dependence of the population (see Marchesseault et al., 1976).
3. Catch-per-unit-effort is proportional to population size.
4. There is constant catchability.
5. Fishing effort is independent of population size and time.
6. There is no appreciable net migration.

Data needs include a sufficient time series of catch and corresponding fishing effort data; these data are usually readily available for commercial fish stocks. However, their interpretation may be difficult and their reliability suspect.

Model parameters that can be modified to incorporate the effects of stress include: r (the intrinsic growth rate), K (the carrying capacity of the environment), and q

(the catchability coefficient). The stress ultimately affects one or more of three basic population processes: mortality, growth (of an individual), and reproduction. These three processes are confounded in the population parameter r (which relates the processes of population increase through reproduction and growth to the processes of population decrease through mortality). However, if only one of the basic processes is affected, then the parameter r, which is equal to the birth rate minus the death rate, can be changed to reflect the effect of a stress on this process.

2.2. Yield Models

Yield models contain explicit terms to represent the three basic population processes. General descriptions and applications of yield models are given by Gulland (1969) and Ricker (1975). The general form of the yield model is

$$\frac{dY(t)}{dt} = F \cdot N(t) \cdot W(t) \tag{6}$$

where: $dY(t)/dt$ is the rate of change in biomass yield, $Y(t)$, to the fishery; F is the instantaneous fishing mortality rate; $N(t)$ is a mathematical function describing population numbers at age t; and $W(t)$ is a mathematical function describing the weight of an average individual at age t. Integrating both sides of Eq. (6) over the lifespan of a cohort permits one to determine the total yield in biomass from the cohort. The effect of stress on a population would probably be accomplished by either an increase in the adult natural mortality rate, a decrease in the individual growth rate, or a reduction in recruitment to the adult stock. Because these factors are treated separately in yield models, in contrast to surplus production models, yield models can more directly utilize laboratory data on the effects of toxic substances on these factors.

Beverton and Holt (1957) employed the exponential decay formula for representing the decline in population numbers with age, that is,

$$N(t) = R \cdot \exp[-(M + F)(t - t_c)] \tag{7}$$

where: $t > t_c$; t_c is the age at recruitment to the exploited phase (or entry to the fishery); R is the number of recruits to the exploited phase, $R = N(t_c)$; and M is the instantaneous natural mortality rate. Weight at age was represented by the von Bertalanffy (1938) growth equation:

$$W(t) = W_\infty\{1 - \exp[-k(t - t_0)]\}^3 \tag{8}$$

where: W_∞ is the maximum (or asymptotic weight of an individual; k governs the rate at which the maximum weight is approached; and t_0 is the hypothetical age when the weight of an individual is zero (t_0 may be either positive or negative).

Substituting Eqs. (7) and (8) into Eq. (6) and integrating from t_c to infinity (t_L, the maximum age of fishable stock is assumed arbitrarily large) produces the following equation for yield in biomass from a cohort over its fishable lifespan:

$$Y = FRW_\infty \sum_{n=0}^{3} \frac{U_n \exp[-nk(t_c - t_0)]}{F + M + nk} \tag{9}$$

where $U_n = 1, -3, 3, -1$, for $n = 0, 1, 2, 3$, respectively. The number of recruits can be factored out of Eq. (9), resulting in the so-called yield-per-recruit model. The traditional end point of a yield or yield-per-recruit analysis is the yield isopleth diagram (Fig. 2). The axes consist of two model parameters that are controllable by the fishery manager: (1) age at entry to the fishery (t_c) on the vertical axis and (2) instantaneous fishing mortality rate (F) on the horizontal axis. The lines on this diagram represent contours of equal yield-per-recruit, with the middle right area representing conditions for obtaining maximum yield to the fishery. Assuming an equilibrium population size (constant recruitment) allows the yield from a cohort over its lifespan to be equivalent to the yield from all cohorts in a single year.

Important assumptions for the yield models include the following:

1. Natural and fishing mortality rates remain constant during the exploited phase and are independent of population density.

2. There is perfect retention of fish by gear on entering the exploited phase.

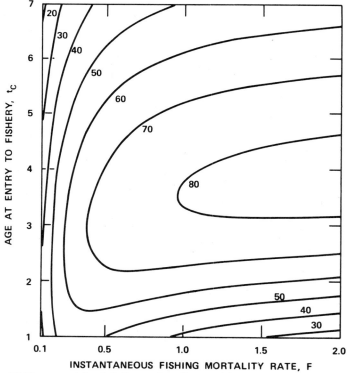

Figure 2. Yield-per-recruit isopleth diagram for yellow perch in western Lake Erie (data from Vaughan, 1981). Yield-per-recruit is in grams, and the parameter values are: $M = 0.26$ per yr; $W_\infty = 340$ g; $k = 0.44$ per yr; and $t_0 = 0.2$ yr.

3. The von Bertalanffy growth equation adequately describes individual growth during the exploited phase
4. Recruitment occurs instantaneously on the same date each year.
5. There is no appreciable net migration.

Data needs for this modeling approach include estimates of the instantaneous natural and fishing mortality rates (M and F), von Bertalanffy growth parameters (W_∞, k, and t_0), age at entry to the fishery (t_c), and the maximum age of the fishable stock (t_L).

This approach allows the effect of stress to be incorporated separately and directly for two of the three population processes: (1) mortality (beyond the age of recruitment to the exploited phase) can be increased, and (2) growth rate (in the von Bertalanffy growth equation) can be decreased. In yield models, any effect of stress on reproduction is confounded with mortality before the age of recruitment. The combined effect could be represented as a decrease in the number of recruits. Percent annual loss in yield-per-recruit to a fishery is illustrated in Fig. 3 with the proportional reduction in growth rate (k') on the vertical axis and the proportional reduction in annual survival (X') on the horizontal axis. The parameter k' varies from 0 (no reduction in the original growth rate) to 1 (100% reduction in the original growth rate), while X' also varies from 0 (no mortality due to the impact of a stress—instantaneous toxic mortality rate X equals 0/year) to 1 (complete annual mortality due to the impact of a stress—X equals infinity). The lower left corner ($k' = 0$, $X' = 0$) is the region of the isopleth diagram for which there is no impact of the stress on the population. The contour lines represent loci of equal percent annual loss in yield-per-recruit to the fishery. The contour lines indicate there is an approximately equal effect due to fractional reductions in either growth rate or survival. The contour lines show that a serious reduction in yield to a fishery can occur for relatively small reductions in growth rate and survival. A proportional reduction in recruitment, say $R' = aR$, would simply rescale the yield-per-recruit function by the same factor (a). This may prove useful for assessing a reduction in the prerecruit population.

2.3. Stock-Recruitment Models

Stock-recruitment models describe the relationship between stock size and the number of subsequent progeny recruited into the stock. Two well-known and widely applied forms are those of Ricker (1954, 1975) and Beverton and Holt (1957), which are special cases of a generalized model studied by Christensen et al. (1977), DeAngelis et al. (1977), and DeAngelis and Christensen (1979). These models can be easily modified to include the effects of environmental factors (Paulik, 1973; Yoshiyama et al., 1981). Applications of stock-recruitment models are given by Cushing (1973) and Cushing and Harris (1973). The general form of the stock-recruitment model is

$$\frac{dN(t)}{dt} = -M \cdot N(t) \tag{10}$$

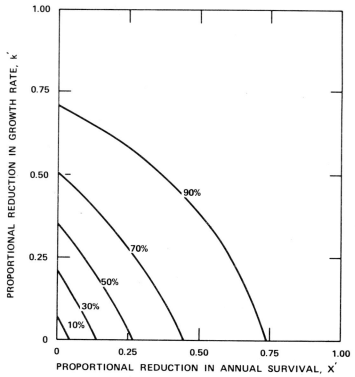

Figure 3. Percent loss in yield-per-recruit isopleth diagram for yellow perch in western Lake Erie (data from Vaughan, 1981). Proportional reduction in growth rate (k') is on the vertical axis, varying from no reduction in original growth rate (k) at $k' = 0$ to complete reduction in growth rate at $k' = 1$. Proportional reduction in annual survival (X') is on the horizontal axis, varying from no reduction in annual survival at $X' = 0$ to complete mortality at $X' = 1$. Parameter values for the unimpacted population are: $M = 0.26$ per yr; $F = 0.85$ per yr; $W_\infty = 340$ g; $k = 0.44$ per yr; $t_0 = 0.2$ yr; and $t_c = 1.6$ yr.

where $dN(t)/dt$ is the rate of change in population numbers [$N(t)$] over time, and M is a mortality parameter. The effect of stress on a population would probably occur through an increase in the mortality parameter, M. As described later, this mortality parameter can be separated into a stock-independent term and a stock-dependent term, either or both of which can be manipulated to reflect stress effects. Unlike with yield models, the mortality parameter for stock-recruitment models considers only the period from spawning to recruitment to the adult stock. A constant or average number of eggs per recruited fish also appears in the stock recruitment model, and this value can also be made to vary in response to stress.

In the Ricker (1954) stock-recruitment model the mortality function, M, is broken into two components: (1) a stock-independent mortality term (m_0), and (2) a stock-dependent mortality term (m_1). Hence,

$$M = m_0 + m_1 \cdot S(t') \tag{11}$$

where $S(t')$ is the stock size at the time of spawning, t'. Letting $R(t' + T)$ be the number of recruits to the stock T units of time (years) later, then integrating Eq. (10) over this time span with Eq. (11) substituted for M in Eq. (10) gives the typical Ricker equation:

$$R(t' + T) = a \cdot S(t') \cdot \exp[-b \cdot S(t')] \qquad (12)$$

where

$$a = E \cdot \exp(-m_0 \cdot T), \qquad b = m_1 \cdot T$$

and E is the mean number of eggs spawned per recruited fish. A constant or average E implies either that the population is in equilibrium or that little change in fecundity occurs with increasing age.

Two Ricker curves are shown in Fig. 4, together with a 45° replacement line. Stock size is shown on the horizontal axis with the subsequent recruits shown on the vertical axis. For single-age spawners having a single sexually mature age-class, recruits and stock size would both be represented by the number of individuals in

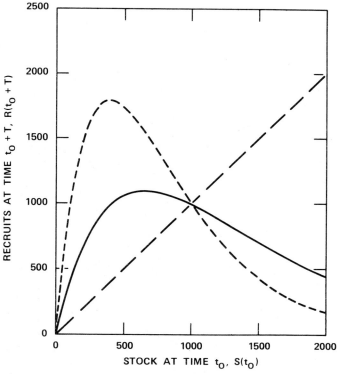

Figure 4. Two hypothetical Ricker stock-recruitment models (solid and dashed curves) with the 45° replacement line. The Ricker curves have the same equilibrium stock (or recruit) size (1000 individuals) but different slopes at the replacement line.

this age-class (with the appropriate lag time for this age-class). A suitable analog is also necessary for multiage spawners. Christensen et al. (1977) suggested converting both recruits and stock to potential egg production. Ricker curves intersect the replacement line at a value of 1000 recruits (or stock size of 1000); then both populations represented by the two Ricker curves will have the same equilibrium population size. However, the dynamics of the population trajectory depends on the slope of the Ricker curve at its intersection with the replacement line (May, 1976). If the absolute value of the slope is less than one (as in the solid curve in Fig. 4), then a stable equilibrium occurs (see Fig. 5a). On the other hand, if the

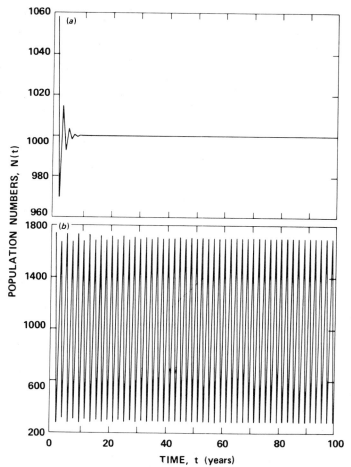

Figure 5. Population trajectories for 100 yr for the two Ricker models in Fig. 4. The dynamics of the trajectories are determined by the slope of the Ricker model at the replacement line. (a) A stable equilibrium is attained when the absolute value of the slope is less than 1 (slope is -0.5 for solid curve in Fig. 4). (b) Two-point bifurcation (or more complex behavior) is attained when the absolute value of the slope is greater than 1 (slope is -1.5 for dashed curve in Fig. 4).

absolute value of the slope is greater than one (as in the dashed curve in Fig. 4), then bifurcations or even chaotic behavior can occur (note the two-point bifurcation shown in Fig. 5b). The slope of the Ricker curve at the replacement line is found by differentiating the Ricker curve (Eq. 12) with respect to $S(t')$ and by setting $S(t')$ equal to the equilibrium stock size $(ln(a)/b)$. The slope (s) is found to be equal to $1 - ln(a)$, or, in terms of E and m_0,

$$s = 1 - ln(E) + m_0 \cdot T \tag{13}$$

In addition to the assumption of a stable age structure that permits the use of an average fecundity value, E, other model assumptions include:

1. all stock-dependent mechanisms occur prior to the age of recruitment,
2. mortality of young is a linear function of the parental stock size at time t',
3. only stock size at the time of spawning affects survivorship of young between time of spawning and age of recruitment, and
4. there is no appreciable net migration. Data needs include an adequate time series of estimates of stock size and subsequent recruits.

This approach allows the effect of stress to be incorporated separately into two of the three population processes: (1) mortality (prior to the age of recruitment) can be increased, especially the stock-independent mortality parameter, m_0; and (2) reproduction (through the average fecundity value, E) can be reduced. The effect of stress on a population can be reflected in two ways. There will be a reduction in the equilibrium yield, and there may be a change in the model dynamics. Because the slope of the Ricker model at the replacement line is typically negative, an increase in m_0 or decrease in E will result in a decrease in the absolute value of the slope (Eq. 13). Hence, the addition of stress to a population assumed regulated by the Ricker model can show a change in dynamics from a two-point bifurcation for the unimpacted stock to a stable equilibrium for the impacted stock. This change in dynamics is a consequence of the Ricker model. Whether it would occur in nature depends on the appropriateness of the underlying Ricker model.

2.4. Leslie Matrix Models

The Leslie matrix, or age-structure, approach was independently developed by Bernardelli (1941), Lewis (1942), and Leslie (1945, 1948) to follow a female population through time. The model, as shown in Fig. 6, relates the age structure at time t, $n(t)$, to the age structure at time $t + 1$, $n(t + 1)$, by multiplication with a square matrix $A(t)$. The first row of this matrix consists of measures of age-class specific fecundities; the subdiagonal contains age-class specific survival probabilities from one age-class to the next and zeros elsewhere. The elements can be assumed constant over time, or they can be made functions of time or population density. Usher (1972) presents an excellent survey of the literature on age-structure models through 1970. Stability of a Leslie matrix with density-dependent (e.g., stock–recruitment relationship) survival is studied by DeAngelis et al. (1980), and an example for yellow perch in western Lake Erie is given in Vaughan (1981). The

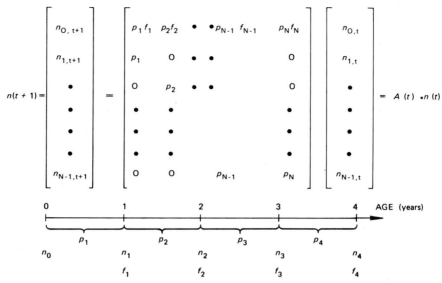

Figure 6.   Leslie matrix formulation of life cycle population model. $n(t)$ and $n(t + 1)$ are the population vectors at times t and $t + 1$, respectively; $A(t)$ is the Leslie matrix; the p_i's are survival rates; and the f_i's are fecundities. The bottom part of this figure is designed to show the relationship of the parameters p_i and f_i to the age-class numbers $n_{i,t}$. For example, p_1 is the survival rate from age-class 0 to age-class 1.

advantage of the Leslie matrix model over previous modeling approaches discussed in this paper is that the effects of stress on mortality and fecundity can be considered in an age-specific manner. If fecundity is determined via a length–age or weight–age relationship, then the effect of stress on growth can also be manipulated.

Emlen (1973) constructed a matrix that assumed that fecundity occurs over a short duration just prior to the end of the year; for example, the age–class specific measures of fecundity are as shown in Fig. 6 $(p_i f_i)$. Fecundity is usually obtained in a three-step process: (1) length-age relationship such as the von Bertalanffy (1938) equation; (2) weight–length relationship or condition factor; and (3) eggs-weight relationship, which is often approximately linear.

As shown in Fig. 6, Leslie matrices are used to project an age structure through time. If the matrix A is independent of time and population density, then the age structure of a population at any time t, $n(t)$, can be obtained given the initial age structure, $n(0)$; that is,

$$n(t) = A^t n(0) \qquad (14)$$

The density-independent Leslie matrix is analogous to the exponential or Malthusian model. The Perron-Frobenius theorem (Demetrius 1971b) implies that the Leslie matrix always has a real, positive eigenvalue (λ) whose modulus is not exceeded by any other eigenvalue. This maximal eigenvalue is related to the Malthusian parameter, r:

$$r = ln(\lambda) \tag{15}$$

If λ exceeds 1 ($r > 0$), then the population will grow without bounds; if λ is less than 1 ($r < 0$), then the population will decay to extinction; and if λ equals 1 ($r = 0$), then the population is perfectly balanced in equilibrium. An additional important property of many Leslie matrices is that they are ergodic; that is, they approach a stable age structure (related to the eigenvector of the maximal eigenvalue) that is independent of the initial age structure (Demetrius 1971a).

Important assumptions in applying the density-independent (or linear) Leslie matrix include:

1. life history parameters are independent of time and density,
2. parameters are constant within an age-class,
3. spawning occurs over a short duration of time on the same date each year, and
4. there occurs no appreciable net migration.

Furthermore, survival through the first year of life (p_1) is often extremely difficult to estimate from field data. This parameter can be estimated from the Leslie matrix by assuming an equilibrium population size ($\lambda = 1$) if one has estimates of the remaining p's and f's (Vaughan and Saila, 1976). Data needs include age-class-specific fecundity per mature female, age-class-specific proportion of females that are mature, age-class-specific sex ratios, first and last age-classes having mature females, and age-class-specific mortality rates (both natural and fishing).

This approach allows the effects of stress to be incorporated separately into all three population processes and in an age-class-specific manner. Mortality (all age-classes), individual growth (via growth rate parameter), and reproduction (via condition factor and egg production and viability) can all be altered to reflect the effects of stress on these processes. In the case of the density-independent Leslie matrix, the population will decline to extinction unless the effects of the stress are removed prior to extinction. For the density-dependent Leslie matrix, a lower equilibrium population size will be attained and a possible change in dynamics may result (Levin and Goodyear, 1980).

2.5. Bioenergetics Models

Bioenergetics models examine the factors affecting the growth of an individual fish (Ursin, 1967, 1979; Elliott, 1976; Andersen and Ursin, 1977; Kitchell et al., 1977; Kitchell and Breck, 1980; Stewart, 1980). The rate of change in average individual biomass (dB/dt) is related to the weight-specific rates of physiological processes of consumption (C), respiration (R), egestion (F), excretion (U), and reproductive loss (G) (Elliott, 1976; Kitchell et al., 1977); that is,

$$\frac{dB}{B\,dt} = C - (R + F + U + G) \tag{16}$$

These physiological rates are described mathematically as functions of other variables.

Consumption rate is typically related to fish biomass or body size (B), ambient water temperature, and food availability. Respiration rate is related to fish biomass, ambient water temperature, activity level, and apparent specific dynamic action. Egestion and excretion rates are related to consumption rate, ambient water temperature, and food quality and availability. Finally, rate of reproductive loss is related to fish biomass and season.

Bioenergetics models are often used to simulate the growth of a representative individual fish from each cohort over its lifetime. When applying the bioenergetics approach to an entire fish population, an accounting must be done for the numbers of fish in each cohort. A population model based on the bioenergetics approach must, therefore, include the process of mortality (see Anderson and Ursin, 1977; Stewart, 1980). This allows the possibility of modeling mortality as dependent on fish size as well as age.

Bioenergetics models generally have a finer temporal resolution than the modeling approaches discussed above. The models discussed previously are usually concerned with the annual increments to a fish population, while bioenergetics models often predict growth on a daily basis. For this reason bioenergetics models can be useful for describing the within-year dynamics of a population's response to stress.

Some important assumptions implicit in the use of these models include these:

1. Major factors influencing the growth rate are body size, ambient water temperature, and food quality and availability.

2. Growth of an average individual represents growth of the cohort.

3. Daily averages adequately represent biological rates.

4. Relative food availability is constant between observations of growth.

5. Temperature-dependent rates, as measured in a series of constant temperature experiments, will apply for the variable temperature in the field (i.e., instantaneous acclimation to temperature and no seasonal adjustments in rates other than those due to temperature).

6. Caloric density is a constant function of body size (or season) and independent of growth and consumption rates.

Data needs include a sufficient time series of ambient water temperatures and concomitant estimates of either body size or daily consumption. In addition, estimates are needed of the physiological parameters describing rates of consumption, respiration, egestion, excretion, and reproductive loss as functions of body size, temperature, and other variables.

One advantage of the bioenergetics approach is that the response to stress can reflect the particular mode of action of the stress. Stress, such as a chemical toxicant, can alter growth rate via changes in food availability, capture efficiency, preferred temperature, or such physiological processes as respiration, assimilation efficiency, excretion, or spawning success and egg production. Because body size is the dependent variable, mortality and fecundity (reproductive loss) rates can be made size dependent. Thus all three population processes can be addressed using this approach. Because these processes are represented in detail, the effects of stress on these processes can be mimicked to a similar level of resolution. This approach is also useful for analyzing the population's response to multiple stresses, which may each act in a different way. One could use this approach to compare and contrast the predicted response to individual stresses or to analyze the overall response to particular combinations of stresses. In either case, this approach enables the study of responses at the individual level that result from a variety of alterations in the mechanisms that produce these responses.

3. DISCUSSION

The usefulness of a particular modeling approach for assessing the effects of stress on fish populations is determined by the applicability and suitability of the model, by the data available for developing the basic model, and by the data available for assessing the effects of a stress on basic population processes. Hence the ability to represent the influence of a stress on three basic population processes (mortality, growth, and reproduction) is an important consideration in the selection of a modeling approach.

The processes of mortality, growth, and reproduction are confounded in the model parameters of surplus production models. In particular, the parameter r represents the difference between processes leading to an increase in population biomass (i.e., growth and reproduction) and the process causing a decline in population biomass (i.e., mortality). But the task of defining relationships among populations and estimating parameters would be nearly insurmountable.

Yield models and stock-recruitment models each treat two of the three population processes separately. Yield or yield-per-recruit models describe mortality beyond the age of recruitment and growth explicitly and separately, while stock-recruitment models describe mortality prior to recruitment and an averaged value of reproduction explicitly and separately. Dynamic pool models, which combine yield models with stock-recruitment models, complement each other and permit all three population processes to be considered separately.

Leslie matrix, or age-structure, models also permit all three population processes to be considered separately and in an age-class-specific manner. Bioenergetics models, although focused on the growth of an average individual, permit all three population processes to be considered. Growth is the dependent variable in bioenergetics models and can be incorporated into a Leslie matrix framework. Mortality and reproduction must be considered for the bioenergetics model to apply to populations. We believe that combining the bioenergetics and Leslie matrix approaches would provide a powerful tool for estimating long-term impacts of toxicants on fish populations. This combination, though very data intensive, would make possible the most detailed comparisons of stressed and unstressed populations.

One should not expect too much of current fish population models. They can answer certain questions quite well, but other questions are beyond the current state of the art. In this regard we recommend that these modeling approaches be used for making relative comparisons of stressed to unstressed fish populations rather than for attempting to attain the unrealistic goal of precisely predicting absolute population levels (see Van Winkle, 1977).

Table 1 provides a summary of the population processes modeled, time scale, and biological/physiological resolution for the five population modeling approaches discussed in this paper. The population processes that can be investigated by each approach have been a major theme of this paper. The time scale by which each of these models operates varies from annual or seasonal to daily. Biological/physiological resolution refers to the level of detail or complexity of the model, and consequently to the data needed for the model. The surplus production approach has minimum resolution, since the basic population processes cannot be separated. Yield, stock-recruitment, and Leslie matrix approaches have an intermediate level of resolution, since the basic population processes can be investigated separately (the Leslie matrix addresses these processes in an age-class-specific manner). Finally, the bioenergetics approach has the most detailed level of resolution, since it addresses the processes underlying growth, and hence reproduction, and, ultimately, mortality.

For all of these approaches the focus is on the population level; ecosystem- or community-level effects are not considered. It would certainly be possible mathematically to link several of these population-level models so that, for example, the effects of predation and interspecific competition could be taken into account in predicting response to stress.

The choice of a modeling approach depends, in part, on data availability. Surplus production and stock-recruitment models generally require the existence of a long time series of data, while yield-per-recruit, Leslie matrix, and bioenergetics models require extensive life history and other data, some of which may be available in the literature. If information on the effects of a particular stress on a fish is limiting, then one of the simpler modeling approaches may be adequate (e.g., surplus production models). If, however, more extensive knowledge of the effects of a stress is available, then the use of a more complicated model structure (e.g., bioenergetics models) would generally be preferable.

The usefulness of a particular modeling approach depends, in part, on the applicability and suitability of the model relative to the level of resolution desired and the level of effort (time, money, staff) available for the assessment. If detailed questions are being asked, for example, about the effect of a stress on the age structure of a population or the effect on within-year growth patterns, then one of the more detailed approaches must be used. In certain cases the constraints of time, money, or staff may prohibit the use of the more complex, data-intensive models, and thereby foreclose the ability to answer detailed questions. Simpler, less data-intensive models can be useful for answering less detailed questions; they can be useful for screening a large number of fish populations that may be at risk. Additional research can then be directed toward implementing more complex, data-intensive models to further investigate a reduced number of fish populations that are considered to be at risk.

4. SUMMARY

The cumulative effects of multiple stresses on fish populations may cause the collapse of a population, even though the effects of a single stress may be insignificant.

Table 1. Population Processes Modeled, Time Scale, and Biological/Physiological Resolution for the Five Population Models Discussed in This Paper

Modeling Approach	Processes Modeled[a]		Mortality		Time Scale	Biological/Physiological Resolution
	Growth	Reproduction	Prerecruitment	Postrecruitment		
Surplus production	cf	cf	cf	cf	Annual	Minimal
Yield	X	cf	X	X	Annual/seasonal	Growth/mortality
Stock-recruitment	—	X[b]	X	—	Age at recruitment[c]	Mortality/reproduction
Leslie matrix	(x)	X	X	X	Annual/seasonal	Age specific
Bioenergetics	X	X	(x)	(x)	Daily	Most detailed[d]

[a] X, process is modeled. (x), process has been included in some models.—, process is not modeled. cf, processes are confounded in this approach.

[b] Constant or average fecundity per recruited fish is used.

[c] Time scale is period from spawning to recruitment, which can range from a fraction of a year to several years, depending on the fish species.

[d] Processes include temperature-, ration-, and weight-dependent rates of consumption, respiration, egestion, excretion, and reproductive loss.

Several population modeling approaches are reviewed and compared for their usefulness in quantifying the aggregate effect of multiple sources of stress on populations. Specifically, ways are investigated to insert the effects of stress (multiple or otherwise) into five major modeling approaches: (1) surplus production models, (2) yield or yield-per-recruit models, (3) stock-recruitment models, (4) Leslie matrix models, and (5) bioenergetics models. For each modeling approach, the kinds of data required to implement the basic model and ways that the effects of stress might be incorporated into the model structure are described. The assumptions, advantages, and usefulness of the various modeling approaches are compared.

The usefulness of a particular modeling approach for assessing the effects of stress on fish populations is determined by (1) the applicability and suitability of the model relative to the level of resolution desired and the level of effort (time, money, staff) available for the assessment, (2) the data available for developing the basic model, and (3) the data available for assessing the effects of a stress on the basic population processes. Simpler, less data-intensive models are useful for screening a large number of fish populations that may be at risk. Additional research can then be directed toward implementing more complex, data-intensive models for the reduced number of fish populations that are considered to be at risk. These modeling approaches should be used for making relative comparisons of stressed to unstressed fish populations rather than for attempting to attain the unrealistic goal of precisely predicting absolute population levels.

ACKNOWLEDGMENTS

This study was funded by the National Power Plant Team (NPPT), Office of Biological Services, U.S. Fish and Wildlife Service, U.S. Department of Interior, under Interagency Agreement DOE 40-1107-80 (Contract No. 14-16-0009-80-1039) with the U.S. Department of Energy under contract W-7405-eng-26 with Union Carbide Corporation. The Project Officer was Paul Rago. The authors gratefully acknowledge the critical reviews by S. W. Christensen (ORNL), J. C. Goyert (ORNL), and Paul Rago (USFWS). Publication No. 2123, Environmental Sciences Division, Oak Ridge National Laboratory, Oak Ridge, Tennessee.

REFERENCES

Andersen, K. P. and Ursin, E. (1977). A multispecies extension to the Beverton and Holt theory of fishing with accounts of phosphorus circulation and primary production. *Medd. Dan. Fisk. Havunders. N.S.* **7**, 319–435.

Bernardelli, H. (1941). Population waves. *J. Burma Res. Soc.* **31**, 1–18.

Beverton, R. J. H. and Holt, S. J. (1957). On the dynamics of exploited fish populations. *U.K. Min. Agric. Fish, Fish. Invest. (Ser. 2)* **19**, 533 pp.

Christensen, S. W., DeAngelis, D. L., and Clark, A. G. (1977). Development of a stock-progeny model for assessing power-plant effects on fish populations. In: W. Van Winkle (Ed.), *Assessing the effects of power-plant-induced mortality on fish populations*, Pergamon Press, New York, pp. 196–226.

Cushing, D. H. (1973). Dependence of recruitment on parent stock. *J. Fish. Res. Board. Can.* **30**, 1965–1976.

Cushing, D. H. and Harris, J. G. K. (1973). Stock and recruitment and the problem of density dependence. *Rapp. P.-V. Reun. Cons. Int. Explor. Mer* **164**, 142–155.

DeAngelis, D. L., and Christensen, S. W. (1979). A general stock-recruitment curve. *J. Cons. Int. Explor. Mer* **38**, 324–325.

DeAngelis, D. L., Christensen, S. W., and Clark, A. G. (1977). Responses of a fish population model to young-of-the-year mortality. *J. Fish. Res. Board Can.* **34**, 2124–2132.

DeAngelis, D. L., Svoboda, L. J., Christensen, S. W., and Vaughan, D. S. (1980). Stability and return times of Leslie matrices with density-dependent survival: Applications to fish populations. *Ecol. Model.* **8**, 149–163.

Demetrius, L. (1971a). Primitivity conditions for growth matrices. *Math. Biosci.* **12**, 53–58.

Demetrius, L. (1971b). Multiplicative processes. *Math. Biosci.* **12**, 261–272.

Elliott, J. M. (1976). The energetics of feeding, metabolism, and growth of brown trout (*Salmo trutta* L.) in relation to body weight, water temperature and ration size. *J. Anim. Ecol.* **45**, 923–948.

Emlen, J. M. (1973). *Ecology: An evolutionary approach*. Addison-Wesley, Reading, MA.

Fox, W. W., Jr. (1970). An exponential surplus-yield model for optimizing exploited fish populations. *Trans. Amer. Fish. Soc.* **99**, 80–88.

Goodyear, C. P. (1980). Compensation in fish populations. In: C. H. Hocutt and J. R. Stauffer, Jr. (Eds.), *Biological monitoring of fish*, D. C. Heath and Company, Lexington, MA, pp. 253–280.

Gulland, J. A. (1969). Manual of methods for fish stock assessment. Part I. Fish population analysis. *FAO Man. Fish. Sci.* **4**, 154 pp.

Kitchell, J. F. and Breck, J. E. (1980). Bioenergetics model and foraging hypothesis for sea lamprey (*Petromyzon marinus*). *Can. J. Fish. Aquat. Sci.* **37**, 2159–2168.

Kitchell, J. F., Stewart, D. J., and Weininger, D. (1977). Applications of a bioenergetics model to yellow perch (*Perca flavescens*) and walleye (*Stizostedion vitreum* vitreum). *J. Fish. Res. Board Can.* **34**(10), 1922–1935.

Leslie, P. H. (1945). On the use of matrices in certain population mathematics. *Biometrika* **33**, 183–212.

Leslie, P. H. (1948). Some further notes on the use of matrices in population mathematics. *Biometrika* **35**, 215–245.

Levin, S. A. and Goodyear, C. P. (1980). Analysis of an age-structured fishery model. *J. Math. Biol.* **9**, 245–274.

Lewis, E. G. (1942). On the generation and growth of a population. *Sankhya* **6**, 93–96.

Marchesseault, G. D., Saila, S. B., and Palm, W. J. (1976). Delayed recruitment models and their application to the American lobster (*Homarus americanus*) fishery. *J. Fish. Res. Board Can.* **33**, 1779–1787.

May, R. M. (1976). Simple mathematical models with very complicated dynamics. *Nature* **261**, 459–467.

Paulik, G. J. (1973). Studies of the possible form of the stock-recruitment curve. *Rapp. P.-V. Reun. Cons. Int. Explor. Mer* **164**, 302–315.

Pella, J. J. and Tomlinson, P. K. (1969). A generalized stock production model. *Bull. Inter-Am. Trop. Tuna Comm.* **14**, 420–496.

Ricker, W. E. (1954). Stock and recruitment. *J. Fish. Res. Board Can.* **11**, 559–623.

Ricker, W. E. (1975). Computation and interpretation of biological statistics of fish populations. Fish. Res. Board Can. bull. 191, 382 pp.

Schaefer, M. B. (1954). Some aspects of the dynamics of populations important to the management of the commercial marine fisheries. *Bull. Inter-Am. Trop. Tuna Comm.* **1**, 25–56.

Schaefer, M. B. (1957). A study of the dynamics of the fishery for yellowfin tuna in the eastern tropical Pacific Ocean. *Bull. Inter-Am. Trop. Tuna Comm.* **2**, 245–268.

Schaefer, M. B. (1968). Methods of estimating effects of fishing on fish populations. *Trans. Amer. Fish. Soc.* **97**, 231–241.

Stewart, D. J. (1980). Salmonid predators and their forage base in Lake Michigan: A bioenergetics-modeling synthesis. Ph.D. thesis (Zoology), Univ. of Wisconsin-Madison, Madison, WI.

Usher, M. B. (1972). Developments in the Leslie matrix model. In: J. N. R. Jeffers (Ed.), *Mathematical models in ecology*. Blackwell, London, pp. 29–60.

Ursin, E. (1967). A mathematical model of some aspects of fish growth, respiration, and mortality. *J. Fish. Res. Board Can.* **24**, 2355–2453.

Ursin, E. (1979). Principles of growth in fishes. *Symp. Zool. Soc. London* **44**, 63–87.

Van Winkle, W. (1977). Conclusions and recommendations for assessing the population-level effects of power plants exploitation: The optimist, the pessimist, and the realist. In: W. Van Winkle (Ed.), *Assessing the effects of power-plant-induced mortality on fish populations*. Pergamon Press, New York, pp. 365–372.

Vaughan, D. S. (1981). An age structure model of yellow perch in western Lake Erie. In: D. G. Chapman and V. F. Gallucci (Eds.), *Quantitative population dynamics*. Int. Co-operative Publishing House, Fairland, MD, pp. 189–216.

Vaughan, D. S. and Saila, S. B. (1976). A method for determining mortality rates using the Leslie matrix. *Trans. Amer. Fish. Soc.* **105**, 380–383.

von Bertalanffy, L. (1938). A quantitative theory of organic growth. *Human Biol.* **10**(2), 181–213.

Yoshiyama, R. M., Van Winkle, W., Kirk, B. L., and Stevens, D. E. (1981). Regression analyses of stock-recruitment relationships in three fish populations. ORNL/NUREG/TM-424, Oak Ridge Nat. Lab., Oak Ridge, TN.

18

MEASURING THE HEALTH OF AQUATIC ECOSYSTEMS

S. R. Kerr and L. M. Dickie

Marine Ecology Laboratory
Bedford Institute of Oceanography
Dartmouth, Nova Scotia

1. INTRODUCTION

It is a basic tenet of the biological sciences that interbreeding populations of organisms have become what they are through a process of adaptation to the environmental changes they experience. Adaptation may be to features of the physical or chemical environment, or to the complex predatory and competitive pressures imposed by other organisms. Whatever its cause, adaptation is reflected in organismic changes that work through individuals and their offspring to give rise to alterations in the average characteristics of populations.

Environmental change, whatever its form, therefore acts on individuals or their offspring. In the case of toxic contaminants, for example, effects on individuals can sometimes be immediate and very clear, resulting in death. But an enormous and growing literature attests, in addition, to ubiquitous sublethal effects, varying from behavioral modification through physiological impairment to morphological deformation. There are proven cases of effects on fecundity, fertility, and zygote viability. In short, a virtual plethora of information reflects direct deleterious effects, of varying subtlety, on the health of individual organisms and sometimes on their progeny.

In the face of a welter of disparate effects, the question naturally arises as to the possibility of assessing, in integral form, the varied responses of individual organisms to environmental insult. That is, can practical indices be derived of the effect, at the community level of ecological organization, of factors whose effects are realized upon individual organisms?

A positive answer to the question depends upon the reality of existence of community-level regularities. Evidence for conservative patterns of ecological structure and function is required. In fact, there is ample evidence that such regularities indeed exist (Ryder et al., 1981; Dickie and Kerr, 1982; Kerr, 1982), providing evidence of control at the community level of organization. Given evidence of the existence of ecological control, contemporary systems theory justifies our aim of seeking ways to describe how environmental contaminants, for example, may disrupt ecological control systems, and consequently, modify the normal patterns of community structure and function.

Accordingly, we deal here with the feasibility of assessing ecological health at the community level of biological organization. In particular, we are concerned to demonstrate that a theoretical basis exists to support the possibility of deriving useful indices of community health, and that there is both theoretical and empirical support for the choice of criterion that we identify here.

2.   AN APPROACH TO ASSESSING ECOLOGICAL HEALTH

Unfortunately, it is not a simple matter to translate evident effects on individuals into ecological effects on aggregations. When the aggregation itself is an identifiable, interbreeding group of organisms it may sometimes be possible to perceive the effects of selective alterations to its production system. For example, death of some individuals may also imply impaired growth and fecundity of others, suggesting a quite general deterioration of living conditions, which could be indexed as death of a certain fraction of the population. Alternatively, removal of some individuals may be reflected in compensatory growth or survival of different genetic strains within the same stock, thus complicating the evident "dosage-response" of the first case. When the concept of aggregation is stretched beyond single, homogeneous, interbreeding stocks to include biological predators, competitors, or prey, which operate as part of the same control or community system, the difficulty of assessment of effects becomes obvious.

The demand placed on contemporary biology by environmental alterations caused by humankind is for an effective means of assessing these effects, whether they be rapid and massive or slow, subtle, and persistent. That is, from observations on alterations in individual well-being the science is required to describe, measure, and predict the emergent effects on ecological communities, in such a manner and with such precision that the relevant social and political processes can be debated and the consequences evaluated.

It is our belief that the problems of judging ecological health are amenable to a great variety of approaches at many different levels of generality. But because scientific enquiry has to develop and present a methodology that is amenable to questions posed at many different analytical scales, we have chosen here to approach the problem of assessment of the health of aquatic ecosystems at the most general level that appears feasible. We advocate that assessments of ecological health initially adopt a nontypological approach, based on the observation that fish production systems can be described by characteristic size-spectra that are sensitive to environmental perturbations. The practice of constructing size-composition diagrams of individual species has a strong tradition in fisheries population study and management, a background that gives some assurance that the techniques are familiar and practicable. However, the application of size-composition information to multispecies associations is new, and derives from developments in community ecology that have only become prominent in the last decade. The evidence suggests that the methodology embodies a remarkably powerful tool that offers distinct advantages for application in environmental analysis and monitoring.

3. ECOLOGICAL SIZE-SPECTRA

Sheldon and Parsons (1967) were the first to point out that the most informative and ecologically appropriate expression of the size-spectrum of particulate matter in seawater takes the form of total particle volume plotted against the logarithm of equivalent particle diameter. Sheldon et al. (1972) then showed that, if organisms in the pelagic zone of the sea are similarly grouped into octave-scale body-size classes, the biomasses of successive groups, ranging in size from bacteria to whales, appear to form a simple linear series that shows a relatively small biomass decrease with increasing body size.

Kerr (1974) demonstrated that this relationship depended on the ubiquitous body-weight–metabolic-rate rule in biology, coupled with an apparently constant allometric exponent of the doubling time for body size (i.e., the growth rate) of aquatic organisms. As Kerr noted, the observed constancy of the biomass relations then depends on the constancy of the body-size relationship between predator and prey. Platt and Denman (1977) point out that this latter restriction may be relaxed in the unique case of the use of the octave scale for classification of organism body size. That is, by using an octave size-scaling, and existing information on specific growth and predation rates, and on gross growth efficiency (or its complement, the fraction of energy lost to respiration), it is possible to generalize the rate of change of biomass with body size. Within this generalization, it is clear that one can then consider the significance of the number of octaves involved in particular energy transfers. Silvert and Platt (1978) suggest that a relaxation of the octave scale may be reasonable, provided one can invoke certain restrictions on the range of the energy growth and loss functions of organisms.

Apart from marine pelagic systems, there is evidence that similar size-dependent considerations are appropriate for the pelagic zone of freshwater lakes. Sprules and Holtby (1979) and Sprules (1980) showed that size-dependent community descriptors are more closely related to the efficiency and nature of energy flow through the pelagic zones of lakes, and bear a stronger relation to the morphometric and hydrological properties of the lakes, than do traditional taxonomic descriptions. More recently, Sprules and Knoechel (1984) concluded that four characteristic size-spectra suffice to describe the pelagic communities of a large and diverse set of lakes. Apart from the direct observations of pelagic size-spectra offered by Sprules and colleagues, there is a large limnological literature (c.f. Kerfoot, 1977) related to the processes underlying freshwater community structure, which further attests to its characteristic orderliness and predictable size composition.

Schwinghamer (1981) found that different, but equally characteristic, size-spectra adequately describe various marine benthic communities. The differences in benthic spectra Schwinghamer attributed to a substrate effect. There are three characteristic modal sizes, corresponding to organisms that colonize the surface of the sediment grains, move between them, or occupy the substrate as an integral whole. Despite these differences from pelagic size-spectra, however, it appears that benthic communities manifest equally characteristic, predictable, size distributions of organisms.

The conclusion we reach is that a spectrum of biomasses in relation to body size provides us with a generalization of the dynamics of aquatic organisms that embodies some very general properties of the organization of ecological production, as well as some apparently general features of the energy transfers between community constituents. It appears that the available theory and observations of community size structure provide good reason to suppose that observation of biomass body-size spectra should provide us with a basic index of ecological health.

4. DYNAMICS OF SIZE-SPECTRA

It remains to examine the evidence for dynamical change of characteristic size-spectra, in consequence of the effects of external perturbation. Woodwell (1970) was apparently the first to notice a general tendency for changes in the average sizes of organisms in communities subjected to many exogenous factors, an observation supported by Kerr and Vass (1973) for toxic contaminants.

Greve and Parsons (1977) suggested that external factors as disparate as storm-induced mixing, lowered nutrient availabilities, and the introduction of hydrocarbons or other contaminants could be expected to cause characteristic changes in the trophic organization, hence size-spectra, of marine pelagic systems. Sprules and Knoechel (1984), reasoning that smaller zooplankton organisms tend to be more resistant to low pH, suggest that community size-spectra could provide useful indicators of the effects of acid precipitation on poorly buffered lakes. Regier (1973) has reviewed a dramatic time series of size-related changes in the Great Lakes biota, reflecting the combined effects of a suite of anthropogenic factors, a subject further considered by Ryder et al. (1981).

Taken together, the foregoing observations, which are by no means exhaustive, attest to the ubiquitous sensitivity of community size-spectra to a variety of human-induced and other environmental changes. Accordingly, despite the fact that various environmental factors act on individual organisms and their progeny in a multitude of possible ways, it is clearly possible to generalize from individual effects on organisms to emergent effects on the integral community. We conclude that deformations of characteristic size-spectra may well provide a cheap (relative to conventional taxonomic approaches), and usefully sensitive, early-warning criterion for monitoring and assessing the health of ecological communities. This is not to say that size-spectra necessarily identify the specific factors causing change, but rather that they provide a suitably generalized tool for detecting change. Identification of specific causal factors will normally entail further detailed analysis once the preliminary evidence of change is indexed by size-spectrum analysis. Although we entertain little reservation concerning the utility of size-spectral analysis as an effective indicator of ecological health, we do draw attention to the need for further work to identify the temporal, spatial, and organism-size scales over which it is effective, and the basic community types (other than aquatic pelagic and marine benthic) to which it is appropriate.

REFERENCES

Dickie, L. M. and Kerr, S. R. (1982). Alternative approaches to fisheries management. In: M. C. Mercer (Ed.), *Multispecies approaches to fisheries management advice.* Can. Spec. Publ. Fish. Aquat. Sci. 59.

Greve, W. and Parsons, T. R. (1977). Photosynthesis and fish production: Hypothetical effects of climatic change and pollution. *Helgolander wiss. Meeresunters* 30, 666–672.

Kerfoot, W. C. (1977). Implications of copepod predation. *Limnol. Oceanogr.* 22, 316–325.

Kerr, S. R. (1974). Theory of size distribution in ecological communities. *J. Fish. Res. Board Can.* 31, 1859–1862.

Kerr, S. R. (1982). The role of external analysis in fisheries science. *Trans. Amer. Fish. Soc.* 111, 165–170.

Kerr, S. R. and Vass, W. P. (1973). Pesticide residues in aquatic invertebrates. In: C. A. Edwards (Ed.), *Environmental pollution by pesticides.* Plenum Publishing Co., London, pp. 134–180.

Platt, T. and Denman, K. L. (1977). Organization in the pelagic ecosystem. *Helgol. wiss. Meeresunters* 30, 575–581.

Regier, H. A. (1973). Sequence of exploitation of stocks in multi-species fisheries in the Laurentian Great Lakes. *J. Fish. Res. Board Can.* 30, 1991–1999.

Ryder, R. A., Kerr, S. R., Taylor, W. W., and Larkin, P. A. (1981). Community consequences of fish stock diversity. *Can. J. Fish. Aquat. Sci.* 38, 1856–1866.

Schwinghamer, P. (1981). Characteristic size distributions of integral benthic communities. *Can. J. Fish. Aquat. Sci.* 38, 1255–1263.

Sheldon, R. W. and Parsons, T. R. (1967). A continuous size spectrum for particulate matter in the sea. *J. Fish. Res. Board Can.* 24, 909–915.

Sheldon, R. W., Prakash, A., and Sutcliffe, W. H., Jr. (1972). The size distribution of particles in the ocean. *Limnol. Oceanogr.* **17**, 327–340.

Silvert, W. and Platt, T. (1978). Energy flux in the pelagic ecosystem: A time-dependent equation. *Limnol Oceanogr.* **23**, 813–816.

Sprules, W. G. (1980). Zoogeographic patterns in the size structure of zooplankton communities, with possible applications to lake ecosystem modelling and management. In: W. C. Kerfoot (Ed.), *Evolution and ecology of zooplankton communities.* Univ. Press of New England, pp. 642–656.

Sprules, W. G. and Holtby, L. B. (1979). Body size and feeding ecology as alternatives to taxonomy for the study of limnetic zooplankton community structure. *J. Fish. Res. Board Can.* **36**, 1354–1363.

Sprules, W. G. and Knoechel, R. (1984). Lake ecosystem dynamics based on functional representations of trophic components. In: D. G. Meyers and R. Strickler (Eds.), *Trophic dynamics in aquatic ecosystems.* Amer. Assoc. Adv. Sci. symp. no. 20.

Woodwell, G. M. (1970). Effects of pollution on the structure and physiology of ecosystems. *Science* **168**, 429–433.

19

ASSESSING THE HEALTH OF THE OCEANS: AN INTERNATIONAL PERSPECTIVE

John B. Pearce

National Marine Fisheries Service
Highlands, New Jersey

1. INTRODUCTION

An interest in the health and multiple use of the marine environment has become a major media topic, nationally and internationally. For example, in recent years there has been increasing concern in regard to the possible impacts of offshore oil exploration and development in or near productive fisheries grounds. At the same time, municipalities and industry have indicated a dissatisfaction with the legislated termination of ocean dumping of sewage sludge and various industrial wastes. Proposals have been made to relocate dumpsites from inshore coastal waters to locations well beyond the shelf-slope break (Federal Register, 1982). Many major ports have proposed, or are conducting, new dredging of channels and deepening of slips to accommodate larger tankers and colliers. Finally, the United States and other nations must face the possible problems associated with disposal of various radioactive wastes or by-products from the nuclear power industry.

Some scientists and administrators have believed that that the oceans are capable of receiving and assimilating increased amounts of wastes and contaminants (Goldberg, 1979). Planning documents or feasibility studies have appeared that propose the possibility of increased contaminant loading of marine waters; they also suggested an urgent need for increased monitoring of and research on ocean disposal of various wastes (Munn, 1973; NOAA, 1979, 1981a). Environmental managers, the public, and legislators must have the information from such studies if appropriate decisions are to be made. The need for long-term monitoring and research in relation to the effects of wastes and other anthropogenic activities has been recognized, especially as these have affected the living resources of the continental shelf waters and adjacent estuaries, and the production of foodstuffs from the sea (McIntyre and Pearce, 1980; Pearce, 1980).

During recent years the Marine Environmental Quality Committee (MEQC) of the International Council for the Exploration of the Sea (ICES), Copenhagen, Denmark, has had a variety of interests in the health of the marine environment. A number of working groups that report directly to MEQC have been concerned with topics such as the biological effects of pollution and environmental stress, monitoring the quality of the environment (using biological effects monitoring), developing regional assessments concerned with the effects of humanity's activities, and intercalibration exercises to standardize analytical procedures, especially those involved with measuring inorganic and organic contaminants in the physical components of the environment, as well as in the biota (see ICES, 1981, 1982, 1983).

In February 1979, ICES, in conjunction with the National Oceanic and Atmospheric Administration (NOAA) and the U.S. Environmental Protection Agency (USEPA), sponsored a workshop on biological effects monitoring. The workshop, held at the Duke University Marine Laboratory, brought together over 60 scientists from Europe, Canada, and the United States. The proceedings were ultimately published as volume 179 of the ICES *Rapports et Proces-verbaux des Reunions* (McIntyre and Pearce, 1980). The volume describes and discusses a number of techniques and protocols, some of which are now being used for biological effects monitoring. Also, ICES intercalibration exercises have resulted in a number of ICES Cooperative Research Reports, which provide information on the efficacy of the efforts of participating laboratories and scientists in terms of providing comparable results when standard samples were analyzed for contaminants.

During the 1978 statutory meetings of ICES, it was recommended by MEQC that member nations begin the preparation of assessments or syntheses setting forth what was then known about the health of major estuarine and coastal waters. A series of papers was presented at the ICES statutory meetings in Warsaw in 1979.

These papers indicated the "status" of certain east coast estuaries of the United States (Larsen and Doggett, 1979; Lippson and Lippson, 1979; Maurer, 1979; Pearce, 1979; Phelps, 1979; Reid, 1979). Subsequently, the Advisory Committee on Marine Pollution (ACMP) of ICES reiterated the importance of developing syntheses on the status of marine waters, especially those estuaries and coastal waters deemed to be highly degraded by human activities (Bewers et al., 1982). Several standing committees of ICES have indicated that it is most important to begin to use the extensive information concerned with degraded estuaries and shelf waters, as well as generic information about effects of specific contaminants on a variety of marine life, to develop assessments of the estuarine and coastal waters of the North Atlantic, both in Europe and North America.

Finally, it was a formal resolution of MEQC that member nations begin to implement biological effects techniques in monitoring the health of estuarine, coastal, and oceanic waters. In response to this, the United States augmented a National Marine Fisheries Service (NMFS) program called Ocean Pulse (OP), a major effort to use biological effects techniques in assessing how various stresses have been and are affecting recruitment and survival of important fishery resources (Pearce, 1980,

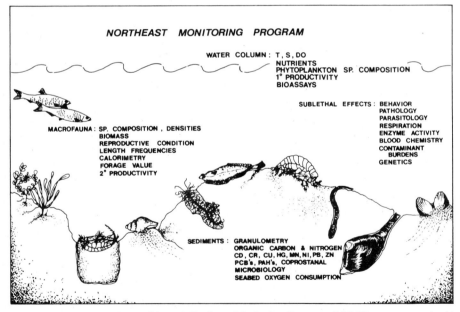

Figure 1. The Ocean Pulse (OP) and Northeast Monitoring Programs (NEMP) are concerned with normal variability *and* the effects of pollution in the physical and chemical environment and the sublethal and lethal effects on resource species and the food chain organisms that support these populations. A variety of monitoring measurements are performed at stations that are polluted or likely to receive pollutants in the future; the principal variables that are measured routinely are indicated in this figure. Other variables, that is, contaminant burdens in biota and water, are measured periodically but with less frequency.

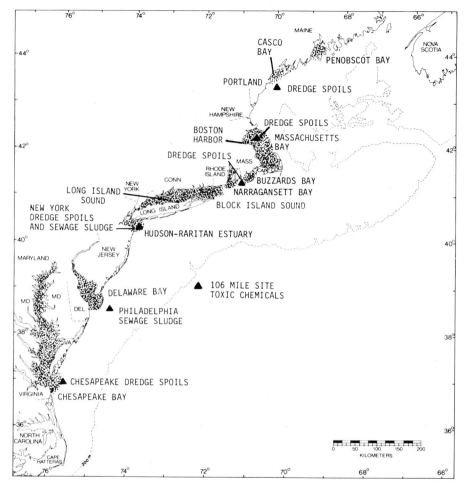

Figure 2. Major estuaries, embayments (stippled), and dumpsites (▲) that are included in the NEMP areas for research and monitoring.

1981). The earlier developed OP program was eventually combined with various other NOAA research and monitoring activities to form a single, unified monitoring program called the Northeast Monitoring Program (NEMP) (see Figs. 1 and 2). The OP program, conducted since 1977 by the Northeast Fisheries Center (NEFC), NMFS, thus formed the basis for the combined monitoring program.

A program development plan (PDP) and the technical development plan (TDP) for the NEMP were completed in September 1980. During the planning phase, however, it was essential to begin to initiate monitoring activities using new, sometimes untested procedures. Numerous cruises and monitoring studies were started in 1980 and these activities culminated in a series of reports submitted to the program manager of NEMP by over 50 principal investigators (PIs). While elements of

NMFS and NOAA were involved in many of the actual field monitoring efforts and ancillary studies, a considerable portion of the work was done through contracts with academic institutions, consulting firms, and other governmental agencies.

The results of these studies were synthesized into the NEMP annual reports for 1981 and 1982 (NOAA, 1981b, 1983). The reports were written for use by environmental managers, conservationists, administrators within several state and Federal agencies, and for the informed public. They indicated the general status of the ecological health of the northeast continental shelf of the United States, and adjunct estuaries, and included statements that reference the considerable data provided in the individual PIs reports. In addition, there were recommendations that evolved from the monitoring activities conducted in 1980 and 1981.

2. RESULTS FROM THE NORTHEAST MONITORING PROGRAM

2.1. Contaminants

Personnel working within the NEMP have begun to develop a series of assessments or statements about what is known of the *sources*, *fates*, and *effects* of contaminants as they are being introduced to estuarine and shelf waters off the northeast coast of the United States. One of the first reports in the series was concerned with the levels of certain organic contaminants in finfish (Boehm and Hirtzer, 1982). This report provided information on the levels of PCBs and petroleum hydrocarbons in several species of demersal and pelagic fish that were taken in trawls made at stations located from Georges Bank south to Cape Hatteras and into the southern U.S. Atlantic Bight and Gulf of Mexico. The results of these studies indicated that, while existing federal action limits were not exceeded in most species of fish, almost all finfish that were collected had trace levels of PCBs and petroleum hydrocarbons in their musculature. In some cases the levels were of the same order of magnitude as the levels that are deemed to be action limits (5 ppm) by federal regulatory agencies. The report therefore suggested that, while there were no detectable violations of action limits in the fish that were analyzed, there was reason for concern. More recently, analyses conducted by the state of New Jersey on certain inshore and estuarine fish taken along the New Jersey coastline indicated that the PCB levels in a number of species, including bluefish and striped bass, were well above the 5 ppm action limit (Belton et al., 1982). These findings have caused considerable concern on the part of the seafood consuming citizenry as well as the fishermen themselves. This has been reflected in widespread newspaper and T.V. accounts (Glading, 1983).

At the same time that fish were being collected for analysis of inorganic and organic contaminants in muscle tissues and certain vital organs, other studies were being conducted in regard to the levels of contaminants in sediments and waters of the New York Bight apex (see Fig. 2) as well as at stations located on the continental shelf of the Middle Atlantic Bight, that part of the shelf between Cape Cod and Cape Hatteras. The findings in regard to the levels of contaminants in sediments

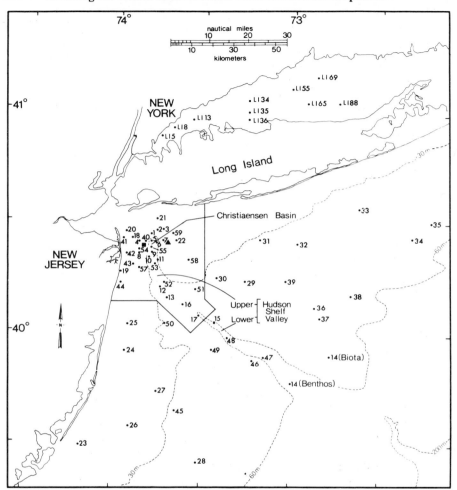

Figure 3.   Station locations in New York Bight and Long Island Sound. Stations within the lined box were a priori considered "contaminated" for comparison with other stations. "Replicate" stations sampled frequently are 4, 6, 7, 15, 26, and 31 (standard NEMP sites), and 40–43. Single grabs were taken at the remainder of stations 1–44. ▲ = sewage sludge disposal site; ■ = dredged material disposal site (Reid et al., 1982).

from the Middle Atlantic Bight and Long Island Sound, and the contaminant effects on the benthos was published as a NEMP report (Reid et al., 1982).

The results of the aforementioned two efforts are indicated, in part, in Figs. 3–11. It can be seen that in areas where there is active dumping of wastes such as sewage sludge and contaminated dredge materials, there are elevations in sediments of the amounts of trace metals, organic contaminants, and certain indicator substances such as coprostanol. Moreover, most information on levels of PCBs in fishes suggests that nearshore areas yield fishes having somewhat higher levels of PCBs and petroleum hydrocarbons; it is important, however, to recognize that these and other contaminants were found in finfish taken at or near the edge of the continental shelf. It has been hypothesized that such offshore fish are migratory forms that move inshore and offshore during certain stages of their life and that the elevated levels of contaminants may be due to exposure to the more contaminated inshore and estuarine waters.

However, there have been speculation and measurements that indicate that contaminants entrained in a dissolved or particulate form may be carried considerable distances from the point of discharge, whether this involves ocean dumping or discharges from pipes in coastal or estuarine situations. It has therefore been deemed important to understand the physical oceanographic processes and features

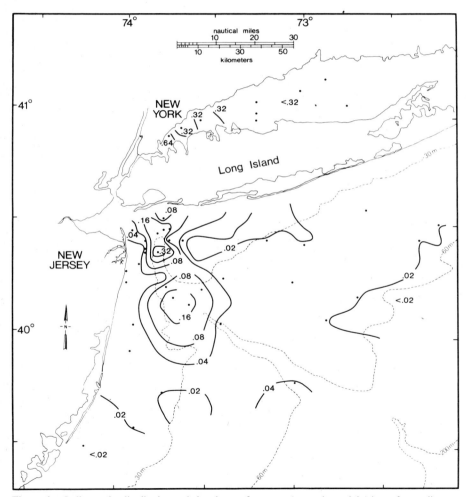

Figure 4.   Indicates the distribution and abundance of mercury (ppm, dry weight) in surface sediments in the New York Bight apex (Reid et al., 1982).

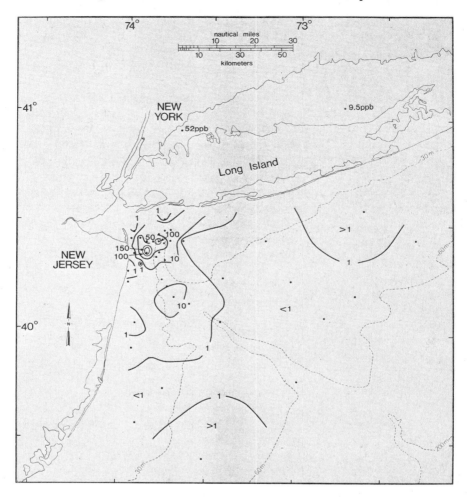

Figure 5. Indicates the distribution of PCB (ppb) in sediments in the New York Bight apex. Principal sources are thought to be sewage sludge and contaminated dredged materials from the Hudson River and several New York and New Jersey dredging sites (Reid et al., 1982).

that might be related to pollution distribution in coastal and offshore waters of the northeastern United States. For this reason, personnel working within the NEMP developed an extensive report on those physical oceanographic processes that might be responsible for distributing contaminants on the northeast continental shelf (Ingham et al., 1982). This document is one of the few comprehensive attempts to draw together all information that is known about the hydrographic processes on the northeast coast of the United States, as these might affect the distribution of contaminants.

2.2. Productivity

While considerable effort has been given to the measurement of benthic community structure (Figs. 10 and 11), contaminants in sediments and waters overlying benthic communities, and other aspects of marine benthic pollution biology, the NEMP also has had a strong interest in measuring the distributions and abundances of nutrients, chlorophyll, primary productivity, and aspects of phytoplankton community structure at stations located between the Canadian border and Cape Hatteras. Recent papers have summarized the findings in regard to the aforementioned variables

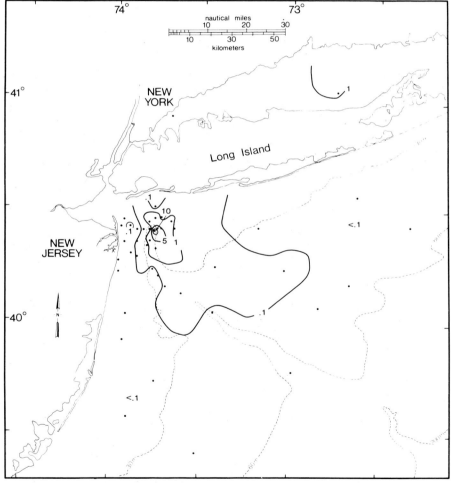

Figure 6. Coprostanol levels (ppm) in sediments. This organic compound is found in mammalian feces and is present in sewage sludge. Its presence in sediments receiving sludges is held as indicative of contamination by human wastes (Reid et al., 1982).

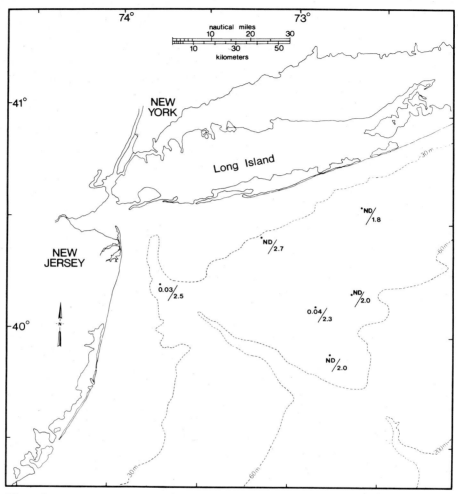

Figure 7. Wet weight values for PCBs (ppm)/% body fat in sea scallops (ND = not detected). This resource species does not appear to have taken up significant quantities of PCBs, whether the samples are collected from the Bight apex or further offshore (Reid et al., 1982).

(Marshall and Cohn, 1981a,b, 1982; O'Reilly and Busch, 1982). Some of the more pertinent results of these studies are shown in Figs. 12–15. The studies have indicated that shelf waters that are influenced by major outwellings from estuaries such as Raritan, Delaware, and Chesapeake Bays have been affected by elevated levels of nutrients. Work ongoing within ICES (1983) suggests that in European waters, particularly those of the more polluted parts of the North Sea and Baltic Sea, there have been unusual plankton blooms and red tides in recent years. While a cause and effect mechanism has not yet been elucidated, it is felt that coastal waters that are undergoing nutrient enrichment due to riverine outflows and waste discharges may be experiencing changes that long-term monitoring programs such as the NMFS's MARMAP and NEMP programs can measure. It has been reported recently that low dissolved oxygen events in the northern Gulf of Mexico might be the result of nutrients that are being added to coastal waters from farm lands and urban areas far removed from the coastal zone (Boesch, 1983). In 1976 there was a very significant event in the Middle Atlantic Bight during which low dissolved oxygen, described as anoxia, occurred during the summer months. An extensive series of papers edited by Swanson and Sindermann (1979) discusses the possible causes of the low dissolved oxygen problem. At the present time the ultimate causes

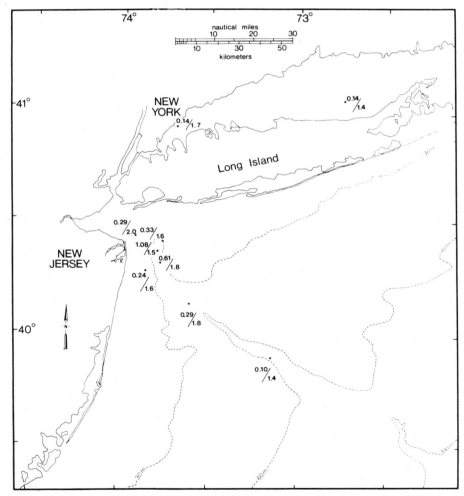

Figure 8.   Wet weight values for PCBs (ppm)/% body fat in the American lobster collected at stations in the heavily contaminated New York Bight apex and further offshore in the Hudson Shelf Valley (Reid et al., 1982).

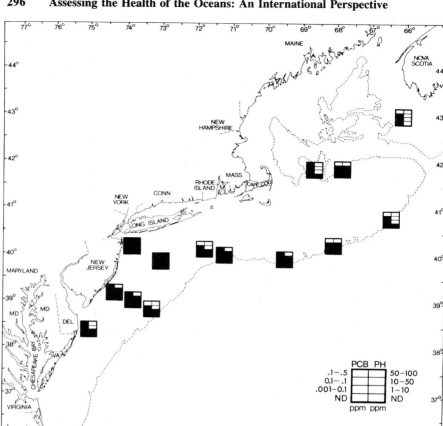

Figure 9. Indicates the amounts of PCBs and petroleum hydrocarbons in silver hake muscle tissue taken at stations on the continental shelf off the northeast coastline. The left vertical column indicates levels of PCBs within the sample and the right column indicates values for petroleum hydrocarbons in ppm. It is apparent that fish taken at the shelf-slope break have "significant" amounts of both contaminants in their musculature. It is presently believed that since these fish are migratory forms, the trace contaminants are taken up during those parts of their life history when they are in inshore areas that contain elevated levels of these contaminants within the physical habitat (after Boehm and Hirtzer, 1982).

are thought to have been unique meteorological and hydrographic conditions and unusual blooms of certain phytoplankton species. The principal species involved, *Ceratium tripos*, was found in great abundance during the summer of 1976. It is known that this species, along with others, was present in such large numbers that available nutrients might have been exhausted with the result that individual plankters died off in large numbers and were consumed by bacteria. This resulted in excessive consumption of oxygen in coastal bottom waters and the development of a condition of true anoxia in some areas, and extensive hypoxia in others. Similar events, as already noted, increasingly have occurred in European and Gulf of Mexico waters. It therefore becomes extremely important to monitor and assess the inputs of nutrients to coastal waters to determine if present land use practices are affecting significantly the coastal zone. Boesch (1983) suggested that these events may be far more important than the effects of local dumping or discharge practices.

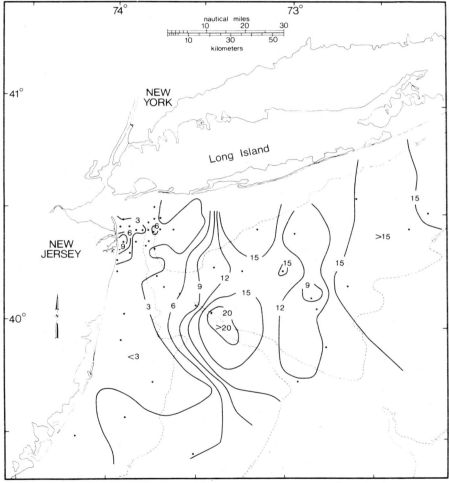

Figure 10.   Numbers of species of benthic amphipod crustaceans per 0.1 m². These small, shrimplike animals are known to be particularly sensitive to contaminant effects. Their total numbers and numbers of species per unit area may be greatly diminished in areas receiving solid wastes such as sewage sludge or contaminants such as petroleum hydrocarbons. Thus benchmarks established for them are valuable in assessing contaminant effects (Reid et al., 1982).

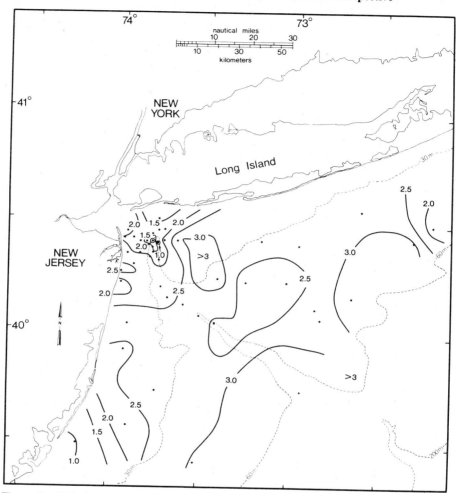

Figure 11. This figure indicates the Shannon-Weaver species diversities (H') per 0.1 m^2 for benthic communities in the New York Bight area of interest to the OP and NEMP programs. Indices such as H', total numbers of species, and total numbers of individuals are important in assessing the effects of stress, including contaminants, on benthic organisms and in interpreting how changes in the benthos may affect fisheries of interest (Reid et al., 1982).

All of the aforementioned tend to relate to those parts of the Middle Atlantic Bight that have been known or recently demonstrated to be highly contaminated with a wide range of materials from terrigenous export, point and nonpoint source discharges, and ocean dumping. One of the principal objectives of the NEMP program has been, however, to establish benchmarks for those areas that are not at the present time significantly contaminated because of dumping, runoff, or discharge of waste materials. We have observed parts of the continental shelf and have been making measurements of levels of contaminants in sediments, waters, and biota to ascertain what the background levels of contaminants may be. The reason for this is that as we enter an era of "multiple-use" of the coastal zone, including mineral exploration and development, transportation, fishing, both recreational and commercial, and increased need for waste disposal, it will be important to have benchmarks or baselines against which the effects of humanity's future activities can be compared. Also, in more remote areas of the region of concern to the NEMP program, there are numerous embayments and estuaries, large and small, which are undergoing increased development. For instance, Portland, Maine is an area that for a variety

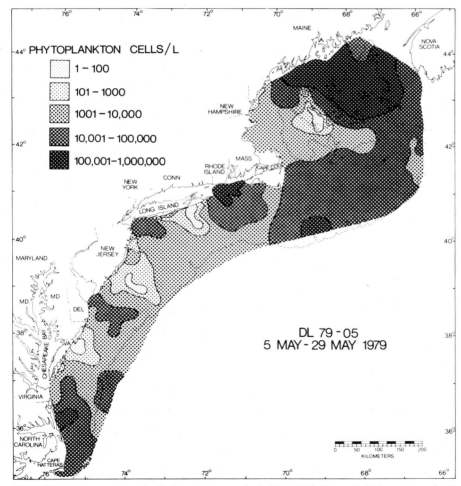

Figure 12.   Concentrations of cells/liter during a cruise in May 1979. Unusually high numbers of cells or changes in numbers of cells with time are important in assessing how natural and human-induced stresses may be affecting plankton populations over broad areas, the food chains, and ultimately, the fisheries (Marshall and Cohn, 1982).

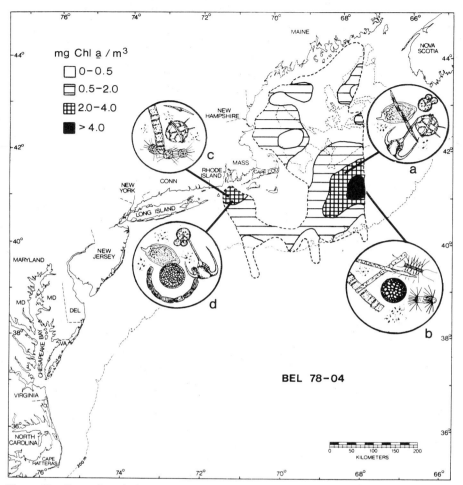

Figure 13. Some phytoplankton species identified in areas of high chlorophyll *a* (determined by C. Evans, NMFS): (*a*) *Nitzschia pungens, Gymnodinium dissimle, Protoperidinium* sp., *Ceratium tripos, Prorocentrum micans*, nanoplankton; (*b*) *Rhizosolenia alata, R. delicatula, R. stolterfothi, Corethron criophilium, Chaetoceros decipiens, Guinardia flaccida, Coscinodiscus radiatus*, nanoplankton; (*c*) *Nitzschia pungens, Protoperidinium* sp., *Thalassiosira* sp., nanoplankton; (*d*) *Gymnodinium dissimile, Ceratium tripos, Rhizosolenia stolterfothi, Coscinodiscus lineatus, Prorocentrum micans*, nanoplankton. Again, by noting changes in the makeup of species assemblages, it becomes possible to assess the effects of nutrient enrichment in estuaries and coastal and shelf waters (Marshall and Cohn, 1981b).

of reasons is increasing rather rapidly in its population density and industrial activities. Personnel from the Bigelow Laboratory, Boothbay, Maine, working within the NEMP program, have established a series of environmental benchmarks for Casco Bay and Penobscot Bay (Figs. 16 and 17). These areas are far removed from major urban areas but constitute estuaries that are projected to have future increased growth in domestic and industrial activities. One of the first NEMP reports on environmental benchmark studies for Casco Bay has recently been published (Larsen et al., 1983a). In addition, Larsen and his co-workers have published papers on levels of metals and petroleum hydrocarbons in sediments of Casco Bay and Penobscot Bay (Larsen et al., 1983b,c). Their findings indicated that indeed there are apparent increases in the amounts of inorganic and organic contaminants in the sediments of these relatively isolated areas. This is not surprising in as much as both domestic and industrial wastes are discharged into such coastal and estuarine waters and undoubtedly associated contaminants accumulate in the sediments. The rate at which this is occurring is a subject of a continuing investigation. Scientists at the Bigelow

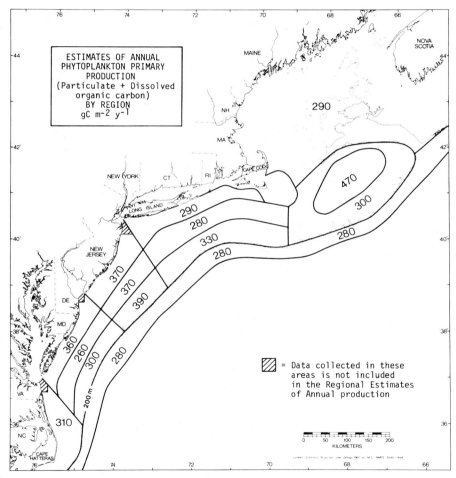

Figure 14. Estimates of annual phytoplankton production (particulate and dissolved organic carbon) by region, g C/m²·yr. Once again, elevated levels of primary production are associated with unique hydrographic phenomena and the introduction of inorganic and organic nutrients from both point and nonpoint sources (see O'Reilly and Busch, 1982, for further discussion; from O'Reilly and Busch, 1982).

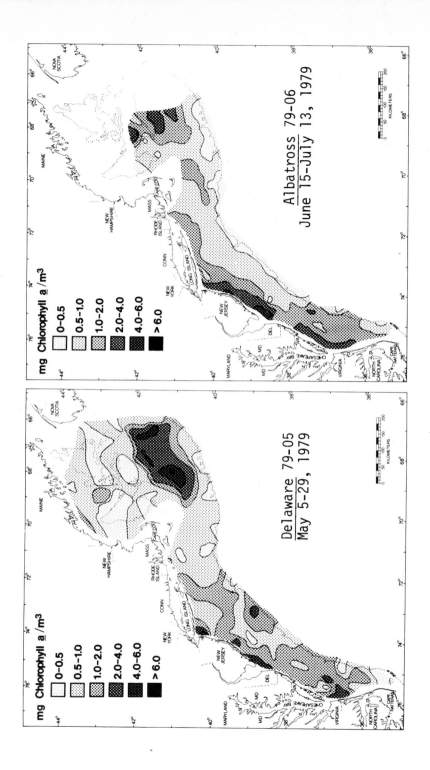

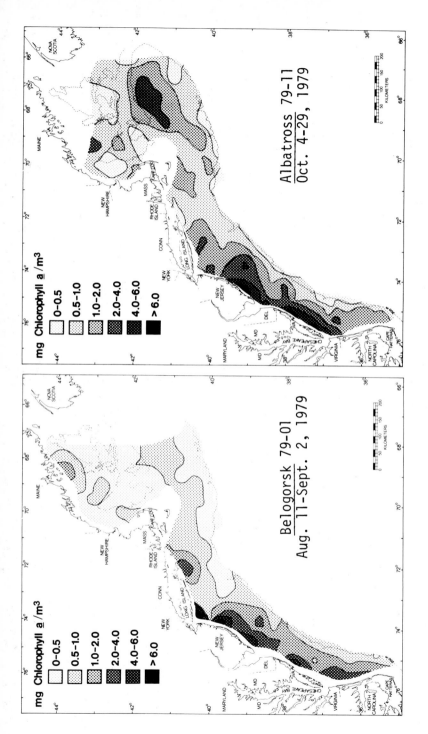

Figure 15. Distribution of average water column concentrations of chlorophyll *a* during four NEFC surveys. Standing stocks of chlorophyll *a* are often indicative of nutrient conditions; unusually high levels of chlorophyll *a* are often associated with greatly elevated levels of nutrients having sources in point and nonpoint sources of organic wastes (O'Reilly and Busch, 1982).

303

Laboratory have collected long sediment cores that will be analyzed at different sediment depths to develop an interpretation of the approximate rate at which materials are accumulating, and to establish the levels of contaminants that might have existed in relatively pristine harbor areas some decades ago. The results of these studies are not yet available.

These studies are thought to be extremely important inasmuch as it is well over a century since embayments adjacent to the New York Harbor area were reported to have become so contaminated with petroleum hydrocarbons from the early refineries that the taking of shellfish and finfish from such waters was precluded (Goode, 1887). We have numerous historical "benchmarks" that suggest that early increases in population and the industrialization of coastal waters resulted in changes that compromised the consumption of seafoods from such areas as early as the time of the U.S. Civil War. It is important now to have an historical perspective about what has happened and is happening in parts of our eastern coastline that have not yet been *significantly* affected by wastes but that will be in the future. If long-term

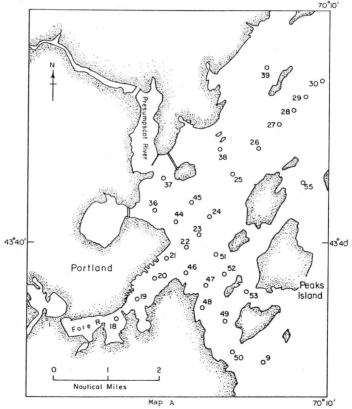

Figure 16. Locations of the 56 benthic stations sampled in Casco Bay, April 1980, for benthos, sediment contaminant levels, and other variables (Larsen et al., 1983a).

CHROMIUM

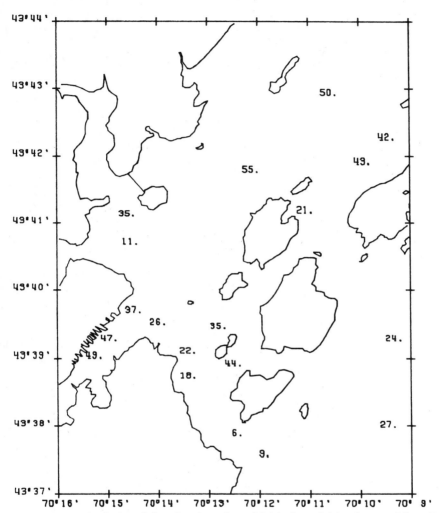

Figure 17. The distribution of chromium (ppm dry weight) in the surficial sediments of Casco Bay, Maine (Larsen et al., 1983a).

monitoring studies are not in place to demonstrate that early development and industrialization in such coastal waters is having an impact, there will be little information available to manage effectively such areas in a multiple-use mode. Obviously, it is not sufficient simply to demonstrate that such changes are taking place; it is very important that information resulting from studies on the effects of contaminants in isolated areas be used in a management program that will allow adjustments to be made in the way various contaminants are released to the environment. Moreover, in areas already heavily impacted we must be able to demonstrate what has happened so as to effect mitigation procedures and pollution abatement.

It should be thoroughly understood that the effects of pollution and habitat degradation can no longer be regarded as being local or even regional problems. Once contaminants are released to marine waters they become truly interstate or even international problems. This obviously has been demonstrated in the Great Lakes and, more recently, in the marine waters of North American and several European nations having common coastal waters.

3.   MONITORING SEQUENCE

Because of the OP and NEMP programs, and similar long-term monitoring and research efforts that exist or are being implemented in North America and Europe, there are, for the first time in history, well-developed, coordinated benchmarks for a range of contaminant variables over extensive stretches of coastal and shelf habitats. Individuals involved in the management of the OP and NEMP programs, based upon information, suggestions, and resolutions from ICES and other international and national organizations, have recently enunciated a sequence of events that are necessary in terms of long-term monitoring and research as these relate to habitat conservation and resource management. These are:

1. Identification of problem areas or "hot spots" using chemistry and biological effects measurements.
2. Identification of trends: spatial and temporal.
3. Quantification of effects:
    (a) Sublethal effects.
    (b) Mortality.
    (c) Effects on stocks.
4. Implementation of risk assessments.
5. Analyses of causation (field and laboratory experiments).
6. Management of fisheries habitats:
    (a) Regulations.
    (b) Legislation.
    (c) Fisheries management plans leading to maintenance of yields or increased yields, and habitat enhancements and mitigation.

4.   BIO-EFFECTS MONITORING

Scientists participating in the OP and NEMP have made comprehensive measurements that have already been used to delineate "hot spots" or problem areas (Reid et al., 1982; Larsen et al., 1983a). Our measurements to date in relation to contaminant buildup have verified that specific areas within the New York Bight and certain embayments (Casco Bay) are characterized by elevated contaminant levels.

These and other benchmarks are now being monitored to determine spatial and temporal trends that might occur because of increased or decreased contaminant loading. The report by Reid et al. (1982) compares the findings from recent measurements with similar research results obtained almost a decade ago. Thus we have already been able to describe temporal change for selected areas.

The OP/NEMP is now into stage 3 (see above). The third stage involves a quantification of measured biological effects. We have been able to demonstrate that certain contaminants, often in a synergistic fashion, result in increased mortality of bottom-dwelling organisms that are important forage species. Biological effects measurements, including hematological studies, have shown that organisms that live along pollution gradients in Long Island Sound and Narragansett Bay reflect environmental stress in blood and blood chemistry measurements, including hematocrit and serum ion change. This work is being done by biochemists and physiologists at the NMFS Laboratory, Milford, Connecticut. The blood measurements also indicate environmental stress.

The biochemists are also studying a range of enzymes and abnormal responses of enzyme systems in organisms that are subjected to stresses, including exposure to contaminants. A number of enzyme activities have been shown to be affected in animals studied in stressed habitats. Included in the enzyme studies have been contaminant analyses for body burdens of inorganic and organic contaminants. Scallops that show enzyme abnormalities have been found to have high levels of cadmium in their viscera. At the present time we are not certain that the enzyme abnormalities are caused, however, by the cadmium.

Other biological effects tests have included examining immunological responses in finfish that have been collected from heavily contaminated and "pristine" habitats. We have found that fish from areas known to be contaminated by pathogenic bacteria do show immunological responses that occur with exposure to such organisms. We are examining these fish from stations located over the entire continental shelf to see if we can "map" the distributions of those fish that show unusual immunological responses.

A range of other biological effects studies have indicated sublethal effects (see Figs. 18 and 19) and related laboratory experiments have shown that mortality can be associated with the various levels of contaminants that have been found in the physical environment; future research will emphasize, however, the effects of contaminants and physical degradation on the resource stocks themselves. In a few limited instances, this has been possible to do, especially for certain shellfish stocks that are sedentary and that tend to bear the brunt of pollution impacts. It can also be stated that effects have been demonstrated on food chain organisms that are important as forage for demersal, bottom feeding fishes (Reid et al., 1982) and that there appear to be good examples of how humans are affecting phytoplankton and primary production, especially in highly enriched estuaries (O'Reilly et al., 1976). There is increasing evidence for a change from dominance by larger diatom species to smaller diatoms and dinoflagellates. Similar changes have been reported in Euro-

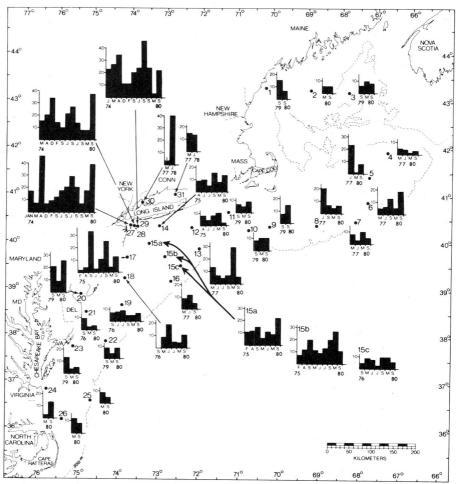

Figure 18. Seabed oxygen consumption values from selected NEMP stations by cruise, in ml/O₂/m²·hr. These data indicate seafloor "metabolism" by season and show heightened oxygen consumption in those areas enriched by organic wastes. Data were developed from samples collected with the Pamatmat multiple corer (NOAA, 1981b).

pean waters that are being enriched by a range of nutrients that have both point and nonpoint sources.

There can be no doubt that future marine and estuarine research and monitoring must quantify effects of pollution and physical degradation on the total fish stocks of interest to humans. This will not be done either quickly or easily. Moreover, in many instances such studies will have to be done on an international basis, often using generic data and information from many sources. Considerable progress has been made, however, in the identification of geographic "hot spots" and problem areas and the development of certain important benchmarks that are crucial to future assessments. Finally, at the same time that scientists are documenting the quantitative effects of pollution and physical degradation on living marine resources, they must be involved in the development of risk assessments that can be used, along with models, to demonstrate how measured effects have or will affect resource populations, both in the specific instance and from the generic point of view. These combined activities: selection of monitoring species, identification of "hot spots," and risk assessment result in the development of habitat management plans. These plans, (Fig. 20) provide a focus for long term resource monitoring and habitat protection.

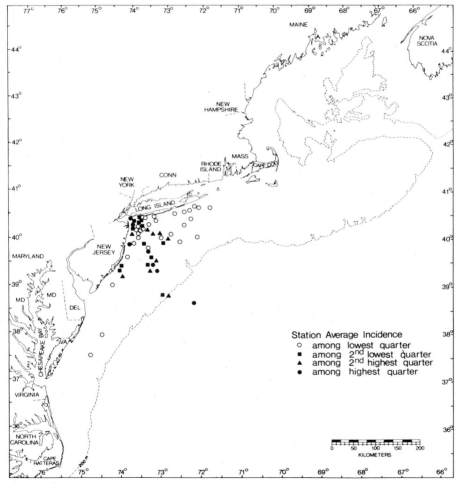

Figure 19.   Station incidence of mitotic telophases with chromosome bridges in gastrulating Atlantic mackerel embryos. Such chromosomal aberrations occur with greater incidence in areas deemed polluted by a range of wastes (see NOAA, 1981b and 1983 for details; from NOAA, 1981b).

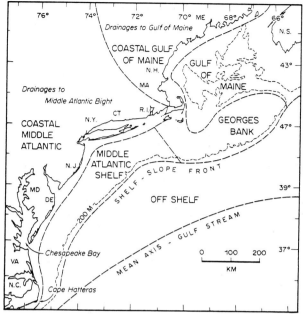

Figure 20. Coherent regions of the northeast continental shelf have been identified as Water Management Units (WMU), which will be the focus for regional assessments of habitat quality. This activity is being done under the Regional Action Plan (RAP) implemented by the NMFS, Northeast Regional Office, and Northeast Fisheries Center.

5. SUMMARY AND RECOMMENDATIONS

Having been involved in long-term monitoring programs and having considered the results discussed in this book, I believe that there are a number of points that should be made for any larger body of water, be it Lake Ontario or the Middle Atlantic Bight. The first step that should be taken is to assess what is known about the area. What are the resources at risk and how have they been affected to date? What are the natural variables that might affect these resources and, perhaps, obscure the impacts of contaminants or physical degradation on the living resources?

A number of nations and commissions have taken steps to develop such assessments. For instance, the Helsinki Commission and the ICES requested and obtained an assessment of the Baltic Sea. This document has been widely used in developing future plans for research and monitoring.

In other areas it should be possible to indicate, at the present time, those areas that have been most degraded and that represent "hot spots" requiring additional research and monitoring.

A next important step is for the scientific community to evaluate various biological effects monitoring techniques and to implement the ones that appear to have greatest efficacy. Moreover, comparative studies should be done using similar techniques but in different geographic areas and ecological settings. The required quality assurance programs will involve input from a wide range of scientists in many nations. National resources are no longer so large that scientists can work alone and in a vacuum. If a scientist working in the Great Lakes develops a technique that can be used in a generic sense in the oceans or fresh waters, there should be a way of bringing the person's findings to the attention of the scientific community.

Finally, once individual scientists are working in concert to solve problems that are common to many aquatic systems, individual agencies and bureaucracies will have to come together to support scientists of several nations that are conducting research that might have generic application. Traditionally, in the United States, scientists working within individual states have often duplicated research on fish that move through waters common to several states. If we are to establish successful fisheries and habitat management, it will have to be done on the basis of interstate and international activities.

REFERENCES

Belton, T. J., Ruppel, B. E., and Lockwood, K. (1982). PCB's (Aroclor 1254) in fish tissues throughout the state of New Jersey: A comprehensive survey. Office of Cancer and Toxic Substances Res. NJ Dep. of Environ. Protec., Trenton, NJ, 36 pp.

Bewers, J. M., Jensen, A., McIntyre, A. D., Parker, M. M., Pearce, J. B., and Portmann, J. E. (1982). Guidelines for the preparation of regional environmental assessments. Hydrography Committee, Int. Council for the Exploration of the Sea, C.M. 1982/E:22. Copenhagen, Denmark, 9 pp.

Boehm, P. D. and Hirtzer, P. (1982). Gulf and Atlantic survey for selected organic pollutants in finfish. NOAA Tech. Memo. NMFS-F/NEC-13. Northeast Fish. Center, Woods Hole, MA, 111 pp.

Boesch, D. F. (1983). Implications of oxygen depletion on the continental shelf of the northern Gulf of Mexico. *Coastal Ocean Pollut. Assess. News* **2** (3), 25–28. Marine Sci. Res. Center, SUNY, Stony Brook, NY.

Federal Register (1982). Ocean dumping; Proposed site designation. Proposed rule by USEPA. vol. 47, no. 244 (56663-56668), Monday, December 20, 1982.

Glading, J. A. (1983). Co-op fishermen: Their side of story. *The Register.* January 30, 1983, Sect. A. Shrewsbury, NJ.

Goldberg, E. D. (Ed.) (1979). *Proceedings of a workshop on scientific problems relating to ocean pollution.* Estes Park, CO, July 10–14, 1978. NOAA Environ. Res. Lab., Boulder, CO, 225 pp.

Goode, G. (1887). *The fisheries and fishery industries of the United States.* Section 2. A geographical review of the fisheries industries and fishery committee for the year 1880. U.S. Government Printing Office, Washington, DC, 384 pp.

Ingham, M. C. (Ed.), Amstrong, R. S., Chamberlin, J. L., Cook, S. K., Mountain, D. G., Schlitz, R. J., Thomas, J. P., Bisagni, J. J., Paul, J. F., and Warsh, C. E. (1982). Summary of the physical oceanographic processes and features pertinent to pollution distribution in the coastal and offshore waters of the northeastern United States, Virginia to Maine. NOAA Tech. Memo. NMFS-F/NEC-17. Northeast Fisheries Center, Woods Hole, MA, 166 pp.

ICES (International Council for the Exploration of the Sea) (1981). *Proces-verbal de la Reunion 1980.* Copenhagen, Denmark, 163 pp.

ICES (International Council for the Exploration of the Sea) (1982). *Proces-verbal de la Reunion 1980.* Copenhagen, Denmark, 163 pp.

ICES (International Council for the Exploration of the Sea) (1983). *Proces-verbal de la Reunion 1980.* Copenhagen, Denmark, 174 pp.

Larsen, P. F. and Doggett, L. F. (1979). An overview of nearshore environmental research in the Gulf of Maine. Mar. Environ. Quality Committee, Int. Council for the Exploration of the Sea, C.M. 1979/E:41, 12 pp.

Larsen, P. F., Johnson, A. C., and Doggett, L. F. (1983a). Environmental benchmark studies in Casco Bay—Portland Harbor, Maine, April 1980. NOAA Tech. Memo. NMFS-F/NEC-19. Northeast Fisheries Center, Woods Hole, MA, 173 pp.

Larsen, P. F., Zdanowicz, V., Johnson, A. C., and Doggett, L. F. (1983b). Trace metals in New England marine sediments: Casco Bay, Maine, in relation to other sites. *Chem. Ecol.* (1), 191–200.

Larsen, P. F., Gadbois, D. F., Johnson, A. C., and Doggett, L. F., (1983c). Distribution of polycyclic aromatic hydrocarbons in the surficial sediments of Casco Bay, Maine. *Bull. Environ. Contam. Toxicol.* **30** (5), May 1983 (accepted for publication).

Lippson, R. L. and Lippson, A. J. (1979). The condition of Chesapeake Bay—An assessment of its present state and its future. Mar. Environ. Quality Committee, Int. Council for the Exploration of the Sea, C.M. 1979/E:42, 24 pp.

Marshall, H. G. and Cohn, M. S. (1981a). Phytoplankton community structure in northeastern coastal waters of the United States. I. October 1978. NOAA Tech. Memo. NMFS-F/NEC-8. Northeast Fisheries Center, Woods Hole, MA, 14 pp.

Marshall, H. G. and Cohn, M. S. (1981b). Phytoplankton community structure in northeastern coastal waters of the United States. II. November 1978. NOAA Tech. Memo. NMFS-F/NEC-9. Northeast Fisheries Center, Woods Hole, MA, 34 pp.

Marshall, H. G. and Cohn, M. S. (1982). Seasonal phytoplankton assemblages in northeastern coastal waters of the United States. NOAA Tech. Memo. NMFS-F/NEC-15. Northeast Fisheries Center, Woods Hole, MA, 31 pp.

Maurer, D. (1979). A brief review of the status of selected pollutants (pesticides, hydrocarbons, trace metals) in relation to benthic invertebrates in Delaware Bay. Marine Environmental Quality Committee, Int. Council for the Exploration of the Sea, C.M. 1979/E:43, 15 pp.

McIntyre, A. D. and J. B. Pearce (Eds.) (1980). *Biological effects of marine pollution and the problems of monitoring.* Proc. from ICES workshop, Beaufort, NC, February 26–March 2, 1979. *Rapp. P.-v. Reun.* **179**. International Council for the Exploration of the Sea, Copenhagen, Denmark. 346 p.

Munn, R. E. (1973). Global Environmental Monitoring System (GEMS) action plan for phase 1. Sci. committee on problems of the environment SCOPE, rep. 3. Toronto, Ont., 130 pp.

NOAA (National Oceanic and Atmospheric Administration) (1979). Federal plan for ocean pollution research, development, and monitoring, fiscal years 1979–83. NOAA Interagency Committee on Ocean Pollution Research, Development, and Monitoring. Washington, DC, 160 pp.

NOAA (National Oceanic and Atmospheric Administration) (1981a). National marine pollution plan; Federal plan for ocean pollution research, development, and monitoring, fiscal years 1981–1985. NOAA Interagency Committee on Ocean Pollution Research, Development, and Monitoring. Washington, DC, 185 pp.

NOAA (National Oceanic and Atmospheric Administration) (1981b). Northeast Monitoring Program first annual NEMP report on the health of the northeast coastal waters of the United States, 1980. NOAA tech. memo. NMFS-F/NEC-10. Northeast Fisheries Center, Woods Hole, MA, 86 pp.

NOAA (National Oceanic and Atmospheric Administration) (1983). Annual NEMP report on the health of the northeast coastal waters of the United States, 1981. NOAA tech. memo. NMFS-F/NEC-20. Northeast Fisheries Center, Woods Hole, MA, 86 pp.

O'Reilly, J. E. and Busch, D. A. (1982). The annual cycle of phytoplankton primary production (netplankton, nannoplankton and release of dissolved organic carbon) for the northwestern Atlantic shelf (Mid-Atlantic Bight, Georges Bank, and Gulf of Maine). In: *Symp. on Biological Productivity of Continental Shelves in the Temperate Zone of the North Atlantic*, March 2–5, 1980, Kiel, Federal Republic of Germany. Int. Council for the Exploration of the Sea, Copenhagen, Denmark, paper 12.

O'Reilly, J. E., Thomas, J. P., and Evans, C. (1976). Annual primary production (nannoplankton, netplankton, dissolved organic matter) in the Lower New York Bay. In: W. H. McKeon and G. J. Lauer (Eds.). Proc. Fourth Symp. on Hudson River Ecology. Hudson River Environ. Soc., Inc., New York, paper 19.

Pearce, J. B. (1979). Raritan Bay—A highly polluted estuarine system. Marine Environmental Quality Committee, Int. Council for the Exploration of the Sea, C.M. 1979/E:45. 16 pp.

Pearce, J. B. (1980). The effects of pollution and the need for long-term monitoring. *Helgolander Meeresuntersuchungen* **34** (2), 207–220.

Pearce, J. B. (1981). Monitoring the health of the northeast continental shelf. *Oceans '81*, pp. 744–751.

Phelps, D. K. (1979). A prognosis for Narragansett Bay. Marine Environmental Quality Committee, Int. Council for the Exploration of the Sea, C.M. 1979/E:46, 6 pp.

Reid, R. N. (1979). Contaminant concentrations and effects in Long Island Sound. Mar. Environ. Quality Committee, Int. Council for the Exploration of the Sea, C.M. 1979/E:47, 15 pp.

Reid, R. N., O'Reilly, J. E., and Zdanowicz, V. S. (Eds.) (1982). Contaminants in New York Bight and Long Island Sound sediments and demersal species, and contaminant effects on benthos, summer 1980. NOAA Tech. Memo. NMFS-F/NEC-16. Northeast Fisheries Center, Woods Hole, MA, 96 pp.

Swanson, R. L. and Sindermann, C. J. (Eds.) (1979). Oxygen depletion and associated benthic mortalities in New York Bight, 1976. NOAA Professional Paper 11. U.S. Dep. of Commerce, Rockville, MD, 345 pp.

20

STATE OF ECOSYSTEM MEDICINE

D. J. Rapport

Statistics Canada
Tunney's Pasture
Ottawa, Ontario

To Agneta, in caring
for the large and the small,
a constant source
of inspiration.

1. INTRODUCTION

In the earliest stages of the development of a science, the preoccupation is the unearthing of "facts," and their cataloging. In the biological sciences, this has given vistas of a landscape of bewildering diversity, and by and large, workers in this field have never quite sprung free from a tendency toward "reductionism," for example, a preoccupation with diversity. In contrast, physics has developed with a more "holistic" view—and evolves around the formulation of generalized theories.

In the environmental sciences, however, particularly the study of the behavior of ecosystems affected by various human activities, there is much to be said for the value of an empirical approach rather than *a priori* generalized theories. There is also the need for a systematic search for pattern in the rapidly accumulating case studies of specific disturbances. One need only look at the history of human medicine to see how centuries of trial and error rooted in the empirical approach have yielded diagnostic protocols for a good many illnesses. If the ecosystem practitioner today has a state of knowledge somewhat comparable to that of the 18th century medical practitioner (Vallentyne, 1974), then developments over the past few centuries in human medicine might offer some guidance for the systematization of signs and causes of ecosystem malfunction, leading to the further development of ecosystem medicine. This essay explores an analogy between medicine and the study of the environment, and indicates how concepts derived from the former might provide the means to perceive a more integrated view of ecosystem behavior under stress.

2. SYMPTOMS OF ECOSYSTEM DISTRESS

Ecosystems, like organisms, depend on the close integration of a number of vital subcomponents. Stoermer (1978) has commented that

> the environmentalist's broad perspective sees an ecosystem such as a large lake as analogous to a living organism and leads him to conclude that Lake Erie is "dead" in the way that an organism that has lost several vital functions is dead. To the environmentalist, survival of large ecosystems depends on the retention of their functional integrity.

The analogy of ecosystem to organism, although much overworked, appears particularly relevant in a discussion of the response of ecosystems to stress. The late Hans Selye (1975) characterized the process by which organisms adapt to stress. A number of different stressors, he pointed out, evoke similar adaptive responses. In mammalian systems, cold, heat, hemorrhage, muscular effort, and nervous irritation give rise to similar symptoms of distress. These typically involve enlargement and hyperactivity of the adrenal cortex, shrinkage of the thymus gland and lymph nodes, and in some cases, the appearance of gastrointestinal ulcers.

The response of ecosystems to stress is less understood. However, from an examination of representative case studies (Rapport et al., 1984), there appears here, too, a surprising degree of similarity in the response syndrome. Whether the stressor be contaminants, fire, floods, harvesting, physical restructuring, or the introduction of exotic species, common symptoms are found frequently enough to suggest an "ecosystem level distress syndrome." For terrestrial systems, these symptoms include decreases in primary productivity, loss of nutrients, reductions in species diversity, shifts in composition of biota to favor more opportunistic species, reduction in size of dominant flora (e.g., the higher mortality of larger life forms), and increased prevalence of disease. Aquatic ecosystems exhibit similar symptoms except these systems generally accumulate nutrients dislodged from surrounding terrestial systems (including urban systems) and, therefore, often exhibit increased primary productivity.

To illustrate the nature of the "ecosystem distress syndrome," I summarize here some of the main features accompanying the intensification of cultural stress on the Laurentian lower Great Lakes. Characteristically, these ecosystems have served as "sinks" for the accumulation of nutrients, resulting in eutrophication of the once oligotrophic waters. This process has continued to the point where large regions of Lake Erie (the Western and Central basins) are highly eutrophic, and other basins (the eastern basin of Erie and the main basin of Ontario) are mesotrophic. These eutrophic waters have an elevated primary productivity (and increased algal biomass), roughly proportional to the increase in population pressure in the surrounding drainage basin.

There are also striking changes in the composition of the fauna. The once abundant and valuable inshore benthics (such as lake trout, blue pike, lake whitefish) have become extinct and the offshore pelagics (in particular the exotics, such as smelt and alewife) have expanded their habitat, covering more of the nearshore zone.

Species diversity in a good many families, at least for the sensitive nearshore zones, has declined. In the more eutrophic zones, near monocultures of blue-green algae (cladophora) have displaced far more diverse diatom communities. The diversity in the large inshore benthic fish has also declined markedly. Similarly, in the *tubificidae*, heavily impacted areas (near urban developments) generally show relatively low diversity with dominance by one or two pollutant-tolerant forms.

The overall transformation of the fisheries community can be characterized as "retrogressive" in the sense that the now dominant forms are more characteristic of the opportunistic species that mount the first wave of the colonization process. These species can be characterized as being relatively short-lived, rapidly reproducing, having short food chains, and relatively independent, that is, showing, little evidence of more symbiotic or mutualistic interactions, characteristic of later stages of colonization.

The destabilization of key populations in stressed ecosystems is often asserted but difficult to document. In the lower Great Lakes, however, especially in the history of the Lake Erie fishery, this is a well-established feature of the transformation process. Nearly all of the important commercial fish stocks that ultimately became extinct in these waters exhibited marked oscillations in the commercial catch, just prior to the final collapse.

Shifts in the size distribution of dominant life forms also characterizes aquatic ecosystems under stress. Here too, in the Laurentian Lower Great Lakes such changes have been striking. Several centuries ago perhaps over 50% of the Lake Erie fishery consisted of fish that weighed in excess of 5 kg. Today, the average weight of an individual is less than 0.5 kg (Christie, 1972; Leach and Nepszy, 1976; Francis et al., 1979).

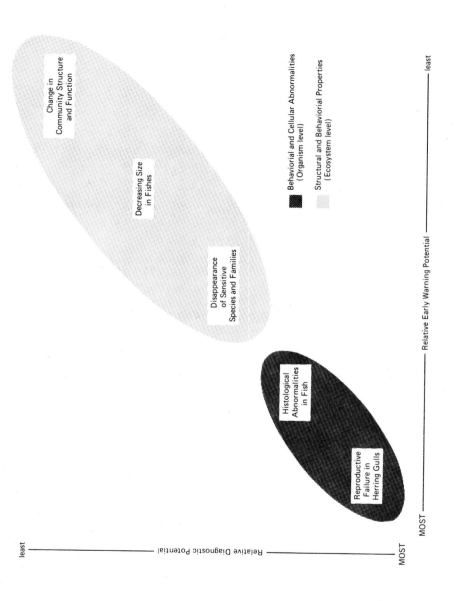

Of course, not all the symptoms of ecosystem distress appear in every case (again, analogous to medical histories). Further, the symptoms of ecosystem distress are usually nonspecific (with respect to the stressor) and "retrospective" rather than "early warning" or "prospective." In this aspect, ecosystem distress signs are somewhat akin to the "vital signs" in human medicine.

A characterization of the state of health at the ecosystem level might include information from lower levels of organization (population, organism, tissue, cellular). In Fig. 1, various classes of environmental indicators used to characterize the state of Lake Erie are arrayed according to their early warning and diagnostic potential. Relatively few indicators are both early warning and diagnostic.

The study of the temporal sequence of signs and symptoms of ecosystem distress is a rather undeveloped area. The earliest signs and symptoms of distress may be subtle physiological changes and/or behavioral changes in sensitive organisms or groups. Such symptoms may easily go undetected, especially if they are transient in nature.

By the time change is manifest in the "ecosystem distress syndrome," the ecosystem is already well advanced in the "breakdown" process. Systemic changes such as alterations of the rates of primary productivity and nutrient recycling generally occur well after the onset of stress. Similarly, in human diseases, alterations in "vital signs" often occur after the disease process has advanced sufficiently so that homeostatic mechanisms no longer suffice to maintain body functions within the normal range.

The search for early warning indicators of ecological transformations is not unlike the search for preclinical signs of disease. In medicine, much useful information can be derived from a single blood sample that yields chemical indicators signaling a wide range of potential disorders. The blood concentration of bilirubin, for example, is useful in the early detection of liver disease and hemolytic anemia. Increased levels of various plasma lipids, singly and in combination, can be indicative of risks of coronary disease. Calcium and phosphorus levels, if abnormal, may be indicative of the onset of hyperparathyroidism. No analagous set of early warning indicators has yet been identified to detect ecosystem degradation.

In medicine, the early warning indicators of disease are often at the subcellular (i.e., biochemical) level. Yet these provide valuable diagnostic information for the functioning of the whole organism. Similarly, monitoring at the lower levels of organization within ecosystems (e.g., at population, organism, and biochemical

Figure 1. Indicators of ecosystem degradation in the Great Lakes. Early warning indicators signal the onset of the process of ecosystem transformation under stress. Characteristically, these symptoms appear at subecosystem levels such as changes in physiology and behavior of sensitive species. An example might be liver histopathology in benthic fish, or abandonment of preferred spawning or feeding habitat by nearshore fauna. Retrospective (historical) indicators confirm widespread ecosystem transformation. These include changes in nutrient concentrations of waters, changes in community structure (e.g., in the Laurentian Lower Great Lakes pelagic fishes have generally displaced the benthics). In some cases, symptoms are diagnostic of a particular type of stress (e.g., contaminant accumulation in Herring gull eggs). More generally, however, the symptoms of ecosystem distress are nonspecific as to causal agent. Examples in figure are from the Laurentian Lower Great Lakes.

levels) provides clues to the processes affecting the behavior of the whole. For example, early warning of significant contaminant stress in the Great Lakes might come from the studies of failures in herring gull reproduction, and the sensitive chemical determinations of bioaccumulated pesticides in gull eggs. Similar potential exists in monitoring other trophic levels such as invertebrates (e.g., clams or mussels) and forage and top predator fishes. Abnormalities in the reproductive process in fish may predict the susceptibility of the aquatic ecosystem to acidification, when it is at a relatively early stage.

3. DIAGNOSIS OF ECOSYSTEM HEALTH

Once a proper set of vital signs and preclinical symptoms is identified, then diagnostic procedures may run parallel to medical practice. It is clear that the normal range for vital signs (pulse, temperature, blood pressure) change with the age of the patient. For example, the fetal hearbeat rate exceeds that of the adult, while normal blood pressure and blood sugar levels increase with age. Ecosystems also exhibit characteristic changes with "age." For example, species diversity and primary productivity generally rise as secondary succession transforms fields to forests.

The challenge to both medical and ecosystem practitioners is to identify symptoms of ill health and correlate them with possible degenerative or transformation processes. When the cause of transformation is a single factor, the diagnosis may be relatively straightforward, and treatment remedies obvious. For example, sewage discharged in the early 1950s to Lake Washington, near Seattle, reduced lake transparency, increased the phosphate content, increased the oxygen depletion rates in the hypolimnion during the summer, reduced diatom species diversity, and increased the dominance of eutrophic diatom forms. When the sewage was diverted away from the lake, these trends were largely reversed.

Multiply stressed ecosystems pose challenges to the ecosystem physician not dissimiliar to those of patients suffering from illnesses involving multiple causal agents. For example, changes in the fish communities in Lake Erie have been attributed to the simultaneous actions of at least 15 major stressors including overfishing, invasion of exotics, entrainment and impingement due to water intakes, thermal outfalls, eutrophication, stream alteration, toxic wastes and microcontaminants, shoreworks, dredging, dumping, sedimentation and infilling, wetland depletion, sand, gravel, and rock extraction, waves and current due to shipping and boating, and winter navigation. These complex synergisms are poorly understood. Indeed, current understanding of ecosystem dynamics under stress is still rather primitive. Vallentyne (1974), in his book *The Algal Bowl*, put the matter in these terms:

> Our knowledge of ecosystems today is equivalent to that of the human body in the latter part of the 18th century. . . . Based on this analogy, we should not be too surprised if, in the latter part of the 20th century, we are inflicted with environmental ills equivalent to epidemics of typhoid fever, cholera and bubonic plague that characterized earlier times. . . . But to the environmental physicians who will look back on us from the vantage point of the 23rd

century, they will rightly attribute our misstatements and misunderstandings to ignorance of the causes of environmental problems.

4.   TREATMENT PROTOCOLS FOR REHABILITATION

What treatment options are available to ecosystem physicians? Once sources of stress are removed, ecosystems may possess sufficient hemeostatic mechanisms to recover unassisted. For example, fishing closures on Lake St. Clair walleye stocks have permitted their recovery. Sewage diversion in Lake Washington resulted in virtual restoration of the lake.

However, in many cases removal of the sources of stress is not sufficient for ecosystem recovery. Acidification of lakes, for example, can yield irreversible changes once the bottom sediments are separated from the hypolimnion by a sphagnum cover, or once key fish stocks disappear. Similarly, terrestrial desertification from overgrazing may progress to the point that removal of grazing pressure cannot arrest further erosion and, ultimately, dune formation.

Risks to ecosystem health from various treatment procedures need more attention and study. Chemicals now used to control spruce budworm outbreaks in eastern Canada have been implicated in the losses of natural predators of budworm as well as in risks to human health from direct chemical exposure.

Treatment may mask some symptoms yet permit continued degradation. Aeration of the hypolimnetic waters to reduce fish die-offs (for example, in Lake Kolbotnvatn, Norway) allows further degradation of water quality (reflected in increases in algal biomass and nutrient levels) from unabated municipal waste loading. This practice also may result in the remobilization of previously sediment-bound contaminants into an aerobic medium.

Physical restructuring (ecosystem surgery) also poses risks and uncertainties. Uncertainties as to outcome are illustrated by comparing the recovery of lakes dredged to remove nutrient laden sediments. Relatively similar lakes, dredged in the same year, have shown very different responses. Washington Park Lake (New York) showed significant improvement in water quality the year following dredging, while Buckingham Lake (New York) showed no sign of improvement. Studies in the Environmental Protection Agency's (U.S.A.) Corvallis Environmental Research Laboratory have been directed toward determining which lake restoration techniques have been most successful. This work involves an analysis of the data base for about 20 different lake types under various stress regimes. As is standard in medical practice, the design of ecosystem treatment protocols will increasingly be based on a systematic review of previous successes and failures with respect to recovery of stressed systems under alternative treatment options.

Ecosystem surgery also may have severe side effects. The clean-up efforts in the aftermath of the Amoco Cadiz oil spill off Brest, France, involved the use of high pressure hoses, steam cleaning, and bulldozing of beaches. These methods destroyed the very habitats that were to be preserved. Commented one investigative team:

In the emotional heat of these disasters, we rarely stop to evaluate the effects of any clean-up technique on the ecosystem suffering the insult. Occasionally, we find the pressure of public opionion so strong that any clean-up may be used just to show the enraged public that some progress is happening.

5.   PREVENTIVE ECOSYSTEM MEDICINE

While the development of ecosystem medicine has progressed to the point of becoming prescriptive, in some instances, far too little attention has been directed toward preventive medicine. While various pieces of legislation (e.g., the Clean Air Act, the Clean Water Act, and the Environmental Protection Act (U.S.A.)) are preventive in intent, more specific guidelines and regulatory procedures are required to make these effective. It is disappointing to note, for example, that the 1978 Water Quality Agreement between the United States and Canada provides for "limited use" zones where raw sewage may be allowed to enter the Great Lakes. Under this plan, local areas will undoubtedly be degraded in terms of their potential to support fisheries and/or recreation.

Inadequate attention to preventive medicine has given rise to ecosystem breakdown or premature "death" (or "aging"). Transformations resulting from cultural eutrophication and desertification are preventable. Further, just as in human medicine, the costs of prevention are (in most instances) far less than the costs of cures after degradation has set in.

Medical science is replete with observations that certain parts of the system are crucial to the functioning and well-being of the whole. If one stresses sensitive parts such as the pituitary gland, then a small perturbation may yield a large system change. The identification of similar sensitive zones or components of ecosystems is less well advanced. In the lower Great Lakes the importance of the nearshore zone with its "critical nodes" for functioning of the entire ecosystem is becoming better understood (Francis et al., 1979). Destruction of these zones, particularly the marshes and shallow gravel and sand bars, has had a significant impact on the composition of the biota, owing to a reduction in the spawning habitat for fishes.

Epidemiological models are now used in forecasting the transmission of forest pathogens such as the bark beetle, spruce budworm, and Dutch elm disease. However, the development of appropriate methodologies for the containment of ecosystem epidemics is less well advanced. One possibility might be to develop an ecosystem analogue of vaccines.

For example, innoculating an ecosytem with relatively harmless forms of the pest (e.g., by the use of lethal genes, sterile males, or temperature-sensitive mutants), might provide some lead time for an ecosystem to mobilize its defense mechanisms (e.g., parasites and predators) to better neutralize the invading species. Should this approach prove feasible, it would offer the prospect of preventing outbreaks of exotics, rather than resorting to more costly and risky extermination procedures after the outbreak. Naturally, in using this approach the introduction of exotics must be strictly controlled (through the above mentioned genetic technique, for example);

otherwise, there is always the risk of destabilizing the system throught the uncontrolled spread of exotics.

While there is a broad appreciation of the importance of synergism among stressors, there are few methods to systematically identify these interactions and evolve effective and efficient strategies for reducing overall stress levels. The general principle that stressed systems become more vulnerable to the effects of additional stressors (true in many medical situations) needs clarification as to its range of applicability in the context of the ecosystem. Empirical studies that identify the most highly interactive stressor complexes (in multiply stressed systems) might provide a first step in development of an efficient treatment protocol for such systems.

Clearly the practice of ecosystem medicine is at a very primitive stage. A good many case studies have now been completed, and these form a rich source of data from which some general patterns of ecosystem breakdown and recovery might be detected. Progress is likely to be painfully slow and to require more of brute force empiricism than preconceived general theories. Further elucidation of the "vital signs" of ecosystem health, and identification of "early warning" indicators of ecosystem breakdown will contribute to a much less costly program of safeguarding the environment.

6. SUMMARY

Ecosystem medicine is in its infancy. While the signs and symptoms of severely disturbed environments are now reasonably well established, few indicators have been found that provide early warning and diagnostic potential. Diagnosis is further made difficult since many stresses produce the same set of symptoms, and ecosystems are often impacted upon by multiple stresses acting in complex synergistic and occasionally antagonistic modes. In the development of ecosystem practice, much can be learned from medical procedures, especially in the design of treatment protocols, and in recognition of the dangers and risks of various treatment options.

ACKNOWLEDGMENTS

I thank J. R. Vallentyne, V. Cairns, A. M. Friend, and H. J. Adler for comments on an earlier draft. I am very grateful to P. Fong for the design and drafting of Fig. 1, and as always to M. E. Scott for painstaking work in preparing the text.

REFERENCES

Christie, W. J. (1972). Lake Ontario: Effects of exploitation, introductions and eutrophication on the salmonid community. *J. Fish. Res. Board Can.* **29**(6), 913–929.

Francis, G. R., Magnuson, J. J. Regier, H. A., and Talhelm, D. R. (1979). Rehabilitating Great Lakes ecosystems, Great Lakes Fish. Commission tech. rep. no. 37, 99 pp.

Gorham, E. and Gordon, A. G. (1960). Some effects of smelter pollution northeast of Falconbridge, Ontario, *Can. J. Bot.* **38**(3), 307–312.

Leach, J. H. and Nepszy, S. J. (1976). The fish community in Lake Erie, *J. Fish. Res. Board Can.*, **33**(3) 622–638.

Rapport, D. J., Regier, H. A., and Hutchinson, T. C. (1984). Ecosystem behaviour under stress, manuscript in preparation.

Raunkiaer, C. C. (1910). Formationsundersogelse Og Formationstatistik. *Bot. Tidsskr.* **30**, 20–43.

Selye, H. (1975). *Stress without distress.* The New American Library of Canada Ltd., 193 pp.

Stoermer, E. (1978). The blue green algae keep coming. *Nat. Hist.* **87**(7) (Aug.–Sept.), 59–61.

Vallentyne, J. R. (1974). *The algae bowl—Lakes and man.* Dep. of Environment Fisheries and Marine Services. Miscellaneous special publ. 22.

Woodwell, G. M. (1970). Effects of pollution on the structure and physiology of ecosystems. *Science* **168**(3930), 429–433.

INDEX